Sea Surface Roughness Observed by High Resolution Radar

Sea Surface Roughness Observed by High Resolution Radar

Special Issue Editors

Atsushi Fujimura
Susanne Lehner
Alexander Soloviev
Xiaofeng Li

MDPI • Basel • Beijing • Wuhan • Barcelona • Belgrade

MDPI

Special Issue Editors

Atsushi Fujimura
University of Guam
USA

Susanne Lehner
German Aerospace Center
Germany

Alexander Soloviev
Nova Southeastern University
USA

Xiaofeng Li
NCWCP - E/RA3
USA

Editorial Office
MDPI
St. Alban-Anlage 66
4052 Basel, Switzerland

This is a reprint of articles from the Special Issue published online in the open access journal *Remote Sensing* (ISSN 2072-4292) from 2018 to 2019 (available at: https://www.mdpi.com/journal/remotesensing/special_issues/ssr_rs)

For citation purposes, cite each article independently as indicated on the article page online and as indicated below:

LastName, A.A.; LastName, B.B.; LastName, C.C. Article Title. *Journal Name* **Year**, *Article Number*, Page Range.

ISBN 978-3-03921-746-5 (Pbk)
ISBN 978-3-03921-747-2 (PDF)

Contents

Xiao-Ming Li, Tianyu Zhang, Bingqing Huang and Tong Jia

Capabilities of Chinese Gaofen-3 Synthetic Aperture Radar in Selected Topics for Coastal and Ocean Observations

About the Special Issue Editors

Atsushi Fujimura, Ph.D. Atsushi Fujimura received a B.Sc. degree in Biology from the University of California, Riverside, in 2006; an M.Sc. degree in Marine Biology from Nova Southeastern University, Dania Beach, FL, in 2010; and a Ph.D. degree in Applied Marine Physics from the University of Miami, FL, in 2015. He has been a Postdoctoral Scholar at the Okinawa Institute of Science and Technology Graduate School, Okinawa, Japan. He is currently an Assistant Professor at the University of Guam Marine Laboratory and an Adjunct Professor at Nova Southeastern University. His research interests include remote sensing, ocean surface waves, coastal oceanography, air–sea interaction, ship hydrodynamics, computational fluid dynamics, ecology, physiology, and biophysical interactions.

Susanne Lehner, Ph.D. Susanne Lehner studied mathematics and physics at the University of Hamburg, Hamburg, Germany. She received an M.Sc. degree in applied mathematics from Brunel University, Uxbridge, U.K., in 1979, and a Ph.D. degree in geophysics from the University of Hamburg, Hamburg, Germany, in 1984. During her Ph.D., she worked as a Research Scientist at the Max Planck Institute for Meteorology, Hamburg. In 1996, she joined the German Aerospace Center DLR/DFD, Wessling, Germany, where she became the head of the Radar Oceanography team at the Institute for Remote Sensing Technology. She holds a faculty position at Nova Southeastern University, Port Everglades, FL, USA. She became an Affiliated Faculty Member in 2013. Her SAR oceanography research focused on developing algorithms to extract information on wind fields, sea state, currents, and underwater topography from SAR images. In addition to global sea state measurements, her recent research interests have been in the high-resolution coastal SAR oceanography, especially TerraSAR-X oceanography, meteo-marine observations, and maritime traffic surveillance in near real-time.

Alex Soloviev, Ph.D. Alexander Soloviev received an M.Sc. degree in physics from the Moscow Institute of Physics and Technology, Moscow, Russia, in 1976; Ph.D. and D.Sc. degrees in physics and mathematics from the former Soviet Academy of Sciences, Moscow, in 1979 and 1992, respectively; and an MBA degree from the University of Florida, Gainesville, FL, USA, in 2010. He has been a Visiting Scientist with the University of Hawaii, Honolulu, HI, USA, and the University of Hamburg, Hamburg, Germany. He has also been a Scientist in the two leading institutions of the former Soviet Academy of Sciences: the P. P. Shirshov Institute of Oceanology and the A. M. Oboukhov Institute of Atmospheric Physics. He is currently a Professor with the Halmos College of Natural Sciences and Oceanography, Nova Southeastern University, Dania Beach, FL. He is also an Adjunct Professor with the Rosenstiel School of Marine and Atmospheric Science, University of Miami, Miami, FL. He has participated in several major oceanographic (POLYMODE, JASIN, FGGE, TOGA COARE, GASEX) and oil spill (GoMRI) experiments. He has been a Principal Investigator (PI) and a co-PI on a number of research projects funded by the U.S. Federal Government and private industries. He has authored and coauthored over 70 peer-reviewed research articles. In coauthorship with Prof. Roger Lukas from the University of Hawaii, he wrote a monograph entitled "The Near-Surface Layer of the Ocean: Structure, Dynamics, and Applications" published by Springer. His research interests include turbulence and microstructure in the near-surface layer of the ocean, upper and bottom ocean boundary layers, satellite oceanography, biophysical interactions in the ocean, carbon dioxide uptake by the ocean, oil spill transport and dispersion, hurricane physics, coastal ocean circulation, and ocean and climate engineering. His major field of study is physical oceanography.

Xiaofeng Li, Ph.D. Xiaofeng Li received a B.Sc. degree in optical engineering from Zhejiang University, Hangzhou, China, in 1985; an M.Sc. degree in physical oceanography from the First Institute of Oceanography, State Oceanic Administration, Qingdao, China, in 1992; and a Ph.D. degree in physical oceanography from North Carolina State University, Raleigh, NC, USA, in 1997. During his M.Sc. program, he completed the graduate course work from the Department of Physics, University of Science and Technology of China, Hefei, China. Since 1997, he has been with the National Environmental Satellite, Data, and Information Service (NESDIS), NOAA, College Park, MD, USA, where he is involved in developing many operational satellite ocean remote sensing products. He has authored over 100 peer-reviewed publications and edited three books. His research interests include remote sensing oceanography and marine meteorology, satellite image processing, oil spill and coastal zone classification with multipolarization synthetic aperture radar, and the development of sea surface temperature algorithms. Dr. Li has received the Individual Award for Science from the NOAA/NESDIS Center for Satellite Applications and Research, the Len Curtis Award from the Remote Sensing and Photogrammetry Society, and the Overseas Expert title from the Chinese Academy of Sciences. He is an Associate Editor of IEEE Transactions on Geoscience and Remote Sensing and the International Journal of Remote Sensing, the Ocean Section Editor-in-Chief of Remote Sensing, and an Editorial Board Member of the *International Journal of Digital Earth and Big Earth Data* from the Chinese Academy of Sciences. He is an Associate Editor of the *IEEE Transactions on Geoscience and Remote Sensing* and the *International Journal of Remote Sensing*, the Ocean Section Editor-in-Chief of the *Remote Sensing*, and an Editorial Board Member of the *International Journal of Digital Earth and Big Earth Data*.

Editorial

Sea Surface Roughness Observed by High Resolution Radar

Atsushi Fujimura [1,*], Susanne Lehner [2], Alexander Soloviev [3] and Xiaofeng Li [4]

[1] Marine Laboratory, University of Guam, Mangilao, Guam 96923, USA
[2] German Aerospace Center, Muenchner Strasse 20, 82234 Wessling, Germany
[3] Nova Southeastern University, 8000 N. Ocean Dr., Dania Beach, FL 33029, USA
[4] NCWCP-E/RA3, 5830 University Research Court, College Park, MD 20740, USA
*   Correspondence: fujimuraa@triton.uog.edu

Received: 9 August 2019; Accepted: 15 August 2019; Published: 28 August 2019

Changes in the sea surface roughness are usually associated with a change in the sea surface wind field. This interaction has been exploited to measure the sea surface wind speed by scatterometry. A number of features on the sea surface associated with changes in roughness can be observed by synthetic aperture radar (SAR) because of the change in Bragg backscatter of the radar signal by damping of the resonant ocean capillary waves. With various radar frequencies, resolutions, and modes of polarization, sea surface features have been analyzed in numerous campaigns, bringing various datasets together, thus allowing for new insights in small-scale processes at a larger areal coverage. This Special Issue aims at investigating sea surface features detected by high spatial resolution radars, such as SAR.

Overview of Contributions

Rikka et al. [1] demonstrate an empirical method for estimating meteo-marine parameters over the Baltic Sea. The empirical function CWAVE_S1-IW combines spectral analysis of Sentinel-1A/B Interferometric Wide swath subscenes with wind data derived with common C-Band Geophysical Model Functions. The estimated wave heights and wind speed agree with the wave model (WAM) and in-situ data, respectively. Their methods are implemented in near-real-time service in the German Aerospace Center's ground station, Neustrelitz.

SAR is applied to tropical storm conditions in several contributed papers. Zhang and Perrie [2] retrieve the wind field of Hurricane Bertha (2008) from the RADARSAR-2 cross-polarized SAR images using the C-3PO (C-band Cross-Polarization Coupled-Parameters Ocean) hurricane wind retrieval model and extract an axisymmetric double-eye structure from an idealized vortex model called Symmetric Hurricane Estimates for Wind. Adding data from airborne measurements using a stepped-frequency microwave radiometer reveal the hurricane's internal dynamic process related to the double-eye structure, which is consistent with past studies.

Zhang et al. [3] examine fetch- and duration-limited parametric models (H-models) using the SAR-retrieved wind speed, Sentinel-1 SAR wave mode, and buoy data to estimate wind wave parameters (wave height and period) generated by hurricanes or typhoons. The models provide an effective method to obtain the wave parameters inside storms.

A paper by Shen et al. [4] introduces a hurricane wind quality index to evaluate SAR wind retrievals from cross-polarization and co-polarization observations for storm conditions. The index shows rain-contaminated wind cells, and it is used for wind correction under heavy rain-contaminated areas. The proposed method improves the SAR-derived wind field under hurricane conditions.

Another wind retrieval model is proposed for the European Space Agency (ESA) Sentinel-1A (S-1A) Extra-Wide swath mode VH-polarized images [5]. The new model is validated by comparing the wind speeds retrieved from S-1A images with the wind speeds measured by Soil Moisture Active

Passive (SMAP) radiometer under tropical cyclone conditions. The results suggest that the proposed model can be used to retrieve wind speeds up to 35 m/s for sub-bands 1 to 4 and 25 m/s for sub-band 5.

Sun et al. [6] develop ocean wind retrieval models for right circular-vertical and right circular-horizontal polarizations from the compact-polarimetry mode of the RADARSAT Constellation Mission (RCM), a set of three satellites just launched in 2019. The wind retrieval models are validated and contribute to temporal oceanography or atmosphere dynamic research based on RCM SAR data.

A hybrid wind retrieval model is proposed by using two models: C-2PO (C-band cross-polarized ocean backscatter) and CMOD4 (C-band model) [7]. Sets of SAR images over the Northwest Pacific off the coast of China are used to establish a wind speed threshold (9.4 m/s). Ocean surface wind speeds are retrieved by the C-2PO model as VH-polarized images when the wind speeds are higher than the threshold, while the CMOD4 geophysical model function for VV-polarized images is used when the wind speeds are less than or equal to the threshold.

Kammerer and Hackett [8] show that phase-resolved ocean wave fields are reconstructed from X-band Doppler radar measurements of the ocean surface by proper orthogonal decomposition (POD) more accurately than the conventional FFT-based dispersion curve filtering. The results indicate that the group line (a linear feature at frequencies lower than the first order dispersion relationship in wavenumber-frequency spectra of the ocean surface) influences the phase-resolved wave field.

Buono et al. [9] discover that under low-to-moderate wind conditions ($\approx$ 3–12 m/s), SAR imaging parameters have a stronger effect on the standard deviation of the co-polarized phase difference than meteo-marine parameters; they use a theoretical model based on the tilted-Bragg scattering. The results can support the improvement of the SAR algorithms for a variety of ocean applications including object detection.

Tings et al. [10] propose an extension of their ship-wake detectability model by using a non-linear basis that allows consideration of all the influencing parameters simultaneously. The parameters affecting wake detectability include environmental conditions (wind speed, wind direction, sea state height, sea state direction, and sea state wave length), ship properties (size, heading, and velocity), and image acquisition settings (incidence angle, beam looking direction). The detectability model can be applied to control an automatic wake-detection system.

An overview of the GeoFen-3 (GF-3), a Chinese C-band SAR satellite launched in August 2016, is provided by Li et al. [11]. They demonstrate the capabilities of the GF-3 SAR in ocean and coastal observations by presenting selected features (i.e., intertidal flats, offshore tidal turbulent wakes, oceanic internal waves, sea surface winds, and waves). For more details and other applications of GF-3, see MDPI journal Sensors Special Issue "First Experiences with Chinese Gaofen-3 SAR Sensor" (https://www.mdpi.com/journal/sensors/special_issues/gaofen_3_SAR_sensor).

Funding: AF was supported by NASA (80NSSC17M0052) "GEOCORE: Geospatial Studies of Reef Ecology and Health using Satellite and Airborne Data" and NASA Guam EPSCoR.

Acknowledgments: We are grateful to the authors who contributed to this Special Issue and the reviewers for their valuable and constructive feedback. We also thank the Remote Sensing editorial team for their support.

Conflicts of Interest: The authors declare no conflict of interest.

References

1. Rikka, S.; Pleskachevsky, A.; Jacobsen, S.; Alari, V.; Uiboupin, R. Meteo-Marine Parameters from Sentinel-1 SAR Imagery: Towards Near Real-Time Services for the Baltic Sea. *Remote Sens.* **2018**, *10*, 757. [CrossRef]
2. Zhang, G.; Perrie, W. Symmetric Double-Eye Structure in Hurricane *Bertha* (2008) Imaged by SAR. *Remote Sens.* **2018**, *10*, 1292. [CrossRef]
3. Zhang, L.; Liu, G.; Perrie, W.; He, Y.; Zhang, G. Typhoon/Hurricane-Generated Wind Waves Inferred from SAR Imagery. *Remote Sens.* **2018**, *10*, 1605. [CrossRef]
4. Shen, H.; Seitz, C.; Perrie, W.; He, Y.; Powell, M. Developing a Quality Index Associated with Rain for Hurricane Winds from SAR. *Remote Sens.* **2018**, *10*, 1783. [CrossRef]

5.   Gao, Y.; Guan, C.; Sun, J.; Xie, L. A Wind Speed Retrieval Model for Sentinel-1A EW Mode Cross-Polarization Images. *Remote Sens.* **2019**, *11*, 153. [CrossRef]

6.   Sun, T.; Zhang, G.; Perrie, W.; Zhang, B.; Guan, C.; Khurshid, S.; Warner, K.; Sun, J. Ocean Wind Retrieval Models for RADARSAT Constellation Mission Compact Polarimetry SAR. *Remote Sens.* **2018**, *10*, 1938. [CrossRef]

7.   Fang, H.; Xie, T.; Perrie, W.; Zhang, G.; Yang, J.; He, Y. Comparison of C-Band Quad-Polarization Synthetic Aperture Radar Wind Retrieval Models. *Remote Sens.* **2018**, *10*, 1448. [CrossRef]

8.   Kammerer, A.J.; Hackett, E.E. Group Line Energy in Phase-Resolved Ocean Surface Wave Orbital Velocity Reconstructions from X-band Doppler Radar Measurements of the Sea Surface. *Remote Sens.* **2019**, *11*, 71. [CrossRef]

9.   Buono, A.; De Macedo, C.R.; Nunziata, F.; Velotto, D.; Migliaccio, M. Analysis on the Effects of SAR Imaging Parameters and Environmental Conditions on the Standard Deviation of the Co-Polarized Phase Difference Measured over Sea Surface. *Remote Sens.* **2019**, *11*, 18. [CrossRef]

10. Tings, B.; Pleskachevsky, A.; Velotto, D.; Jacobsen, S. Extension of Ship Wake Detectability Model for Non-Linear Influences of Parameters Using Satellite Based X-Band Synthetic Aperture Radar. *Remote Sens.* **2019**, *11*, 563. [CrossRef]

11. Li, X.-M.; Zhang, T.; Huang, B.; Jia, T. Capabilities of Chinese Gaofen-3 Synthetic Aperture Radar in Selected Topics for Coastal and Ocean Observations. *Remote Sens.* **2018**, *10*, 1929. [CrossRef]

remote sensing

MDPI

Article

Meteo-Marine Parameters from Sentinel-1 SAR Imagery: Towards Near Real-Time Services for the Baltic Sea

Sander Rikka [1],[*], Andrey Pleskachevsky [2], Sven Jacobsen [2], Victor Alari [1] and Rivo Uiboupin [1]

[1] Department of Marine Systems at Tallinn University of Technology, Tallinn 12618, Estonia;
victor.alari@ttu.ee (V.A.); rivo.uiboupin@ttu.ee (R.U.)
[2] German Aerospace Center (DLR), Remote Sensing Technology Institute, Bremen 28199, Germany;
andrey.pleskachevsky@dlr.de (A.P.); sven.jacobsen@dlr.de (S.J.)
[*] Correspondence: sander.rikka@ttu.ee; Tel.: +372-5342-8135

Received: 18 April 2018; Accepted: 9 May 2018; Published: 15 May 2018

Abstract: A method for estimating meteo-marine parameters from satellite Synthetic Aperture Radar (SAR) data, intended for near-real-time (NRT) service over the Baltic Sea, is presented and validated. Total significant wave height data are retrieved with an empirical function CWAVE_S1-IW, which combines spectral analysis of Sentinel-1A/B Interferometric Wide swath (IW) subscenes with wind data derived with common C-Band Geophysical Model Functions (GMFs). In total, 15 Sentinel-1A/B scenes (116 acquisitions) over the Baltic Sea were processed for comparison with off-shore sea state measurements (52 collocations) and coastal wind measurements (357 collocations). Sentinel-1 wave height was spatially compared with WAM wave model results (Copernicus Marine Environment Monitoring Service (CMEMS). The comparison of SAR-derived wave heights shows good agreement with measured wave heights correlation r of 0.88 and with WAM model ($r = 0.85$). The wind speed estimated from SAR images yields good agreement with in situ data ($r = 0.91$). The study demonstrates that the wave retrievals from Sentinel-1 IW data provide valuable information for operational and statistical monitoring of wave conditions in the Baltic Sea. The data is valuable for model validation and interpretation in regions where, and during periods when, in situ measurements are missing. The Sentinel-1 A/B wave retrievals provide more detailed information about spatial variability of the wave field in the coastal zone compared to in situ measurements, altimetry wave products and model forecast. Thus, SAR data enables estimation of storm locations and areal coverage. Methods shown in the study are implemented in NRT service in German Aerospace Center's (DLR) ground station Neustrelitz.

Keywords: SAR; Sentinel-1; wave height; wind speed; Copernicus; CMEMS; Baltic Sea

1. Introduction

1.1. Meteo-Marine Parameters in the Baltic Sea in Relation to Synthetic Aperture Radar

Space-borne Synthetic Aperture Radar (SAR), known for its independence of daylight and weather, can provide two-dimensional (2D) information about the ocean surface with global coverage [1,2]. It is due to the Bragg scattering of the short capillary waves in the dimension of centimeters, produced by wind stress, which allows extraction of wave and wind parameters from radar imagery [3–5].

The investigation of SAR ocean surface imaging mechanisms and the extraction of wave and wind parameters started with the launch of L-band SAR onboard SEASAT in 1978 [3,4]. Since then, numerous different algorithms have been developed over time to estimate oceanographic parameters and ocean wave spectra from SAR imagery [6–8].

The methods for sea state estimation are largely divided into two main groups; first one being functions where image spectra are transferred into wave spectra using transfer functions (e.g., [6,7,9,10]). These methods are suitable for estimations of swell's spectra, and its output can be assimilated into spectral wave models. The key to success is to understand the nonlinear SAR imaging of the moving sea surface waves that can be incorporated in "transfer functions" [9]. This approach requires SAR acquisitions with clearly visible wave patterns (e.g., Sentinel-1 Wave Mode (WM) data, high resolution Stripmap Mode TerraSAR-X data). Otherwise, the waves are substantially distorted and are not visible in the SAR images and thus are not represented in the image spectra.

The second group of sea state estimation algorithms use a direct estimation of the wave parameters from the image spectrum with empirical functions (e.g., [11–14]). Although empirical methods for C-band SAR exist, e.g., CWAVE_ERS and CWAVE_ENVI [12,14], they are only applicable to ERS-2 and Envisat-ASAR WM data. The most recent method for Sentinel-1 WM data by Stopa et al. [15] uses neural network techniques to retrieve wave parameters. However, since Sentinel-1A/B WM data is not available over the coastal areas of world ocean (including the Baltic Sea), moderate resolution Interferometric Wide (IW) swath mode images are used for sea state parameter retrieval. Short windsea waves produce unclear wave pattern in Sentinel-1 IW mode and are hardly distinguishable from ocean clutter. The SAR images are being affected by strong non-linear distortions due to the defocusing effects. Empirical functions, deduced from large sets of representative data, are proven to be more suitable for the short windsea waves and noisy images. The direct estimation of wave parameters from subscene spectra allows fast, straightforward, and reliable near-real-time (NRT) processing of satellite scene while excluding only a fragment of the data [13,16].

For the semi-enclosed micro-tidal Baltic Sea with the absence of long swell waves and short wave "memory" [17], and the significant wave heights remaining mostly between 0 and 2 m (rarely exceeds 4 m [18,19]), the second type of mentioned methods is recommended [13,20,21]. Windsea waves are short-crested and represent a considerable number of small, nonstable, fast, and erratically moving targets for a SAR sensor. Such sea state is typically imaged similar to noise with radar echoes of every scatterer blurred in azimuth and shifted randomly in range direction due to the individual Doppler contribution. The resulting pattern is hardly recognized as a wave pattern. A strong windsea contribution to the total wave height is therefore equivalent to more substantial uncertainties in SAR imaging.

With the launch of C-band Sentinel-1A/B constellation, different methods to estimate meteo-marine parameters, software realizations, and infrastructure open possibilities for NRT services for oceanographic applications [16]. As shortly as 10 min after image downlink, information about wave height, wind speed, as well as ice coverage, oil spills, and ship detection can be transferred to interested institutions or weather services [13,20,22–24].

Sentinel-1A/B data is already used worldwide for different applications. For example, estimating wave-induced orbital velocities from which elevation spectra is derived over ice-covered regions [25], calculating significant wave height and mean wave period from Sentinel-1A/B StripMap images using semi-empirical methods [26], or using neural network techniques on Sentinel-1 data to retrieve wave height [15].

SAR-based wave products have also proven to be valuable in the open ocean applications for swell tracking (e.g., [27,28]). In operational wave monitoring and forecasting, several organizations provide relevant information on wave conditions in the Baltic Sea: Baltic Operational Oceanographic System (BOOS), Copernicus Marine Environment Monitoring Service (CMEMS). However, the inclusion of Sentinel-1 wave products over the Baltic Sea into the CMEMS product portfolio would improve the service quality which currently provides only model wave forecast, altimetry wave products, and in situ data [29–31].

1.2. Sentinel-1A/B Data over the Baltic Sea

The Baltic Sea is situated in temperate latitudes between 53°N to 66°N and from 9°E to 30°E which makes it one of the most frequently imaged locations by the Sentinel-1 satellites. Various parts of the Baltic Sea are imaged by Sentinel-1A/B daily and often even twice a day by ascending and descending orbits in the morning and in the evening correspondingly. The most suitable Sentinel-1A/B relative orbit numbers are 22 paired with 29, 51 with 58, and 124 with 131 (Figure 1).

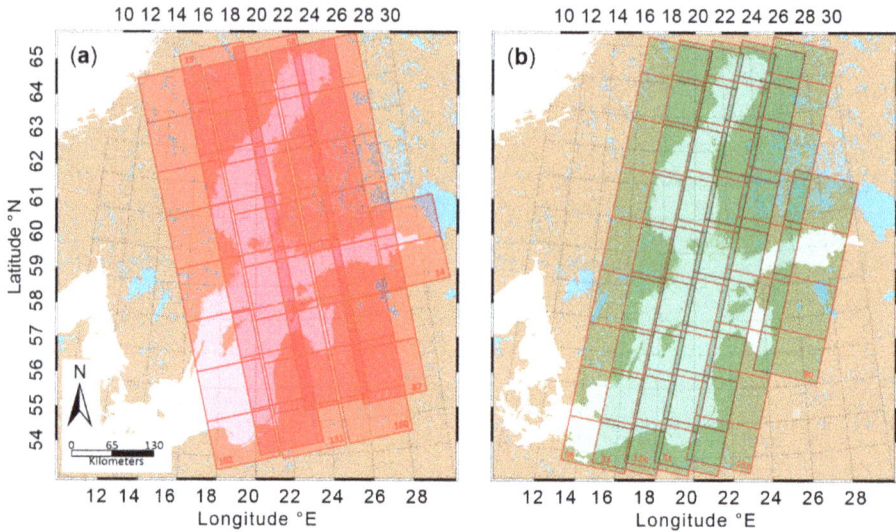

Figure 1. Sentinel-1A/B IW relative orbit overlays and corresponding orbit numbers over the Baltic Sea. (**a**) ascending/morning orbits and (**b**) descending orbits in the evenings.

Similar usability of satellite SAR data in the Baltic Sea was available when Envisat/ASAR (Advanced Synthetic Aperture Radar) was operational. With the launch of Sentinel-1A/B constellation and the freely available data on the Copernicus Open Access Hub, all the services can be continued. Different methods can be applied on the images to estimate meteo-marine parameters in the Baltic Sea for operational maritime awareness applications. The Sentinel-1 IW level-1 products have 250 km wide swath with 10 m pixel resolution to cover the length of the Baltic Sea with sequential SAR acquisitions.

1.3. Aim of the Study

The main purpose of this study is to assess current state-of-the-art method in estimating meteo-marine parameters, such as wind speed or total significant wave height, in the Baltic Sea from medium resolution Sentinel-1A/B IW swath mode satellite radar imagery. The main advantages of the method as well as challenges are also brought out. The study focuses on the possibilities of making the method available as a near-real-time service over the Baltic Sea using three examples of different sea state in comparison to spectral wave model and available in situ measurements.

The specific objectives of the study are: (i) to validate CWAVE_S1-IW wave retrievals in the Baltic Sea; (ii) to validate CMOD wind speed retrievals in the coastal zone of the Baltic Sea; (iii) to demonstrate potential of Sentinel-1A/B SAR wave retrievals with CWAVE_S1-IW algorithm for operational monitoring in coastal area.

2. Data

2.1. In Situ Data

Wind measurement data from 39 coastal stations (357 collocations with SAR data) around the Baltic Sea were used for statistical validation of Sentinel-1 wind retrievals (Figure 2). Sentinel-1 SAR sea state retrievals were validated with in situ wave measurements from 5 offshore stations (52 collocations with SAR data) (Table 1 and Figure 2).

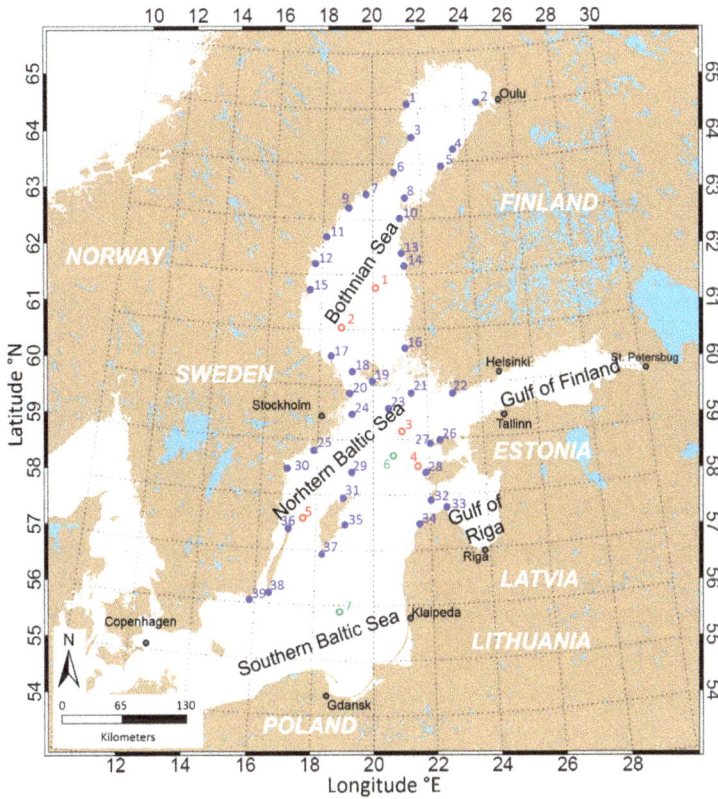

Figure 2. The map of the Baltic Sea and locations of measurement stations used in the study. The location of wave measurements—significant wave height, wave propagation direction, wave period (red), and coastal wind measurements—speed, gusts, direction (blue) are indicated on the map; green marks extra stations (virtual buoys).

Table 1. Overview of wave measurements used in the study. H_S represents total significant wave height.

No. (Origin)	Station	Lat (°N)	Lon (°E)	Sensor	Data Used
1 (FIN)	Selkämeri	61.8001	20.2327	Waverider	H_S
2 (SWE)	Finngrundet	61.0000	18.6667	Waverider	H_S
3 (FIN)	NBP	59.2500	20.9968	Waverider	H_S
4 (EST)	Vilsandi	58.4889	21.6333	Waverider	H_S
5 (SWE)	Knolls grund	57.5167	17.6167	Waverider	H_S
6	NBP Extra	58.7500	20.8271	Virtual buoy	H_S
7	Södra Östersjön	55.9167	18.7833	Virtual buoy	H_S

2.2. Sentinel-1A/B Data

The C-band SAR data from Sentinel-1A/B, namely IW mode data, are used for estimation of meteo-marine parameters in this study. The IW mode allows combining large swath width of 250 km in range direction with moderate geometric (5 × 20 m) resolution. Sentinel-1A/B products are available in single (HH or VV) or dual polarizations (HH+HV, VV+VH). For the meteo-marine parameter estimation, either one of the single polarization data is used. The Normalized Radar Cross Section (NRCS) σ_0 is firstly processed from image pixel digital number (DN):

$$\sigma_0 = \frac{DN^2}{k_s^2} \tag{1}$$

where k_s is the calibration factor given in the products metadata. The process of estimating sea state parameters is based on FFT (Fast Fourier Transform) of the subscene. Before the analysis, each pixel value $\sigma_0(x, y)$ of the subscene is normalized resulting in a value $\sigma_n(x, y)$:

$$\sigma_n = \frac{\sigma_0(x, y) - \sigma_0}{\sigma_0} \tag{2}$$

where σ_0 is the mean value of σ_0 in the subscene.

Images were processed with a 3 nautical mile grid with the FFT window of 1024 × 1024 pixels with four-factor resampling and Gaussian smoothing. The processing was implemented for latitudes up to 65°N.

Although all the Sentinel-1A/B IW scenes (460 scenes) over the Baltic Sea from the beginning of 2015 until the end of 2016 were processed, only 15 overpasses (number of acquisitions per satellite overpass ranged from 5 to 9) were selected for validation of the meteo-marine parameter retrieval method as well as for analysis and comparison. All the selected data in Table 2 were acquired in VV polarization. The SAR data for validation were selected to have equal representation of different meteo-marine conditions (i.e., high and low sea states) (Table 2).

Table 2. Sentinel-1A/B acquisitions used for the study. Relative orbit numbers with acquisitions per scene are listed. Mean and maximum significant wave height and wind speed calculated with the methods described in Section 3 are shown with the number of collocations per overpasses.

Sentinel-1 UTC	Relative Orbit no.	Images in Scene	Mean/Max H_S per Scene	Mean/Max U_{10} per Scene	Collocations (Wave/Wind)
11 January 2015 16:19	29	6	2.4/7.5	9.0/18.7	3/11
22 April 2015 16:28	102	6	0.3/1.7	2.6/11.8	2/3
04 June 2015 05:04	22	9	0.9/2.8	5.3/14.0	4/32
11 June 2015 04:56	124	9	0.6/2.1	4.1/11.5	5/28
25 June 2015 04:56	124	8	0.8/1.8	5.1/13.8	5/27
28 June 2015 05:04	22	9	0.6/2.2	3.4/8.7	4/25
05 July 2015 04:56	124	9	0.7/2.5	4.7/12.4	4/22
28 July 2015 04:56	124	9	0.9/2.4	6.6/14.4	5/25
08 August 2015 05:04	22	9	1.7/2.9	10.8/16.0	4/31
08 September 2015 16:19	29	6	1.3/2.7	8.5/17.3	3/19
02 October 2015 05:04	22	9	1.8/3.6	11.6/19.1	4/32
02 October 2015 16:19	29	5	2.5/4.8	13.2/18.2	1/16
02 November 2015 04:56	124	9	1.6/2.4	10.7/17.1	5/32
09 August 2016 16:19	29	6	1.9/3.7	9.7/13.8	1/28
14 December 2016 04:56	124	7	1.4/2.6	8.6/13.9	2/26
15		116			52/357

2.3. Spectral Wave Model

The wave model WAM [32] is a third-generation wave model which solves the action balance equation without any a priori restriction to the evolution of spectrum. The action density spectrum N is considered instead of the energy density spectrum E because in the presence of ambient currents, action density is conserved, but energy density is not. Action density is related to energy density through the relative frequency [33]:

$$N(\sigma, \theta) = \frac{E(\sigma, \theta)}{\sigma} \tag{3}$$

The variable σ is the relative frequency (as observed in a frame of reference moving with the current velocity) and θ is the wave direction (the direction normal to the wave crest of each spectral component). The action balance equation in Cartesian coordinates reads:

$$\frac{\partial N}{\partial t} + \left(\vec{c}_g + \vec{u} \right) \nabla_{x,y} N + \frac{\partial c_\sigma N}{\partial \sigma} + \frac{\partial c_\theta N}{\partial \theta} = \frac{S_{wind} + S_{nl4} + S_{wc} + S_{bot}}{\sigma} \tag{4}$$

On the left-hand side of Equation (4) the first term represents the local rate of change of action density in time; the second term denotes the propagation of wave energy in two-dimensional geographical space, where $\vec{c}_g$ is the group velocity and $\vec{u}$ is the ambient current. The third term represents shifting of the relative frequency due to variations in depths and currents (with propagation velocity c_σ in σ space). The fourth term represents depth-induced and current-induced refraction (with propagation velocity c_θ in θ space). At the right-hand side of the action balance equation is the source term that represents physical processes which generate, redistribute, or dissipate wave energy in the WAM model. These terms denote, respectively, wave growth by the wind S_{wind}, non-linear transfer of wave energy through four-wave interactions S_{nl4} and wave dissipation due to whitecapping S_{wc} and bottom friction S_{bot}.

A pre-operational version of the WAM model which is since April 2017 used for the production of CMEMS wave forecast over the Baltic Sea was used [29]. The model domain covers the Baltic Sea with a grid resolution of one nautical mile, yielding 800 × 775 model grid points. The model was forced with High Resolution Limited Area Model (HIRLAM) winds with a spatial resolution of 11 km and temporal resolution of one hour. In winter, ice concentration data from the Finnish Meteorological Institute's Ice Service was used. Model grid points in which the ice concentration exceeds 30% are excluded from the calculation. Data assimilation was not used in the wave model.

3. Methods

3.1. Wind

Sea state is strongly dependent on local wind characteristics which SAR data can provide. By analyzing the roughness of the sea, wind speed is received using Geophysical Model Functions (GMF) which relate the local wind conditions and sensor geometry to radar cross section values.

For Sentinel-1 IW data, separate GMFs are used for HH or VV polarizations. For HH, CMOD4 function, developed by Stoffelen et al. [34] is used, and for VV polarization CMOD5.N algorithms shows the best results [35]. The selection of the respective GMF is based on an extensive comparison of GMF performance in comparison with an advanced scatterometer (ASCAT), METOP-A, and METOP-B satellite data performed by [36]. As stated in [36], Thompson polarization ratio [37] with $\alpha = 1$ is applied to HH polarized data. Also following the authors' suggestion, a bias of 0.004 is subtracted from VV polarized data, to achieve an overall better agreement with scatterometer data. In total, an accuracy of approximately 1.5 m s^{-1} has been found in the comparison with the ASCAT data within the validity range of 2–25 m s^{-1} of the two GMFs [36]. In the current processing procedure, no information from the cross-polar channel is exploited, although a future application of a respective GMF as e.g., proposed by [38] is foreseen to improve wind data reliability in storm situations. The data analyzed in this paper is entirely in the validity range of the applied GMFs for co-polar channels.

In the common procedure, GMFs in general and thus also the CMOD algorithms are inversion methods and require the local wind direction to reduce the number of free parameters in the forward calculation. For the work presented in this paper, wind direction from Weather Research and Forecasting Model (WRF) is used [39]. The model is run for the given area and time of the data acquisition. Initial and boundary conditions are adopted from the corresponding National Oceanic and Atmospheric Administration Global Forecast System (NOAA GFS) analysis model values. For NRT applications, NOAA GFS Forecast values are used instead and the model is run shortly prior to satellite data downlink with a configuration based on the scene parameters (region and time) available in the data processing system schedule. Finally, WRF model values for the wind direction are interpolated to the sea state calculation grid and wind speeds are calculated directly within the sea state algorithm procedure for a given subcell.

3.2. Sea State

An empirical algorithm CWAVE_S1-IW, developed by Pleskachevsky et al. [20], is used to estimate integrated sea state parameters straight from SAR image spectra without transformation into wave spectra. The method is chosen since traditional functions (image spectrum transfer to wave spectrum) are not able to calculate total significant wave height from Sentinel-1 IW mode imagery in the Baltic Sea. The main reasons are the relatively coarse resolution of Sentinel-1A/B and generally lower sea state without long swell compared to the open ocean.

In comparison to e.g., TerraSAR-X/TanDEM-X StripMap scenes with about 3 m resolution, the Sentinel-1A/B IW mode resolution is by an order of magnitude larger. In case of such Sentinel-1 SAR imaging setting the wave structures, if visible, are disturbed by the vast amount of noise. In addition, a standard FFT window of 1024 × 1024 pixels covers a relatively large area of 10240 × 10240 m. To overcome the limitation, four-factor resampling and Gaussian smoothing were applied to selected subscenes. The modified resolution becomes to 2.5 m with areal coverage of 2560 × 2560 m [20].

An important part of sea state estimation is pre-filtering of any natural or man-made objects from subscene which yields to inaccuracies in wave height estimation. Such spectral perturbations result in an integrated value which leads to the total image energy not connected to the sea state. The radar signal disturbances can be divided into two main groups:

- radar signal much stronger than background backscatter from sea state produced mainly by ships or offshore constructions. In these cases, the subscene is additionally analyzed with 100 × 100 m sliding window. The statistics of each window σ_0^{win} is compared with σ_0 of the subscene. In a case of $\sigma_0^{win} > q_{ship}\sigma_0$ with tuned q_{ship} value of 2.3 (for 100 × 100 m window), the outliers in the current window are replaced with the mean value of the subscene σ_0 [20];
- radar signal much weaker than background backscatter from sea state produced, for example, by oil spills, or commonly occurring algae blooms in the Baltic Sea [20]. In those cases, the filtering algorithm was extended by employing $\sigma_0^{win} > q_{spills}\sigma_0$ with tuned threshold coefficient q_{spills}.

To obtain integrated wave parameters, FFT operation is applied to the radiometrically calibrated subscene. Image power Spectrum $IS(k_x, k_y)$ is calculated by integration over 2D wavenumber domain:

$$E_{IS} = \int_{k_x^{min}}^{k_x^{max}} \int_{k_y^{min}}^{k_y^{max}} IS(k_x, k_y)dk_x dk_y \tag{5}$$

The integration over wavenumber domain is limited by $k_{max} = 0.003$ and $k_{min} = 0.201$ which correspond to wavelength of 2000 to 30 m, where wavenumber $k = \sqrt{k_x^2 + k_y^2}$. In the Sentinel-1A/B image spectra the wavenumber domain ~0.201 < k < 0.060 represents the clutter produced by waves shorter than about 100 m. The domain ~0.060 < k < 0.010 represents long waves with wavelength of ~100 < L_p < 600 m, and the domain ~0.010 < k < 0.003 represents the longest structures such as wind streaks [20].

During the algorithm's development, it became clear that estimating sea state parameters based only on image spectral properties is not accurate enough. Additional information about each subscene is therefore acquired by using Grey Level Co-occurrence Matrix (GLCM) [40]. By using image texture analysis, accuracy in low and high sea state was improved [20].

The resulting function CWAVE_S1-IW for Sentinel-1A/B imagery to calculate total significant wave height is expressed as:

$$H_S^{XWAVE\_C} = a_0 \sqrt{B_0 E_{IS} tan(\theta)} + \sum_{i=1}^{n} a_i B_i \tag{6}$$

where θ is local incidence angle, a_i are coefficients, and B_i are functions of spectral parameters, wind and GLCM results.

The first term in Equation (6) connects the sea state and image spectra energy which contributes the most in the case of long prominent waves with over 100 m wavelengths. The non-linearity of the imaging mechanism is represented by the B_0, which represents noise scaling of total image spectrum energy E_{IS}. The relation $B_0 = KE_{IS}^{100}/E_{IS}^{600}$, where K serves as a constant found by collocating buoy data, connects the spectrum energy between the wavelength domain of 30–100 m (noisy part of the image spectrum) with the wavelength domain of 100–600 m (the area where wave-looking patterns can be observed). The rest of the terms in Equation (6) represent a series of corrections and filtering of different origins. For example, to consider the wind speed, the term $a_1 B_1$, where $B_1 = U_{10}$, is used. Full information about the function development, tuning and results can be found in [20].

3.3. Comparison Methods

The total significant wave height H_S and wind speed U_{10} derived from SAR are used for comparisons with collocated in situ measurements. The Sentinel-1A/B scenes were processed with 3×3 nautical miles posting with ~30×45 = ~1350 subscenes per IW image. The collocations were done for five Sentinel-1A/B scenes with a time window of $\pm$20 min and almost 30 min for one case. For the rest of the nine cases, the time difference between the measurements or WAM model data and SAR-derived values is less than 5 min (Table 2). For the spatial collocation, the closest SAR-subscenes are used with a mean value between subscene centre and measurement equipment location or WAM wave model grid point being 4.1 km and 0.7 km, correspondingly. In case the buoy location remains outside the image, the results from the closest subscene to the SAR acquisition edge in the range of up to 10 km was incorporated.

In the case of the wind speed comparison, the average distance between in situ measurement location and the closest subscene centre is 7.7 km. The reason is that the majority of the stations are at the coast (Figure 2) and the SAR subscenes which are close to the shore (contaminated by land backscatter) are filtered out. The time difference remains the same as for wave height comparison, mostly below 5 min.

The Root Mean Square Error (RMSE), Pearson correlation coefficient r, and Scatter Index (SI) (where SI = RMSE/(average of observations)) are calculated for each collocated dataset for the statistical comparisons. Standard deviation (STD) is used to measure variabilities of datasets. All collocated data are presented in scatterplots for wave height and wind speed.

4. Results

4.1. Validation

The inter-comparison and the scatter plots in Figure 3 show a good general agreement of SAR wave retrievals and WAM model fields with in situ wave measurements. The corresponding correlation coefficients are 0.88 to 0.89 (Table 3). Also, the RMSE of SAR-derived wave heights and WAM model wave heights are very similar, 0.40 m and 0.39 m correspondingly (Table 3). Slightly poorer statistics ($r = 0.81$ and RMSE = 0.47 m) are observed when the SAR wave is compared with WAM model data (based on 52 collocated observations), which indicates that SAR and model data resolve distinct aspects of the observed wave parameters. Therefore, SAR and model data could both provide complementary information for accurate description of the wave field. The benefits of multiple data sources for understanding wave field variations are discussed in Sections 4.2 and 5.1 based on characteristic examples of wave conditions.

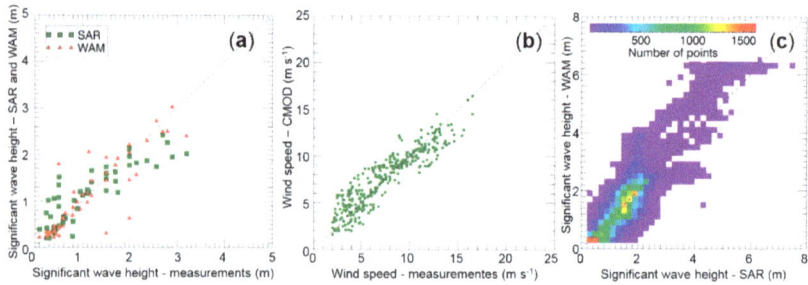

Figure 3. (**a**) Scatterplot for sea state for available collocated data acquired over the Baltic Sea including 15 Sentinel-1A/B scenes (overflights/events/days) with 116 individual Sentinel-1 IW mode images and 52 buoy collocations. The correlation coefficient between SAR and in situ measurements is 0.88, RMSE is 0.40 m, and Scatter Index is 0.37. (**b**) Scatterplot for surface wind speed for all available collocated data acquired over the Baltic Sea. The correlation coefficient *r* is 0.91, RMSE is 1.43 m s^{-1}, and SI is 0.19. (**c**) Histogram plot for all the collocated SAR versus WAM results. The bin size for histogram calculations is 0.2 m. The statistics between the datasets are as follows: *r* = 0.86, RMSE = 0.47 m, and SI = 0.33.

Table 3. Overview of inter-comparison of significant wave height and wind speed: correlation coefficient (*r*), root mean square error (RMSE), scatter index (SI), and number of collocations (*n*). The values in brackets in the 3rd column represent the statistics when all collocated data of Synthetic Aperture Radar (SAR) and wave model (WAM) wave fields were used (49,315 colocations).

Parameter	SAR vs. In Situ Wave Height	SAR vs. WAM Wave Height	SAR vs. In Situ Wind Speed	WAM vs. In Situ Wave Height
r	0.88	0.81 (0.86)	0.91	0.89
RMSE	0.40	0.47 (0.47)	1.43	0.39
SI	0.37	0.42 (0.33)	0.19	0.36
n	52	52 (49314)	357	52

Scatter plot on Figure 3b shows the collocated in situ data comparison with estimated wind speed results from the corresponding CMOD algorithm. The wind speed varied from 2 m s^{-1} to 17 m s^{-1} with the mean wind speed value of all 357 collocations being 7.53 m s^{-1}. The correlation coefficient between SAR wind retrievals and coastal wind speed measurements was 0.91 (Table 3).

Figure 3c shows the wave height histogram plot of all 49314 SAR-derived values and the corresponding WAM results. Figure 3c clearly indicates that most values are around 1 m. The statistics between the two methods in the case of a larger dataset (49,314 collocations) is slightly better compared to the dataset that was collocated with 52 observations—*r* = 0.86 and RMSE = 0.47 m (Table 3).

4.2. Case Studies: High, Medium, and Low Sea State

A high sea state example from 11 January 2015 (16:19 UTC) is presented in Figure 4a–c. Considering the general Baltic Sea wave conditions, high significant wave height values (up to 7.5 m) were observed along the Polish and Lithuanian coasts. Both the SAR-derived results and WAM model field show good general agreement in the wave height values and location of maximum (*r* = 0.91). The area of the storm on the SAR image is smaller and does not spread as much to the north as in the WAM results. The maximum significant wave height from SAR is about 0.5 m higher

(Figure 4b, Table 4). Another region with some differences between the wave fields retrieved with the two different methods is seen in the Bothnia Sea area, where SAR-derived wave height along the Swedish coast is about two meters lower compared to the model data.

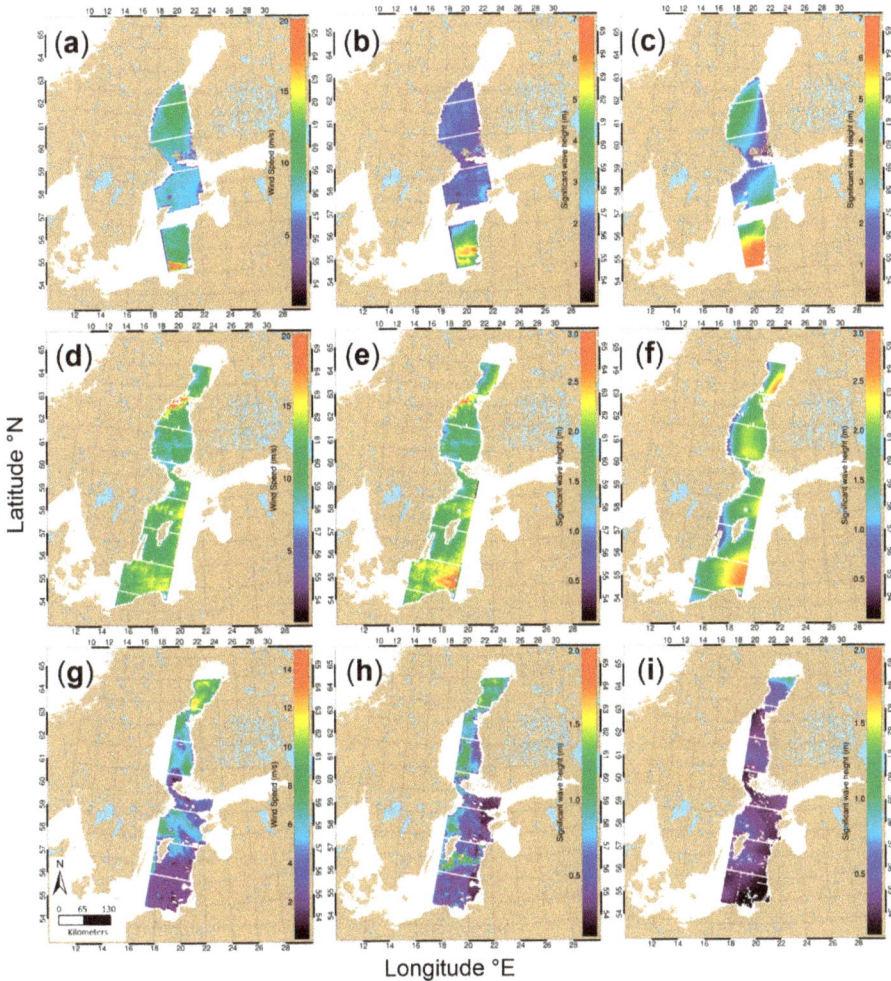

Figure 4. Examples of spatially collocated SAR wind (**a,d,g**) fields, SAR wave fields (**b,e,h**) and WAM wave fields (**c,f,i**) during three characteristic situations over the Baltic Sea: high sea state on 11 January 2015 at 16:19 (**a–c**), medium sea state on 2 October 2015 at 04:56 (**d–f**) and low sea state on 5 July 2015 at 04:56 (**g–i**)).

The second example depicts medium sea state conditions in the Baltic Proper and the Gulf of Bothnia on 2 October 2015 (Figure 4d–f). Again, a good general match between wave fields (considering the wave height and spatial pattern) estimated from SAR data and the WAM model outcome can be observed. SAR-derived wave height field is more variable (STD from 1.14 m to 1.51 m) than the model field (STD from 0.17 m to 1.48 m), which is the case for all examples (Table 4). Also, there are some differences in the wave field pattern along the Swedish coast where the wave height is underestimated by WAM model data compared to SAR-derived fields.

Table 4. Statistics between Sentinel-1 H_S retrievals and WAM numerical model outputs.

Time UTC	Variable	Sentinel-1	WAM
11 January 2015 16:19:22 High sea state	Mean (m)	2.41	3.02
	Maximum (m)	7.47	6.97
	STD (m)	1.51	1.48
	r		0.91
	RMSE (m)		1.02
02 October 2015 05:04:47 Medium sea state	Mean (m)	1.82	1.68
	Maximum (m)	3.62	2.65
	STD (m)	1.32	0.57
	r		0.51
	RMSE (m)		0.39
05 May 2015 04:56:28 Low sea state	Mean (m)	0.57	0.33
	Maximum (m)	1.84	1.02
	STD (m)	1.14	0.17
	r		0.51
	RMSE (m)		0.41

Most commonly occurring, the low sea state [19] example on 5 July 2015 over the Baltic Sea (H_S is around 1 m) is presented in Figure 4g–i. Although WAM wave model results are smoother and lower than SAR-derived values, they represent a very similar large-scale pattern. One can notice the increased wave height values to the north and to the south of Gotland Island. A similar pattern from both datasets is also observed in Bothnia Sea region. The low sea state conditions might not be the most relevant from a safe navigation point of view and operational monitoring/forecasting of the wave conditions is not as critical as during storm conditions. Nevertheless, it is still relevant for routine environmental monitoring and therefore noteworthy that during the low sea state, most of the wave field variability is lost in the model outcome compared to SAR-derived values. A similar example from TerraSAR-X satellite data is presented in Rikka et al. [21], where during low sea state conditions the local wave height increases by 0.5–1 m in kilometre-size "islands" (small local area with elevated wave height values). In Sentinel-1A/B (Figure 4g–i), the size of the observed "island" is larger due to larger SAR resolution and processing grid step which does not allow retrieving such fine scale variations as in the case of TerraSAR-X data. Similarly, Romeiser et al. [41] showed that in hurricane situations, the wavelength is analogously retrieved in island-like fashion from C-band satellite radar.

The case studies showed good general agreement between the SAR-derived and WAM model wave fields. However, there are some differences between the results obtained with the two methods: (i) the area and the location of the storm might be different; (ii) the wave height variability of WAM model fields is lower compared to the SAR-derived fields. The variation in WAM model fields is lost mostly due to wind forcing fields (HIRLAM) used in the wave modelling which have 1 h temporal resolution and 11 km spatial resolution. Therefore, the forcing fields do not include local fine-scale wind field variations and gusts that influence the radar backscatter and related wave field pattern on SAR imagery.

5. Discussion

The current study, as well as previous studies [13,15,21–25] demonstrate the advantages of SAR data in general and Sentinel-1 A/B IW data in particular for the operational sea state monitoring (downstream) services. The meteo-marine parameters derived from Sentinel-1 A/B IW data provide added value to operational monitoring/forecasting services (NRT open source data with high spatial resolution and large spatial coverage; frequency of acquisitions) and statistical analysis (large dataset with sufficient spatial coverage in the coastal zone). Together with the unrestricted access to operational in situ data collected by various Baltic Sea countries and model forecast (e.g., CMEMS, BOOS), the SAR-derived meteo-marine parameters form a basis for improving maritime situation awareness. Furthermore, other applications in the Baltic Sea region, e.g., oil spill detection (impact of wave-wind conditions on detection accuracy), sea ice monitoring (waves under ice), wave-circulation coupling [42], etc. will benefit from incorporation of the Sentinel-1 A/B sea state products in these service chains [29,43].

5.1. Benefits of Sentinel-1A/B IW Wave Field Data for Operational Services

An independent time series from 1 August 2016 until the end of 2016 from four separate locations demonstrate the benefits of using SAR-derived significant wave height retrievals (Figure 5).

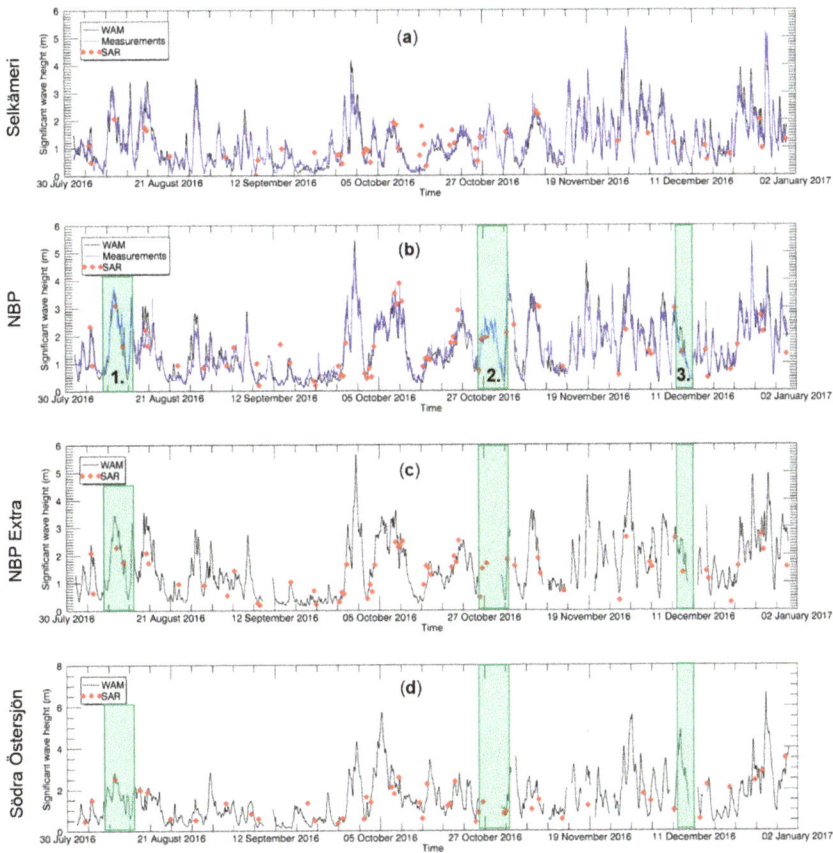

Figure 5. A timeseries from 1 August 2016 until the end of 2016 from four stations. Two stations—Selkämeri and NBP (Figure 2, Table 1)—include all the data: measurement, WAM, and SAR-derived results; other two stations—NBP Extra (58.7500°N, 20.8271°E; no. 6 in Figure 2) and Södra Östersjön (55.9167°N, 18.7833°E; no. 7 in Figure 2) include WAM result and SAR-derived significant wave height. Highlighted areas indicate some benefits of using SAR data over the Baltic Sea: "case 1" and "case 3" bring out the variability aspect of SAR-derived values whereas "case 2" shows missing measurements that can be replaced with SAR data.

In areas where average significant wave height is very low, for example, Selkämeri station in Figure 5a, the SAR-derived results ($r = 0.79$) are not as accurate as the results over the open part of the

Baltic Sea in the NBP station (r = 0.92) on Figure 5b. It is known from previous studies that dominant wave height in the Baltic Sea is around 1 m [19] and relatively low spatial resolution of Sentinel-1 IW mode data might complicate the accurate wave height estimation.

In Figure 5b,c three cases highlighted in green are brought out to explain the benefits of using SAR data. In "case 1" of Figure 5b, one can observe that both WAM wave model results and SAR-derived results match closely with the in situ measurements of the NBP station. However, in Figure 5c which represents a location 60 km away from NBP, a mismatch between SAR and WAM results can be seen in the "case 1" region. The reason could be that since SAR represents better detailed spatial variability/pattern, the actual significant wave height was lower than WAM had predicted at the specific time and location. "Case 3" in Figure 5b,c shows good general match between in situ measurements, SAR-derived wave height, and WAM output in separate places, suggesting that the wave field was spatially more uniform. In general, SAR-derived results could be used as validation data for wave models.

Since the Baltic Sea is seasonally ice-covered, in situ measurement devices are removed for the winter period. Similarly, when the buoys have technical problems (e.g., no data connection) or during their maintenance, highly valuable information is lost. Moreover, wave models may also have short periods with technical problems when no wave forecast is provided. These situations can be observed in "case 2" in Figure 5c, where SAR-derived results become the only source of wave information.

Figure 5d demonstrates the added benefit of using SAR data to retrieve wave information over the poorly sampled area. Although Södra Östersjön station (55.9167°N, 18.7833°E) is included into BOOS measurement stations, the last unrestricted access measurement data was received in 2011. However, Southern Baltic Sea is a location where the highest waves occur [18,19]. As no in situ measurements are carried out in the region, the SAR-derived results would be highly valuable for model validation and/or assimilation into the wave model.

5.2. Statistical Mapping of Coastal/Regional Wave Field: Comparison with Altimetry

Although Sentinel-1A/B are not able to cover the extent of the Baltic Sea (or any sea in that matter) as frequently as wave models can, the SAR data can be as valuable as any other satellite-based wave product (e.g., altimetry products). Altimetry products validations have shown reliable performance (RMSE less than 0.5 m) in the open ocean [44–47] and in the coastal sea (RMSE up to 0.37 m) [48–52]. However, the spatial coverage of altimetry products is limited and restricted to offshore areas (30–70 km from coast) [53]. The low-resolution altimetry wave products/algorithms and open ocean SAR wave mode products (not available for the coastal areas, including Baltic Sea) are not sufficient for local and regional applications in the complex coastal environment, such as Baltic Sea. The sea state products derived from Sentinel-1 SAR IW data provide information over a large area, including the coastal zone with similar product accuracy (r = 0.88, RMSE = 0.40 m, Table 3) to the altimetry products. Thus, the high-resolution SAR wave data would provide added value for user communities dealing with coastal processes. Moreover, SAR wave products enable to resolve detailed spatial variability while in situ data describes detailed temporal variability in a limited number of locations (Figures 4 and 5).

Besides NRT, an example of SAR data benefits is the statistical analysis of wave conditions (e.g., wave climate). Figure 6 represents the average wind speed and significant wave height values from Sentinel-1A/B IW data over the 2015 and 2016 interpolated onto WAM wave model grid. The average significant wave height values over the two-year period (Figure 6c) generally represent similar values to previous studies that used either model data reanalysis or altimetry products over a longer period (up to 23 years) (e.g., Figure 6 in Tuomi et al. [19]; Figure 2 in Kudryavtseva et al. [54]). There are clearly higher average wave height values in the open parts of the Baltic Sea (around 1.8 m) and lower values in the Gulf of Riga (up to 1.0 m in the open part; below 0.8 m in the coastal areas) or the Bothnian Sea (from 0.7 m to 1.2 m). However, from Figure 6a we can conclude that more than 100 points would be necessary for calculating average values since a limited number of samples may cause artificial features and improbable wind speed/wave height fields (e.g., in Southern Baltic Sea

(Figure 6b). Considering the long-term objectives of the Copernicus program and the revisit cycle of the Sentinel-1 mission, the statistical bases for wave mapping will improve over time.

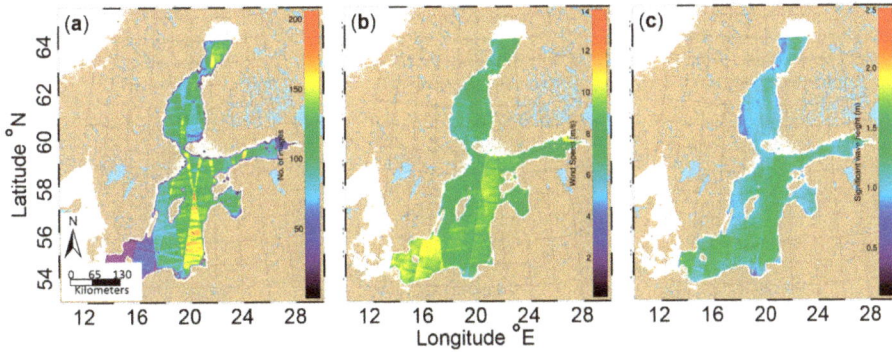

Figure 6. (a) Number of SAR points; (b) average wind speed; and (c) average significant wave height over 2015–2016 interpolated to WAM model grid.

Compared to altimetry, SAR data have a benefit of much higher resolution and larger coverage. For example, Kudryavtseva et al. [54] calculated average significant wave height maps for the Baltic Sea over the period of 23 years using approximately 660,000 data points with the end-resolution of $0.2 \times 0.1°$. To retrieve the analogous map from SAR data, over 3 billion data points can be obtained from a two year period (using the 3 nm processing step). Furthermore, SAR data provides much greater detail, especially in the coastal zones where the vicinity of the coastline influences the altimetry signal and the related wave height retrievals.

6. Conclusions

A method for sea state parameter and marine wind estimation from Sentinel-1 IW SAR imagery in the Baltic Sea (proposed by Pleskachevsky et al. [20]) was validated. The sea state parameters were retrieved from image spectrum using an empirical algorithm CWAVE_S1-IW to estimate integrated sea state parameters directly from SAR image spectra without transformation into wave spectra. The study shows that wave field retrievals from Sentinel-1 IW SAR data correlate with in situ data ($r = 0.88$, RMSE = 0.40) as well as with WAM wave model ($r = 0.86$, RMSE = 0.47). Furthermore, the wind speed retrievals that were derived with the CMOD algorithm correlated with the values recorded at the coastal meteorological stations ($r = 0.91$, RMSE = 1.43).

The advantages of Sentinel-1 SAR IW mode wave products in Baltic Sea were demonstrated. The free/open data, high spatial resolution, large spatial coverage, and frequent acquisitions of Sentinel-1 A/B IW images leads to the following improvements that Sentinel-1 can offer to operational wave field monitoring and forecasting: (i) improved description of spatial variability of significant wave height; (ii) improved estimation of the area and the location of the storms during high sea state.

Considering the advantages, the operational wave product retrieved from Sentinel-1 A/B IW mode data by a dedicated algorithm for the coastal ocean (including Baltic Sea) would be valuable for many communities dealing with wave modelling, operational monitoring, and forecasting, etc. The SAR wave field retrievals would improve the downstream of monitoring services by improving the forecast accuracy, thus enabling a better understanding of coastal processes.

This work contributes to the uptake of Sentinel-1 A/B IW data in the fully automated operational service for meteo-marine parameter retrieval in the Baltic Sea. Implementation of Sentinel-1 sea state products for assimilation into an operational wave model and usage for model forecast quality checking would improve general marine awareness. All the SAR processing methods presented in the study are running as NRT services in the German Aerospace Center's (DLR) ground station, Neustrelitz.

Author Contributions: S.R. was responsible for designing the study, processing and analysis of SAR data, interpretation of the results and writing the manuscript; A.P. was responsible for CWAVE_S1-IW method development and contributed in data analysis; S.J. was responsible for method development and contributed in writing the manuscript; V.A. performed wave model experiments and contributed in writing the paper; R.U. contributed in interpretation of results and writing the paper.

Acknowledgments: The authors express the gratitude towards ESA for making Sentinel constellation data freely available. Special thanks to Finnish Meteorological Institute (FMI), Swedish Meteorological and Hydrological Institute (SMHI), Estonian Environmental Agency (KAUR), and Latvian Environment, Geology and Meteorology Centre that have open data policy for wind and wave measurement data. We would also like to thank colleagues T. Kõuts, U. Lips, and K. Vahter at Department of Marine Systems at TUT for providing wave and wind measurement data. The authors are grateful for the DLR ground station Neustrelitz team for continuous cooperation and organization of the NRT services for the users. The corresponding author gratefully acknowledges the European Regional Development Fund for financial support for the stay in DLR's SAR Oceanography group. The author would also like to thank for the warm welcome by the SAR Oceanography group and the exceedingly qualified knowledge shared by them in various SAR-related subjects. The study was supported by institutional research funding IUT (19-6), by Personal Research Funding PUT1378 of the Estonian Ministry of Education and Research, by the European Regional Development Fund and through CMEMS Copernicus grant WAVE2NEMO.

Conflicts of Interest: The authors declare no conflict of interest.

References

1. Lehner, S.; Schulz-Stellenfleth, J.; Brusch, S.; Li, X.M. Use of TerraSAR-X data for oceanography. In Proceedings of the 2008 7th European Conference on Synthetic Aperture Radar (EUSAR), Rome, Italy, 26–30 May 2008; pp. 1–4.

2. Li, X.; Lehner, S.; Rosenthal, W. Investigation of ocean surface wave refraction using TerraSAR-X data. *IEEE Trans. Geosci. Remote Sens.* **2010**, *48*, 830–840.

3. Beal, R.C.; Tilley, D.G.; Monaldo, F.M. Large-And Small-Scale Spatial Evolution of Digitally Processed Ocean Wave Spectra from SEASAT Synthetic Aperture Radar. *J. Geophys. Res. Oceans* **1983**, *88*, 1761–1778. [CrossRef]

4. Masuko, H.; Okamoto, K.I.; Shimada, M.; Niwa, S. Measurement of Microwave Backscattering Signatures of the Ocean Surface Using X Band and Ka Band Airborne Scatterometers. *J. Geophys. Res. Oceans* **1986**, *91*, 13065–13083. [CrossRef]

5. Schulz-Stellenfleth, J. *Ocean Wave Measurements Using Complex Synthetic Aperture Radar Data*; University of Hamburg: Hamburg, Germany, 2004.

6. Hasselmann, K.; Hasselmann, S. On the Nonlinear Mapping of an Ocean Wave Spectrum into a Synthetic Aperture Radar Image Spectrum and Its Inversion. *J. Geophys. Res. Oceans* **1991**, *96*, 10713–10729. [CrossRef]

7. Hasselmann, S.; Brüning, C.; Hasselmann, K.; Heimbach, P. An Improved Algorithm for the Retrieval of Ocean Wave Spectra from Synthetic Aperture Radar Image Spectra. *J. Geophys. Res. Oceans* **1996**, *101*, 16615–16629. [CrossRef]

8. Schulz-Stellenfleth, J.; Lehner, S.; Hoja, D. A Parametric Scheme for the Retrieval of Two-Dimensional Ocean Wave Spectra from Synthetic Aperture Radar Look Cross Spectra. *J. Geophys. Res. Oceans* **2005**, *110*, C05004. [CrossRef]

9. Alpers, W.R.; Ross, D.B.; Rufenach, C.L. On the Detectability of Ocean Surface Waves by Real and Synthetic Aperture Radar. *J. Geophys. Res. Oceans* **1981**, *86*, 6481–6498. [CrossRef]

10. Lyzenga, D.R. Unconstrained Inversion of Waveheight Spectra from SAR Images. *IEEE Trans. Geosci. Remote Sens.* **2002**, *40*, 261–270. [CrossRef]

11. Bruck, M. *Sea State Measurements Using TerraSAR-X/TanDEM-X Data*; Christian-Albrechts-Universität zu Kiel: Kiel, Germany, 2015.

12. Li, X.M.; Lehner, S.; Bruns, T. Ocean Wave Integral Parameter Measurements Using ENVISAT ASAR Wave Mode Data. *IEEE Trans. Geosci. Remote Sens.* **2011**, *49*, 155–174. [CrossRef]

13. Pleskachevsky, A.L.; Rosenthal, W.; Lehner, S. Meteo-Marine Parameters for Highly Variable Environment in Coastal Regions from Satellite Radar Images. *ISPRS J. Photogr. Remote Sens.* **2016**, *119*, 464–484. [CrossRef]

14. Schulz-Stellenfleth, J.; König, T.; Lehner, S. An Empirical Approach for the Retrieval of Integral Ocean Wave Parameters from Synthetic Aperture Radar Data. *J. Geophys. Res. Oceans* **2007**, *112*, C03019. [CrossRef]

15. Stopa, J.E.; Mouche, A. Significant wave heights from Sentinel-1 SAR: Validation and applications. *J. Geophys. Res. Oceans* **2017**, *122*, 1827–1848. [CrossRef]

16. Schwarz, E.; Krause, D.; Berg, M.; Daedelow, H.; Maass, H. Near Real Time Applications for Maritime Situational Awareness. *Int. Arch. Photogr. Remote Sens. Spat. Inf. Sci.* **2015**, *40*, 999–1003. [CrossRef]

17. Soomere, T.; Räämet, A. Spatial Patterns of the Wave Climate in the Baltic Proper and the Gulf of Finland. *Oceanologia* **2011**, *53*, 335–371. [CrossRef]

18. Björkqvist, J.-V.; Lukas, I.; Alari, V.; van Vledder, G.P.; Hulst, S.; Pettersson, H.; Behrens, A.; Männik, A. Comparing a 41-year model hindcast with decades of wave measurements from the Baltic Sea. *Ocean Eng.* **2018**, *152*, 57–71. [CrossRef]

19. Tuomi, L.; Kahma, K.K.; Pettersson, H. Wave hindcast statistics in the seasonally ice-covered Baltic sea. *Boreal Environ. Res.* **2011**, *16*, 451–472.

20. Pleskachevsky, A.; Jacobsen, S.; Tings, B.; Schwarz, E. Sea State from Sentinel-1 Synthetic Aperture Radar Imagery for Maritime Situation Awareness. *Int. J. Remote Sens.*. submitted.

21. Rikka, S.; Pleskachevsky, A.; Uiboupin, R.; Jacobsen, S. Sea state in the Baltic Sea from space-borne high-resolution synthetic aperture radar imagery. *Int. J. Remote Sens.* **2018**, *39*, 1256–1284. [CrossRef]

22. Ressel, R.; Singha, S.; Lehner, S.; Rösel, A.; Spreen, G. Investigation into Different Polarimetric Features for Sea Ice Classification Using X-Band Synthetic Aperture Radar. *IEEE J. Sel. Top. Appl. Earth Obs. Remote Sens.* **2016**, *9*, 3131–3143. [CrossRef]

23. Singha, S.; Velotto, D.; Lehner, S. Dual-polarimetric feature extraction and evaluation for oil spill detection: A near real time perspective. In Proceedings of the 2015 IEEE International on Geoscience and Remote Sensing Symposium (IGARSS), Milan, Italy, 26–31 July 2015; pp. 3235–3238.

24. Velotto, D.; Bentes, C.; Tings, B.; Lehner, S. First Comparison of Sentinel-1 and TerraSAR-X Data in the Framework of Maritime Targets Detection: South Italy Case. *IEEE J. Ocean. Eng.* **2016**, *41*, 993–1006. [CrossRef]

25. Ardhuin, F.; Stopa, J.; Chapron, B.; Collard, F.; Smith, M.; Thomson, J.; Doble, M.; Blomquist, B.; Persson, O.; Collins III, C.O. Measuring ocean waves in sea ice using SAR imagery: A quasi-deterministic approach evaluated with Sentinel-1 and in situ data. *Remote Sens. Environ.* **2017**, *189*, 211–222. [CrossRef]

26. Shao, W.; Zhang, Z.; Li, X.; Li, H. Ocean wave parameters retrieval from Sentinel-1 SAR imagery. *Remote Sens.* **2016**, *8*, 707. [CrossRef]

27. Ardhuin, F.; Chapron, B.; Collard, F. Observation of swell dissipation across oceans. *Geophys. Res. Lett.* **2009**, *36*. [CrossRef]

28. Collard, F.; Ardhuin, F.; Chapron, B. Monitoring and analysis of ocean swell fields from space: New methods for routine observations. *J. Geophys. Res. Oceans* **2009**, *114*. [CrossRef]

29. Tuomi, L.; Vähä-Piikkiö, O.; Alari, V. *Baltic Sea Wave Analysis and Forecasting Product BALTICSEA_ANALYSIS_FORECAST_WAV_003_010*; Issue 1.0, CMEMS, 2017.

30. Tuomi, L.; Vähä-Piikkiö, O.; Alari, V. CMEMS Baltic Monitoring and Forecasting Centre: High-resolution wave forecast in the seasonally ice-covered Baltic Sea. In Proceedings of the 8th International EuroGOOS Conference, Bergen, Norway, 3–5 October 2017.

31. Von Schuckmann, K.; Le Traon, P.-Y.; Alvarez-Fanjul, E.; Axell, L.; Balmaseda, M.; Breivik, L.-A.; Brewin, R.J.; Bricaud, C.; Drevillon, M.; Drillet, Y.; et al. The Copernicus marine environment monitoring service ocean state report. *J. Oper. Ocean.* **2016**, *9*, s235–s320. [CrossRef]

32. Group, T.W. The WAM model—A third generation ocean wave prediction model. *J. Phys. Ocean.* **1988**, *18*, 1775–1810. [CrossRef]

33. Whitham, G.B. *Linear and Nonlinear Waves*; John Wiley & Sons: New York, NY, USA, 1974.

34. Stoffelen, A.; Anderson, D. Scatterometer data interpretation: Estimation and validation of the transfer function CMOD4. *J. Geophys. Res. Oceans* **1997**, *102*, 5767–5780. [CrossRef]

35. Hersbach, H.; Stoffelen, A.; de Haan, S. An improved C-band scatterometer ocean geophysical model function: CMOD5. *J. Geophys. Res. Oceans* **2007**, *112*. [CrossRef]

36. Monaldo, F.; Jackson, C.; Li, X.; Pichel, W.G. Preliminary evaluation of Sentinel-1A wind speed retrievals. *IEEE J. Sel. Top. Appl. Earth Obs. Remote Sens.* **2016**, *9*, 2638–2642. [CrossRef]

37. Thompson, D.R.; Elfouhaily, T.M.; Chapron, B. Polarization ratio for microwave backscattering from the ocean surface at low to moderate incidence angles. In Proceedings of the 1998 IEEE International on Geoscience and Remote Sensing Symposium (IGARSSS'98), Seattle, WA, USA, 6–10 July 1998; pp. 1671–1673.

38. Horstmann, J.; Falchetti, S.; Wackerman, C.; Maresca, S.; Caruso, M.J.; Graber, H.C. Tropical cyclone winds retrieved from C-band cross-polarized synthetic aperture radar. *IEEE Trans. Geosci. Remote Sens.* **2015**, *53*, 2887–2898. [CrossRef]
39. Skamarock, W.C.; Klemp, J.B.; Dudhia, J.; Gill, D.O.; Barker, D.M.; Wang, W.; Powers, J.G. *A Description of the Advanced Research WRF Version 2 (No. NCAR/TN-468+ STR)*; National Center For Atmospheric Research Boulder Co Mesoscale and Microscale Meteorology Div: Boulder, CO, USA, 2005.
40. Haralick, R.M.; Shanmugam, K. Textural Features for Image Classification. *IEEE Trans. Syst. Man Cyber.* **1973**, *3*, 610–621. [CrossRef]
41. Romeiser, R.; Graber, H.C.; Caruso, M.J.; Jensen, R.E.; Walker, D.T.; Cox, A.T. A new approach to ocean wave parameter estimates from C-band ScanSAR images. *IEEE Trans. Geosci. Remote Sens.* **2015**, *53*, 1320–1345. [CrossRef]
42. Alari, V.; Staneva, J.; Breivik, Ø.; Bidlot, J.-R.; Mogensen, K.; Janssen, P. Response of water temperature to surface wave effects in the Baltic Sea: Simulations with the coupled NEMO-WAM model. In Proceedings of the EGU General Assembly Conference Abstracts, Vienna Austria, 17–22 April 2016; p. 4363.
43. European Maritime Safety Agency. *EMSA CleanSeaNet First Generation Report*; EMSA: Lisboa, Portugal, 2011.
44. Ducet, N.; Le Traon, P.-Y.; Reverdin, G. Global high-resolution mapping of ocean circulation from TOPEX/Poseidon and ERS-1 and-2. *J. Geophys. Res. Oceans* **2000**, *105*, 19477–19498. [CrossRef]
45. Pascual, A.; Faugère, Y.; Larnicol, G.; Le Traon, P.Y. Improved description of the ocean mesoscale variability by combining four satellite altimeters. *Geophys. Res. Lett.* **2006**, *33*. [CrossRef]
46. Pujol, M.-I.; Faugère, Y.; Taburet, G.; Dupuy, S.; Pelloquin, C.; Ablain, M.; Picot, N. DUACS DT2014: The new multi-mission altimeter data set reprocessed over 20 years. *Ocean Sci.* **2016**, *12*, 1067–1090. [CrossRef]
47. Ray, R.D.; Beckley, B. Simultaneous ocean wave measurements by the Jason and Topex satellites, with buoy and model comparisons special issue: Jason-1 calibration/Validation. *Mar. Geod.* **2003**, *26*, 367–382. [CrossRef]
48. Bouffard, J.; Vignudelli, S.; Cipollini, P.; Menard, Y. Exploiting the potential of an improved multimission altimetric data set over the coastal ocean. *Geophys. Res. Lett.* **2008**, *35*. [CrossRef]
49. Cazenave, A.; Bonnefond, P.; Mercier, F.; Dominh, K.; Toumazou, V. Sea level variations in the Mediterranean Sea and Black Sea from satellite altimetry and tide gauges. *Glob. Plan. Chang.* **2002**, *34*, 59–86. [CrossRef]
50. Kudryavtseva, N.A.; Soomere, T. Validation of the multi-mission altimeter wave height data for the Baltic Sea region. *arXiv*, 2016; arXiv:1603.08698.
51. Madsen, K.S.; Høyer, J.; Tscherning, C.C. Near-coastal satellite altimetry: Sea surface height variability in the North Sea–Baltic Sea area. *Geophys. Res. Lett.* **2007**, *34*. [CrossRef]
52. Vignudelli, S.; Cipollini, P.; Roblou, L.; Lyard, F.; Gasparini, G.; Manzella, G.; Astraldi, M. Improved satellite altimetry in coastal systems: Case study of the Corsica Channel (Mediterranean Sea). *Geophys. Res. Lett.* **2005**, *32*. [CrossRef]
53. Høyer, J.L.; Nielsen, J. Satellite significant wave height observations in coastal and shelf seas. In Proceedings of the Symposium on 15 Years of Progress in Radar Altimetry, Venice, Italy, 13–18 March 2006.
54. Kudryavtseva, N.; Soomere, T. Satellite altimetry reveals spatial patterns of variations in the Baltic Sea wave climate. *arXiv*, **2017**, arXiv:1705.01307.

remote sensing

MDPI

Article

Symmetric Double-Eye Structure in Hurricane *Bertha* (2008) Imaged by SAR

Guosheng Zhang * and William Perrie

Fisheries and Oceans Canada, Bedford Institute of Oceanography, Dartmouth, NS B2Y 4A2, Canada;
William.Perrie@dfo-mpo.gc.ca
* Correspondence: zgsheng001@gmail.com; Tel.: +1-902-426-7797

Received: 15 July 2018; Accepted: 13 August 2018; Published: 15 August 2018

Abstract: Internal dynamical processes play a critical role in hurricane intensity variability. However, our understanding of internal storm processes is less well established, partly because of fewer observations. In this study, we present an analysis of the hurricane double-eye structure imaged by the RADARSAT-2 cross-polarized synthetic aperture radar (SAR) over Hurricane *Bertha* (2008). SAR has the capability of hurricane monitoring because of the ocean surface roughness induced by surface wind stress. Recently, the C-band cross-polarized SAR measurements appear to be unsaturated for the high wind speeds, which makes SAR suitable for studies of the hurricane internal dynamic processes, including the double-eye structure. We retrieve the wind field of Hurricane *Bertha* (2008), and then extract the closest axisymmetric double-eye structure from the wind field using an idealized vortex model. Comparisons between the axisymmetric model extracted wind field and SAR observed winds demonstrate that the double-eye structure imaged by SAR is relatively axisymmetric. Associated with airborne measurements using a stepped-frequency microwave radiometer, we investigate the hurricane internal dynamic process related to the double-eye structure, which is known as the eyewall replacement cycle (ERC). The classic ERC theory was proposed by assuming an axisymmetric storm structure. The ERC internal dynamic process of Hurricane *Bertha* (2008) related to the symmetric double-eye structure here, which is consistent with the classic theory, is observed by SAR and aircraft.

Keywords: hurricane internal dynamical process; synthetic aperture radar (SAR); eyewall replacement cycles; ocean winds

1. Introduction

Accurate observations of surface winds of a tropical cyclone (hurricane or typhoon), particularly the high-resolution structures, play a critical role in improving hurricane dynamic readiness and understanding of its evolution process. Over the past 25 years, prediction skill for hurricane intensity has had comparatively few improvements because of limited knowledge regarding the hurricane internal dynamical processes, whereas its track forecasts errors have steadily declined [1]. Moreover, determinations of hurricane inner core structures and surface winds remain considerable operational challenges to the hurricane dynamic studies [2]. Routinely, surface winds of hurricanes are measured by the airborne Stepped Frequency Microwave Radiometer (SFMR) along flying track. Then, hurricanes are assumed to be axisymmetric. Based on this axisymmetric assumption, hurricane internal dynamics have been analyzed using the SFMR measurements, i.e., the vortex Rossby wave dynamics [3], eyewall replacement cycles (ERCs) [4], and hurricane pressure-wind model [5]. However, the actual hurricane structure, whether axisymmetric or not, is difficult to discern, only based on the aircraft reconnaissance low-level SFMR data. The surface wind fields of hurricanes have high azimuthal asymmetries, and these asymmetries are hard to measure by the aircraft, which typically flies at roughly fixed azimuths with time legs [5–7].

Compared with optical satellite sensors, the spaceborne synthetic aperture radar (SAR) is suitable for two-dimensional ocean surface wind field observations, for its advantages of a high spatial resolution, relatively large spatial swath, and its ability to work day-and-night under almost all-weather conditions. Additionally, hurricanes have been frequently observed by spaceborne SAR images, since the first spaceborne SAR image became available in 1978 [8]. Over the last few decades, SAR observations have been adopted in much research on hurricane readiness [9], morphology [8,10], precipitation [11], ocean surface current [12], and hurricane internal dynamic processes [13]. In this study, we present an analysis of one cross-polarized ScanSAR image from C-band RADARSAT-2 over Hurricane *Bertha* (2008) containing a double-eye structure.

To estimate the hurricane double-eye core structure from observed SAR surface winds, the idealized surface wind structure model, known as the Symmetric Hurricane Estimates for Wind (SHEW) model, is extended from a single-eye to double-eye structure. The double-eye SHEW model is the successor of approaches that we proposed in 2014 and 2017 [10,14]. To simplify the process, we present the SHEW model for the double-eye structure based on three assumptions: (1) both the hurricane eyewalls containing the maximum winds are circular-shaped; (2) the maximum wind speeds on either eyewall are axisymmetric; and (3) the wind speed within a hurricane is a function of the radius.

The theory for ERCs related to the hurricane double-eye structure was originally presented by Willoughby et al. (1982) [15]. This theory suggests that any secondary eyewall will shrink and contract due to the presence of annular convective heating. The circulation in a hurricane is changed by the axisymmetric outer eyewall heating, which results in cutting off the boundary layer inflow to the primary eyewall. Then, the hurricane is weakened by the outage of boundary layer inflow [15,16]. Based on an analysis of 79 Atlantic basin hurricanes observed by aircraft reconnaissance, Sitkowski et al. (2011) [4] demonstrated that the maximum wind speed of the primary core decreases significantly after the formation of the outer eyewall in a general ERC process. The RADARSAT-2 VH-polarized ScanSAR image over Hurricane *Bertha* (2008) provides us with a chance to study the sea wind field of a double-eye structure and its relationship to the ERC internal dynamic process.

The modified SHEW model for a double-eye structure is based on the one-dimensional modified Rankine vortex functions [3,4,17] and double circles for the eyewalls containing maximum wind speeds. We organize the remainder of this paper as follows. The data sets of cross-polarized SAR observations from C-band RADARSAT-2 and the respective aircraft SFMR measurements, as well as the double-eye SHEW model, are presented in Section 2. Then, we show the extracted double-eye structure of Hurricane *Bertha* (2008) in the SHEW model in Section 3 and discuss the respective ERC process with the aircraft measurements in Section 4. Finally, we present conclusions in Section 5.

2. Materials and Methods

Bertha (2008) was an early season category 3 hurricane and the longest-lived Atlantic July tropical cyclone on record. In this study, a hurricane wind speed retrieval model is employed and an idealized vortex model for a symmetric double-eye hurricane structure is proposed.

2.1. Data Sets

To study the role of the symmetric secondary eyewall, we adopt one Cross-polarized (VH-polarized) ScanSAR image from C-band RADARSAT-2 in dual-polarized (VV and VH) mode (acquired at 10:14 UTC, 12 July 2008) over Hurricane *Bertha* (2008). To show the respective ERC process, the aircraft SFMR (stepped-frequency microwave radiometer) measurements acquired at times close to the SAR image are also employed. We show the SAR image, aircraft tracks, and the SAR location with respect to the Best Track of Hurricane *Bertha* (2008) in Figure 1.

The VH-polarization SAR image from C-band RADARSAT-2 used here is in ScanSAR wide swath mode. Its medium resolution is 50 m and the swath width is about 450 km. To reduce the image speckle noise, we calibrated the SAR image and then downscaled the spatial resolution to 1 km using the boxcar averaging method [18]. As shown in Figure 1, the SAR image here captured the whole core structure of Hurricane *Bertha* (2008). We retrieved the wind speeds directly from the SAR image using the C-3PO (C-band Cross-Polarization Coupled-Parameters Ocean) hurricane wind retrieval model, which was developed for cross-polarization SAR wind speed retrieval [19]. The operational surface winds measured by the SFMR radar on board the research aircrafts of NOAA (National Oceanic and Atmospheric Administration) WP-3D and U.S. Air Force are adopted. The SFMR radar can potentially provide along-track mapping of wind speeds at relatively high temporal (1 Hz) and spatial (~120 m) resolutions. The root-mean-square errors of the SFMR winds are less than 4 m/s~5 m/s [17,20], with validations of the measurements from dropwindsonde and an *in situ* instrument.

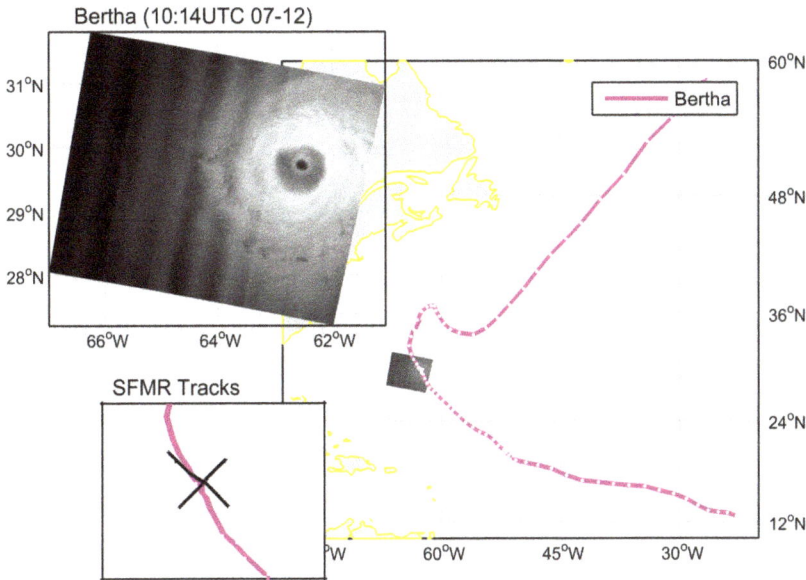

Figure 1. VH-polarized SAR image from C-band RADARSAT-2 over Hurricane *Bertha* (10:14 UTC, 12 July 2008), as well as flight tracks of SFMR measurements relative to the same hurricane internal dynamic processes and the Best Track (BT) of Hurricane *Bertha* (2008). RADARSAT-2 Data and Product MacDonald, Dettwiler, and Associates Ltd., All Rights Reserved.

2.2. C-3PO Hurricane Wind Retrieval Model

We retrieved the sea surface wind from the knowledge of normalized radar cross section (NRCS) imaged by SAR, using the C-3PO hurricane wind retrieval model. The C-3PO hurricane high-wind-speed retrieval model is [19]:

$$\sigma_0[dB] = [0.2983 \cdot u_{10} - 29.4708] \cdot \left[1 + 0.07 \cdot \frac{\theta_{ind} - 34.5}{34.5}\right] \tag{1}$$

where u_{10} is the surface wind speed (at 10 m reference height), σ_0 is the NRCS of the radar signal with the units of dB, and θ_{ind} is the SAR incidence angle.

The C-3PO model was developed from a theoretical analysis and a database including 650 sets of wind vectors, VH-polarized NRCSs, and the associated incidence angles [19]. This database was built based on five RADARSAT-2 VH-polarized SAR images covering five different hurricanes, as well as the collocated wind vectors measured by the aircraft SFMR. The SFMR wind vectors were selected during the 30-mintute windows with respect to the SAR acquired time. As the aircraft measured the rain rate simultaneously with the wind vector, we removed the measurements with rain rates higher than 10 mm/hr. Then, we collocated the VH-polarized NRCSs with the wind vectors by considering the hurricane movements and rotations.

The incidence angles should play a role in the cross-polarized (VH/HV) SAR wind retrieval. However, the dependences of the wind induced cross-polarized NRCSs on the incidence angles are difficult to extract because of a lack of observational data. Based on this limited database, we [19] developed a wind retrieval model (C-3PO), including the incidence angles, by a theoretical analysis. In fact, in that study, we proposed a theoretical model for the C-band Cross-polarization based on two existing theoretical models developed for the Co-polarizations (VV and HH). We simulated a relationship between the VH-polarized NRCSs and the incidence angles for various wind speeds. We show the C-3PO hurricane wind retrieval model and the database in Figure 2.

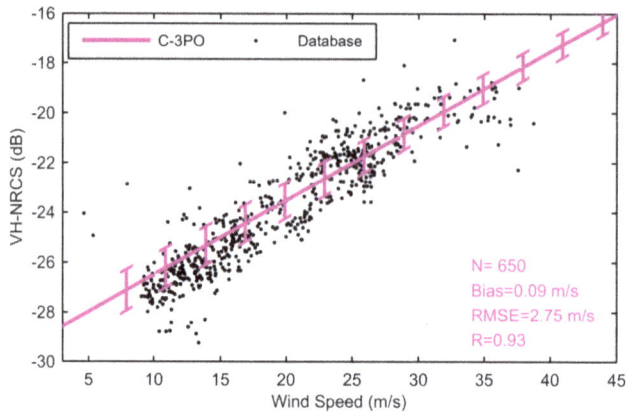

Figure 2. C-3PO hurricane wind retrieval model. The magenta line is for the middle incidence angle (34.5°), as well as the bars for the incidence angle ranges (from 19.5° to 49.5°) of the RADARSAT-2 ScanSAR mode.

2.3. SHEW Idealized Vortex model

For the symmetric double-eye structure estimation, our SHEW model [10] is modified using the double modified Rankine vortex functions [4,17]. Using a given hurricane center (HE) location, we *firstly* transfer the geographical coordinates to a polar coordinate system in the modified SHEW model:

$$\begin{cases} r = distance(grid, HE) \\ \theta = azimuth(grid, HE) \end{cases} \tag{2}$$

In this model, $\theta = 0$ means east and rises counter-clockwise. *Then*, we propose the symmetric double-eye SHEW model as:

$$u_f(r,\theta) = \begin{cases} u_1(\theta) \cdot \left[\dfrac{r}{r_1(\theta)}\right], & (r \le r_1(\theta)) \\[3mm] u_1(\theta) \cdot \left[\dfrac{r_1(\theta)}{r}\right]^{\alpha_1}, & (r_1(\theta) < r \le r_{moat}) \\[3mm] u_1(\theta) \cdot \left[\dfrac{r_1(\theta)}{r_{moat}}\right]^{\alpha_1} \\[1mm] \quad + \left\{u_2(\theta) - u_1(\theta) \cdot \left[\dfrac{r_1(\theta)}{r_{moat}}\right]^{\alpha_1}\right\} \cdot \left[\dfrac{r - r_{moat}}{r_2(\theta) - r_{moat}}\right], & (r_{moat} < r \le r_2(\theta)) \\[3mm] u_2(\theta) \cdot \left[\dfrac{r_2(\theta)}{r}\right]^{\alpha_2}, & (r_2(\theta) < r \le 150 \text{ km}) \end{cases} \qquad (3)$$

where $u_1(\theta)$ and $u_2(\theta)$ are the maximum wind speeds for the *inner* and *outer* vortexes, respectively; $r_1(\theta)$ and $r_2(\theta)$ are the radius of the maximum wind (RMW) for the *inner* and *outer* vortexes, respectively; and α_1 and α_2 are the decay parameters for the two vortexes, respectively. Moreover, a specific pivot position between the two vortexes is labeled (u_{moat}, r_{moat}). For symmetric hurricane structures, the maximum wind speeds and RMWs are set as constants, (u_1, u_2) and (r_1, r_2), respectively. In Figure 3, we provide an example for the symmetric double-eye hurricane structure using the SHEW model with the following values of $(u_1 = 35 \text{ m/s}, r_1 = 15 \text{ km}, \alpha_1 = 0.5)$ for the inner vortex, and $(u_2 = 35 \text{ m/s}, r_2 = 45 \text{ km}, \alpha_2 = 0.5)$ for the outer vortex, as well as $(u_{moat} = 25 \text{ m/s}, r_{moat} = 33 \text{ km})$ for the pivot position.

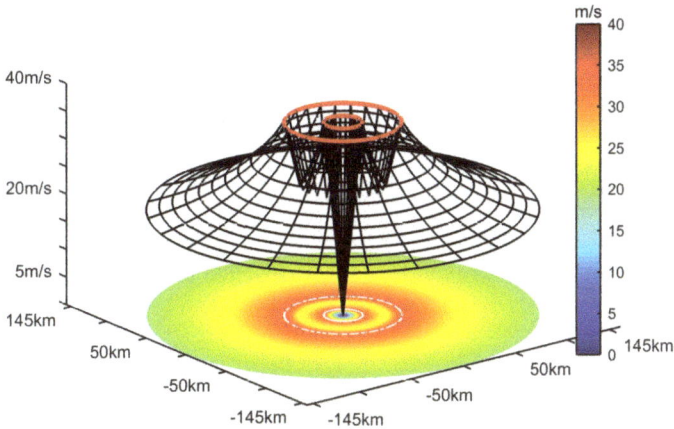

Figure 3. An example of the SHEW symmetric double-eye hurricane structure. The maxima wind speeds for two eyes are both 35 m/s, the RMW for the inner core is 15 km and for the secondary core is 45 km, and the two decay parameters are both 0.5.

To estimate the double-eye structure of Hurricane *Bertha* (2008), we display the procedures in the flowchart (Figure 4) needed to use the modified SHEW. Firstly, we give an initial hurricane center to start the procedures. With various parameters, we fit the winds relative to the radius from the SHEW model to the SAR winds using the least-squares approximation. We note that the parameters of the SHEW model are independent.

25

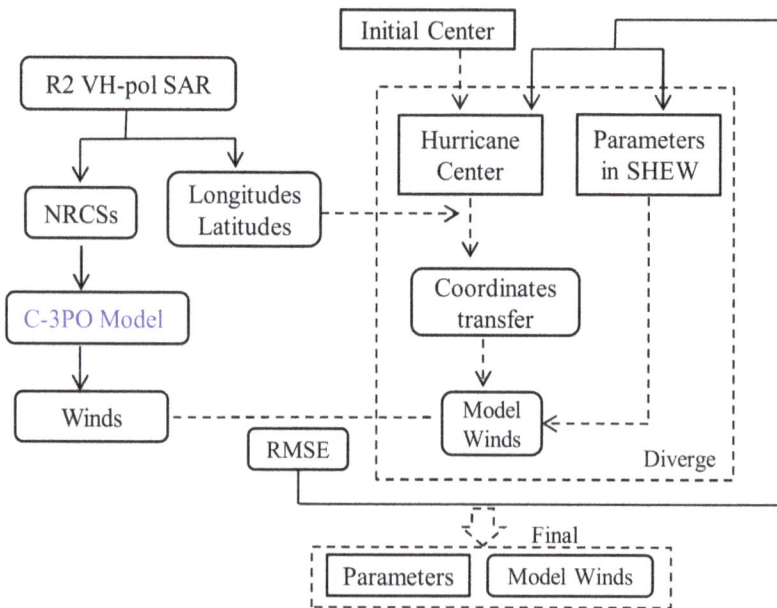

Figure 4. Flowchart for the modified SHEW model to estimate the hurricane double-eye structure.

3. Results

In Figure 5a, we show the time-series of hurricane *Bertha*'s intensity from the Best Track database and the corresponding SAR image. In Figure 5b, we show the wind fields retrieved from the VH-polarized SAR image by employing the C-3PO model (Equation (1)). Relative to the hurricane center, identified by very low wind speeds, two high wind speed rings are obviously separated by relatively low wind areas. Following the flowchart in Figure 4, we estimate the closest idealized axisymmetric hurricane structure, as shown in Figure 5c, by fitting the symmetric double-eye SHEW model (Equations (2) and (3)) to wind field retrieved from SAR. Although the maximum wind speed of the inner eyewall is much smaller than the outer eyewall, the outline of the inner core is very clear in the wind speed field. The detected intensity of the inner core is 20.9 m/s and for the secondary eyewall, 27.9 m/s. The RMW for the *primary* vortex is 13 km and for the *secondary* vortex, 52 km. Within the estimated axisymmetric framework (Figure 5c), we show the storm-centered wind field of Hurricane *Bertha* (2008) observed by SAR in Figure 5d. Using the wind fields reconstructed by the modified SHEW model (Figure 5c) and imaged by SAR (Figure 5d), we compute the standard deviation between the closest idealized axisymmetric structure and the 'real' wind field as 2.45 m/s, as well as the correlation coefficient as 0.827. The confidence interval of the correlation coefficient is higher than 99%. In a related hurricane study, Zhang et al. (2017) [10] demonstrated that correlation coefficients between the idealized structure by the SHEW model and 'real' wind imaged by SAR are routinely between 0.60 to 0.85. By comparison, the correlation coefficient here of 0.827 implies that the double-eye structure of Hurricane *Bertha* (2008) observed by SAR is close to symmetric.

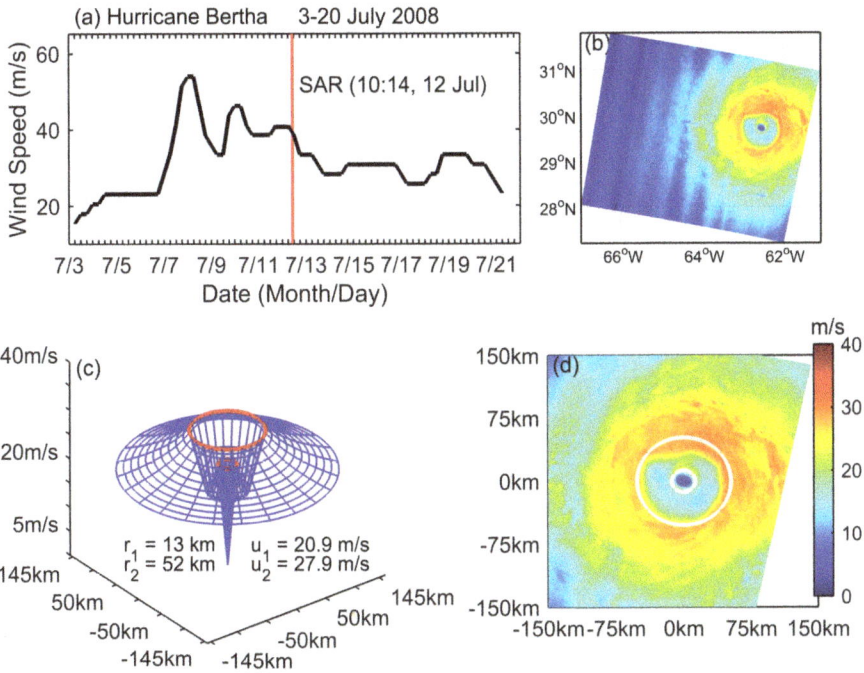

Figure 5. Hurricane *Bertha* (2008): (**a**) SAR captured time respective to the time series of Best Track Intensities; (**b**) retrieved sea surface wind speed from SAR images acquired at 10:14 UTC, 12 July 2008; (**c**) SHEW estimated closest idealized structure; and (**d**) storm-centered wind field of hurricane core.

In the axisymmetric framework for storm internal dynamic processes, Willoughby et al. (1982) [15] suggested that the hurricane intensity decreases as a result of the outage of boundary layer inflow to the primary eyewall, which is cut off by a negative tendency induced by the secondary eyewall heating. According to the classic ERC theory, the symmetric double-eye structure of Hurricane *Bertha* (2008) observed by SAR suggests that the secondary eyewall would contract and replace the primary eyewall.

As the SAR imagery of hurricane *Bertha* (2008) captures a double-eye structure (Figure 5), we address two points consistent with the classic theories for the ERC internal dynamic process: (1) the structure of the sea surface wind field is axisymmetric, and (2) the intensity of the inner eyewall is smaller than the outer eyewall.

4. Discussion

Routinely, tangential wind profiles are used for the aircraft measurements in applications for studies on the hurricane internal dynamic processes [14]. We adopt the four profiles associated with the SAR image which should be seen during one ERC process. Based on an analysis of 79 Atlantic basin hurricanes observed by aircraft reconnaissance, Sitkowski et al. (2011) [4] demonstrated that the process of replacing an inner eyewall with an outer eyewall lasts for an average of 36 h, in a general ERC process. They [4] also suggested that the maximum wind speed of the primary core decreases significantly after the formation of the outer eyewall. The maximum difference between the time of SAR observation and SFMR winds used here is no more than 10 h. Therefore, they are supposed to be part of the same ERC process.

To analyze the axisymmetric double-eye structure, we adopt the definition of the complete profile [13], which contains two continuous normal profiles across the hurricane center (one is flying into the hurricane and the other is flying out). As shown in Figure 6b, there are two complete profiles composited by four normal profiles. Although only four normal profiles were measured by SFMR, these measurements allow us to present a clear display of the hurricane evolution with the axisymmetric double-eye structure in Figure 6. In Figure 6a, we show a time-series composed of intensities for the two eyewalls observed by spaceborne SAR and airborne SFMR. The hurricane intensities from SAR are derived from the SHEW model, and the intensities of SFMR are extracted from the profiles using the maximum wind speeds. An outline of the ERC internal dynamic process of Hurricane *Bertha* (2008) is shown by the two complete profiles with the axisymmetric double-eye structure (Figure 6b–d). In Figure 6b, we show the distributions of SFMR wind speeds along the flight tracks.

In Figure 6c, we show a complete wind profile, which was captured about seven hours after the SAR image (hereafter "the 7 h profile"). Additionally, "the 7 h profile" is the first SFMR measurement of the same secondary eyewall as that detected from the ScanSAR imagery. By comparing the storm-centered SAR wind field (Figure 5d) to the aircraft flying tracks (Figure 6b) of "the 7 h profile", the double-eye structure measured by SFMR is also axisymmetric, which indicates that this structure lasts more than seven hours. In Figure 6d, the other complete profile was measured about 8.5 h after the SAR image (hereafter "the 8.5 h profile"), which should be related to the same ERC process. We cannot distinguish the secondary eyewall in "the 8.5 h profile", which demonstrates that the inner eyewall was replaced by the outer eyewall during the period from 7 h to 8.5 h after the SAR image.

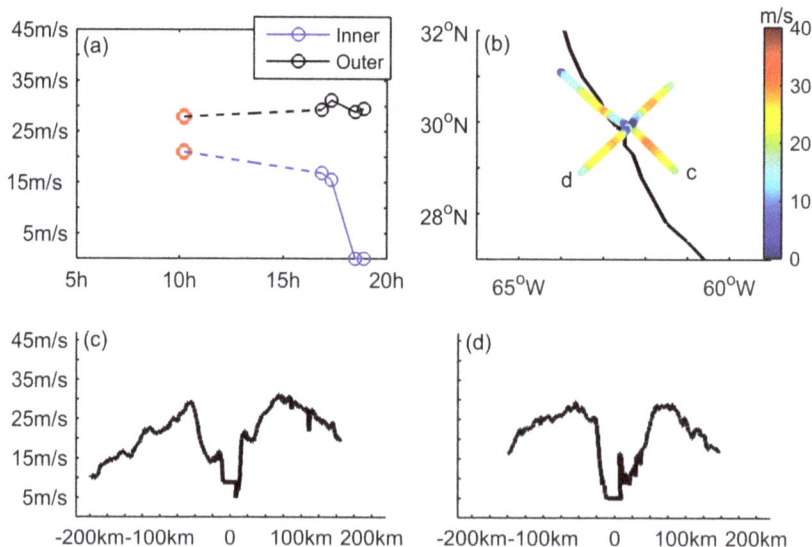

Figure 6. ERC dynamic process of Hurricane *Bertha* measured by aircraft SFMR: (**a**) hurricane intensities of the double-eye structure, (**b**) flight tracks, as well as (**c**), (**d**) the respective complete profiles along radius.

Recent studies [21] have suggested that symmetric heating appears to be the dominant factor for hurricane intensity change, although asymmetries controlled by the large-scale shearing environment may also be important. As is typical of hurricane processes during the axisymmetric framework, an early study of [16] demonstrated that a negative tendency for the tangential winds inside the secondary eyewall can be induced by the symmetric eyewall heating, which cuts off the boundary layer inflow to the primary eyewall and further weakens the hurricane intensity. Finally, towards the end of this progression, the primary eyewall was replaced by the secondary eyewall. Using two-dimensional observations of SAR with a high spatial resolution, one can see that the double-eye structure of

Hurricane *Bertha* (2008) is relatively axisymmetric. Associated with the airborne SFMR measurements, we provide the relative ERC process. Therefore, the role of the axisymmetric double-eye structure in the ERC dynamic process is verified as consistent with the classic theory [15,16]. Moreover, many hurricanes are detected not to significantly weaken after the formation of the secondary eyewall [22], which may be a result of the asymmetric hurricane structure [13]. As Hurricane *Bertha* (2008) is relatively axisymmetric, the SFMR observed in one-dimension is supposed to represent the surface wind.

5. Conclusions

A one cross-polarized (VH) ScanSAR image from C-band RADARSAT-2 over Hurricane *Bertha* (2008) provides us with a chance to research the axisymmetric double-eye structure effects on the hurricane intensity change. In this study, we employ the C-3PO hurricane wind retrieval model and modify the SHEW idealized vortex model for the double-eye structure. By comparing the wind fields simulated by the symmetric double-eye SHEW model and observed by SAR, we compute the correlation coefficient of 0.827 with a confidence interval higher than 99% and the standard deviation of 2.45 m/s. In a related hurricane study, Zhang et al. (2017) [10] demonstrated that correlation coefficients between the idealized structure by the SHEW model and *'real'* wind imaged by SAR are routinely between 0.60 to 0.85, which implies that the double-eye structure of Hurricane *Bertha* (2008) observed by SAR is symmetric. To analyze the ERC process related to the double-eye structure captured by SAR, we employ two complete tangential wind profiles measured by airborne SFMR. After about 7 h from the SAR observation, the profile appears to capture the double-eye structure; but 8.5 h later, the primary eyewall should be replaced by the secondary eyewall.

Compared to the previous studies [4,15,16,22], we find three characteristics consistent with the classic ERC theory as follows: (1) the structure of the sea surface wind field is axisymmetric; (2) the intensity of the primary eyewall is smaller than the secondary eyewall; and (3) the primary eyewall is replaced by the secondary eyewall. We suggest that the axisymmetric structure plays a major role in the ERC dynamic process.

Author Contributions: G.Z. was responsible for designing this study, processing datasets, building the model, and writing-Original Draft Preparation; W.P. was responsible for supervision, writing-review, and editing.

Funding: This research was funded by National Natural Science Youth Foundation of China under Grant 41706193, in part by the Canadian Space Agency SWOT and Office of Energy Research and Development (OERD) Programs, in part by the Canadian Data Utilization and Application Program (DUAP) "Winds from SAR" RCM Readiness Project between ECCC and DFO, and in part by the Open Fund of Key Laboratory of Geographic Information Science (Ministry of Education), East China Normal University (Grant No. KLGIS2017A06).

Acknowledgments: The authors thank the Canadian Space Agency for providing RADARSAT-2 data, NOAA HRD for supplying SFMR data (http://www.aoml.noaa.gov/hrd/), and NOAA NHC for hurricane best track data (HURDAT2) (http://www.nhc.noaa.gov/data/#hurdat).

Conflicts of Interest: The authors declare no conflict of interest.

References

1. Montgomery, M.T.; Smith, R.K. Recent developments in the fluid dynamics of tropical cyclones. *Annu. Rev. Fluid Mech.* **2017**, *49*, 541–574. [CrossRef]
2. Sanabia, E.R.; Barrett, B.S.; Celone, N.P.; Cornelius, Z.D. Satellite and aircraft observations of the eyewall replacement cycle in Typhoon Sinlaku (2008). *Monthly Weather Rev.* **2015**, *143*, 3406–3420. [CrossRef]
3. Mallen, K.J.; Montgomery, M.T.; Wang, B. Reexamining the near-core radial structure of the tropical cyclone primary circulation: Implications for vortex resiliency. *J. Atmos. Sci.* **2005**, *62*, 408–425. [CrossRef]
4. Sitkowski, M.; Kossin, J.P.; Rozoff, C.M. Intensity and structure changes during hurricane eyewall replacement cycles. *Mon. Weather Rev.* **2011**, *139*, 3829–3847. [CrossRef]
5. Kossin, J.P. Hurricane wind–pressure relationship and eyewall replacement cycles. *Weather Forecast.* **2015**, *30*, 177–181. [CrossRef]

Remote Sens. **2018**, *10*, 1292

6. Kossin, J.P.; Eastin, M.D. Two distinct regimes in the kinematic and thermodynamic structure of the hurricane eye and eyewall. *J. Atmos. Sci.* **2001**, *58*, 1079–1090. [CrossRef]

7. Franklin, J.L.; Black, M.L.; Valde, K. GPS dropwindsonde wind profiles in hurricanes and their operational implications. *Weather Forecast.* **2003**, *18*, 32–44. [CrossRef]

8. Li, X.; Zhang, J.A.; Yang, X.; Pichel, W.G.; DeMaria, M.; Long, D.; Li, Z. Tropical cyclone morphology from spaceborne synthetic aperture radar. *Bull. Am. Meteorol. Soc.* **2013**, *94*, 215–230. [CrossRef]

9. Du, Y.; Vachon, P.W. Characterization of hurricane eyes in RADARSAT-1 images with wavelet analysis. *Can. J. Remote Sens.* **2003**, *29*, 491–498. [CrossRef]

10. Zhang, G.; Perrie, W.; Li, X.; Zhang, J.A. A hurricane morphology and sea surface wind vector estimation model based on C-band cross-polarization SAR imagery. *IEEE Trans. Geosci. Remote Sens.* **2017**, *55*, 1743–1751. [CrossRef]

11. Zhang, G.; Li, X.; Perrie, W.; Zhang, B.; Wang, L. Rain effects on the hurricane observations over the ocean by C-band Synthetic Aperture Radar. *J. Geophys. Res.* **2016**, *121*, 14–26. [CrossRef]

12. Zhang, G.; Perrie, W. Dual-Polarized Backscatter Features of Surface Currents in the Open Ocean during Typhoon Lan (2017). *Remote Sens.* **2018**, *10*, 875. [CrossRef]

13. Zhang, G.; Perrie, W. Effects of asymmetric secondary eyewall on tropical cyclone evolution in Hurricane Ike (2008). *Geophys. Res. Lett.* **2018**, *45*, 1676–1683. [CrossRef]

14. Zhang, G.; Zhang, B.; Perrie, W.; Xu, Q.; He, Y. A hurricane tangential wind profile estimation method for C-band cross-polarization SAR. *IEEE Trans. Geosci. Remote Sens.* **2014**, *52*, 7186–7194. [CrossRef]

15. Willoughby, H.E.; Clos, J.A.; Shoreibah, M.G. Concentric eye walls, secondary wind maxima, and the evolution of the hurricane vortex. *J. Atmos. Sci.* **1982**, *39*, 395–411. [CrossRef]

16. Shapiro, L.J.; Willoughby, H.E. The response of balanced hurricanes to local sources of heat and momentum. *J. Atmos. Sci.* **1982**, *3*, 378–394. [CrossRef]

17. Wood, V.T.; White, L.W.; Willoughby, H.E.; Jorgensen, D.P. A new parametric tropical cyclone tangential wind profile model. *Mon. Weather Rev.* **2013**, *141*, 1884–1909. [CrossRef]

18. Shen, H.; Perrie, W.; He, Y.; Liu, G. Wind speed retrieval from VH dual-polarization RADARSAT-2 SAR images. *IEEE Trans. Geosci. Remote Sens.* **2014**, *52*, 5820–5826. [CrossRef]

19. Zhang, G.; Li, X.; Perrie, W.; Hwang, P.A.; Zhang, B.; Yang, X. A hurricane wind speed retrieval model for C-band RADARSAT-2 cross-polarization ScanSAR images. *IEEE Trans. Geosci. Remote Sens.* **2017**, *55*, 4766–4774. [CrossRef]

20. Horstmann, J.; Falchetti, S.; Wackerman, C.; Maresca, S.; Caruso, M.J.; Graber, H.C. Tropical cyclone winds retrieved from C-band cross-polarized synthetic aperture radar. *IEEE Trans. Geosci. Remote Sens.* **2015**, *53*, 2887–2898. [CrossRef]

21. Cotto, A.; Gonzalez, I., III; Willoughby, H.E. Synthesis of vortex Rossby waves. Part I: Episodically forced waves in the inner waveguide. *J. Atmos. Sci.* **2015**, *72*, 3940–3957. [CrossRef]

22. Kuo, H.-C.; Chang, C.-P.; Yang, Y.-T.; Jiang, H.-J. Western North Pacific typhoons with concentric eyewalls. *Mon. Weather Rev.* **2009**, *137*, 3758–3770. [CrossRef]

remote sensing

MDPI

Article

Typhoon/Hurricane-Generated Wind Waves Inferred from SAR Imagery

Lei Zhang [1,2]**, Guoqiang Liu** [1,3,*]**, William Perrie** [2,3,4]**, Yijun He** [1,*] **and Guosheng Zhang** [1,3]

[1] School of Marine Sciences, Nanjing University of Information Science and Technology, Nanjing 210044, China; lei.zhang5657@gmail.com (L.Z.); zgsheng001@gmail.com (G.Z.)
[2] Department of Oceanography, Dalhousie University, Halifax, NS B3H 4R2, Canada; william.perrie@dfo-mpo.gc.ca
[3] Fisheries and Oceans Canada, Bedford Institute of Oceanography, Dartmouth, NS B2Y 4A2, Canada
[4] Department of Engineering Mathematics and Internetworking, Dalhousie University, Halifax, NS B3H 4R2, Canada
* Correspondence: Guoqiang.Liu@dfo-mpo.gc.ca (G.L.); yjhe@nuist.edu.cn (Y.H.)

Received: 29 August 2018; Accepted: 5 October 2018; Published: 9 October 2018

Abstract: The wide-swath mode of synthetic aperture radar (SAR) is a good way of detecting typhoon/hurricane winds with a cross-polarization mode. However, its ability to detect wind waves is restricted because of its spatial resolution and nonlinear imaging mechanisms. In this study, we use the SAR-retrieved wind speed, Sentinel-1 SAR wave mode and buoy data to examine fetch- and duration-limited parametric models (denoted H-models), to estimate the wave parameters (significant wave height H_s, dominant wave period T_p) generated by hurricanes or typhoons. Three sets of H-models, in total 6 models, are involved: The H-3Sec model simulates the wave parameters in 3 sections of a given storm (right, left and back); H-LUT models, including the H-LUTI model and H-LUTB model, provide a better resolution of the azimuthal estimation of wind waves inside the storm by analyzing the dataset from Bonnie 1998 and Ivan 2004; and the third set of models is called the H-Harm models, which consider the effects of the radius of the maximum wind speed r_m on the wave simulation. In the case of typhoon Krovanh, the comparison with wave-mode measurements shows that the duration-limited models *underestimate* the high values for the wind-wave H_s, while the fetch models' results are *more accurate*, especially for the H-LUTI model. By analyzing 86 SAR wave mode images, it is found that the H-LUTI model is the best among the 6 H-models, and can effectively simulate the wind-wave H_s, except in the center area of the typhoon; root mean square errors (*rmse*) can reach 0.88 m, and the coefficient correlation (R^2) is 0.86. The H-Harm models add r_m as an additional factor to be considered, but this does not add significant improvement in performance compared to the others. This limitation is probably due to the fact that the data sets used to develop the H-Harm models have only a limited coverage range, with respect to r_m. Applying H-models to RADARSAT-2 ScanSAR mode data, we compare the retrieved wave parameters to collected buoy measurements, showing good consistency. The H-LUTI model, using a fetch-limited function, does the best among these 6 H-models, whose *rmse* and R^2 are 0.86 m and 0.77 for H_s, and 1.06 s and 0.76 for T_p, respectively. Results indicate the potential for H-models to simulate waves generated by typhoons/hurricanes.

Keywords: synthetic aperture radar (SAR); typhoon/hurricane-generated wind waves; fetch- and duration-limited wave growth relationships

1. Introduction

Wind-waves generated by extreme weather, such as typhoons or hurricanes, are among the most important dynamic elements of the marine environment [1–5]. A well-known example is the Perfect

Storm in October 1991, which is sometimes known by wave researchers as the Halloween Storm, which generated maximum waves in excess of 30 m [6]. The development of remote sensing has greatly motivated the studies of large storm waves [7–9], as in situ observations are spatially sparse and expensive undertakings [10–13]. Because of its ability to make measurements under almost all weather conditions, day or night, synthetic aperture radar (SAR) is an important way to monitor marine winds and waves.

Over the past decades, studies concerning SAR wave detection and the inversion algorithm methodologies have achieved some developments. Originally, Hasselmann et al. (1991) suggested a wave spectrum inversion algorithm for SAR imaging of waves [14]. On the basis of that approach, many successive nonlinear wave spectral algorithms were proposed, such as SPRA (Semi-Parametric Algorithm) [15], PFSM (Parameterized First-guess Spectrum Method) [16–18] and PARSA (Partition Rescaling and Shift Algorithm) [19]. These algorithms need to build a first guess wave spectrum, based on the additional wave information, which might be provided by a numerical wave model or parametric wave spectrum model. However, because SAR imaging of ocean waves is affected by nonlinear imaging mechanisms causing distortion of shorter waves [20,21], it is difficult to observe short wind waves, which is a key problem in observations of storm-generated waves.

Instead of using an image-to-wave spectra inversion scheme, alternative approaches were developed to empirically estimate integral wave parameters, such as significant wave heights or wave periods, from SAR images. CWAVE_ERS [21] and CWAVE_ENVI [22] are empirical models for ERS-2 and Envisat ASAR wave mode images. Based on the Sentinel-1 SAR wave mode, Stopa et al. proposed CWAVE-S1A and Fnn models to retrieve significant wave heights and dominant wave periods by using cut-off wavelength and other SAR image parameters [23]. The imaging range of the SAR wave mode is quite small (~20 km), making it impractical for large-scale monitoring of a wide range of observations of storm generated waves. Alternatively, ScanSAR mode images have a larger swath-width for observations, for example as much as 500 km for RADARSAT-2, making it applicable for monitoring large scale fields of storm winds and waves. However, its low spatial resolution, ranging from 60–100 m for RADARSAT-2, limits its observational ability for short wind waves [24–27].

Despite the complicated spatiotemporal distribution characteristics of the wind fields associated with storms, many studies show that most of measured 1-D wave spectra under storms are monomodal and suggest that the associated surface waves follow the same similarity concept as the waves that grow in response to steady winds in fetch-limited conditions [28–32]. Young [29] suggested that the nonlinear wave-wave interactions play a central role in stabilizing the shape of unimodal spectrum [33,34]. This is achieved by the transfer of wave energy within the spectrum, from high frequencies to low frequencies, and the other way around, from low to high frequencies [35,36], although the wind input and dissipation by breaking may also be important at this stage of wave growth [11,31,37]. Many parametric models were built on the assumption of the fetch-limited condition, particularly in the 1970s and 1980s [38], providing a rapid means to explore the general characteristics of waves generated by storms, especially the asymmetry of the wave field [39–41] and energy [32] inside the storms. Based on the fetch- or duration-limited wind wave growth relations, Hwang et al. [42–44] proposed three sets of wind wave models for storms, denoted as the H-models, which use the wind field to estimate the significant wave heights H_s and dominant wave periods T_p directly for the wind waves. Parameters of the wind waves can be estimated in three separate sectors (left, right, and back) of the storm by the H-3Sec model [42]. After analyzing more wind and wave spectral data generated by hurricanes and collected by the Scanning Radar Altimeter (SRA), Hwang and Walsh [43] proposed the H-LUT model, which provides a better azimuthal resolution for estimating the wind wave parameters inside storms. In addition, by considering the influence of the radius of maximum storm winds on the wave estimation, Hwang and Fan [44] further developed the H-Harm models.

At present, SAR has been widely applied for observations of typhoon wind fields, especially the cross-polarization SAR [45–50] which appears to be able to largely resolve the signal saturation problem at high winds, experienced in co-polarization SAR and scatterometer observations [51]. Therefore, by combing the relatively accurate storm winds determined from cross-polarization SAR measurements, with the H-models, it is possible to obtain the wind waves generated by typhoon and hurricane environments.

The remainder of the paper is organized as follows: Section 2 describes the Sentinel-1 SAR wave mode, RADARSAT-2 ScanSAR mode and the fetch- and duration-limited wind wave models (H-models). Section 3 shows the detailed validations of H-models for estimating significant wave heights for typhoon Krovanh and other typhoons using Sentinel-1 SAR wave mode wind and wind-wave data. By combing the buoy wave data and RADARSAT-2 ScanSAR mode winds, the H-models are applied to estimate significant wave heights and dominant wave periods for six additional hurricanes. Discussion follows in Section 4, and Conclusions in Section 5.

2. Data and Methods

2.1. Sentinel-1 SAR Wave Mode

Sentinel-1 (S1A) Level-2 Ocean Products, OCN, can provide wind speed and significant wave height from SAR wave mode. Significant wave heights for wind waves are calculated from the nonlinear part of the SAR image through the cross spectrum methodology [52] combined with the SPRA inversion algorithm [15]. S1A was launched on April 2014 and was the first satellite with a SAR observation system among the ESA Sentinel series. The S1A SAR wave mode was put into service in July 2015 for observations of waves over the global ocean. This SAR wave mode has 4 m spatial resolution, with a small imaging range (20 km × 20 km). In addition, S1A operates in the C-Band at two incidence angles, 23° and 36°, alternating along the satellite orbital direction at 100 km intervals. Thus, the S1A SAR wave mode can provide almost continuous sampling, collecting abundant simultaneous wind and wave measurements under storm conditions.

2.2. RADARSAT-2 ScanSAR Mode

RADARSAT-2 is a C-band spaceborne SAR, which was launched on 14 December 2007. The satellite has the capability to provide single-, dual- and quad-polarization SAR imaging mode data, day or night, in almost all-weather, with multi-spatial resolution of the sea surface. We focus on measurements from RADARSAT-2 cross-polarization ScanSAR mode, which provides wide swath (450 km) images and has a pixel spacing of 50 m, which is high potential for hurricane/typhoon monitoring over a relatively large spatial scale.

The wind speeds used to calculate waves are retrieved by the C-2POD (C-band cross-polarized ocean surface wind retrieval model) model [46]. The wind speed range of the fit for C-2POD is 3.7–39.7 m/s, which is an important motivation for using C-2POD to retrieve the wind speeds from dual-polarization SAR images. The model was previously validated against SFMR measurements and H*wind, and the comparisons showed good agreement.

2.3. The Fetch- and Duration-Limited Wind Wave Models (H-Models)

According to recent studies, the fetch- and duration-limited wind wave growth relations derived from steady wind forcing conditions appear to be applicable for waves generated by typhoon/hurricane winds [28–41], which conforms the essential role of nonlinear wave-wave interactions in maintaining the shape of wave spectrum. Based on this assumption, Hwang et al. successively proposed three sets of models for wind waves generated by hurricanes [42–44], which are summarized here. The fetch- and duration-limited wind wave growth relations, in terms of the dimensionless variance $\eta_\#$ and dimensionless frequency $\omega_\#$ as functions of significant wave height H_s and wave period T_p, can be represented as equations of dimensionless fetch $x_\#$:

$$\begin{cases} \eta_\# = 6.19 \times 10^{-7} x_\#^{0.81} \\ \omega_\# = 11.86 x_\#^{-0.24} \end{cases} \tag{1}$$

where $\eta_\# = H_s^2 g^2 (16 U_{10}^4)^{-1}$, $\omega_\# = 2\pi U_{10} (T_p g)^{-1}$, dimensionless fetch is $x_\# = x_f g U_{10}^{-2}$ and wind speed at 10 m height is U_{10}. In the same way, the equations of dimensionless duration $t_\#$ are:

$$\begin{cases} \eta_\# = 1.27 \times 10^{-8} t_\#^{1.06} \\ \omega_\# = 2.94 t_\#^{-0.34} \end{cases} \tag{2}$$

where dimensionless duration is $t_\# = t_d g U_{10}^{-1}$. The x_f and t_d are fetch and duration respectively. All of the parameters in (1) and (2) above are the functions of wind-wave triplets: H_s, T_p and U_{10}. Substituting these relations into Equations (1) and (2) leads to

$$\begin{cases} \dfrac{H_s^2 g^2}{16 U_{10}^4} = 6.19 \times 10^{-7} \left(\dfrac{x_f g}{U_{10}^2}\right)^{0.81} \\ \dfrac{2\pi U_{10}}{T_p g} = 11.86 \left(\dfrac{x_f g}{U_{10}^2}\right)^{-0.24} \end{cases} \tag{3}$$

and

$$\begin{cases} \dfrac{H_s^2 g^2}{16 U_{10}^4} = 1.27 \times 10^{-8} \left(\dfrac{t_d g}{U_{10}}\right)^{1.06} \\ \dfrac{2\pi U_{10}}{T_p g} = 2.94 \left(\dfrac{t_d g}{U_{10}}\right)^{-0.34} \end{cases} \tag{4}$$

As shown by these equations, given fetch x_f or duration t_d, H_s and T_p can then be directly calculated from the wind field U_{10}. Wave height and wave period calculated from the fetch-limited growth functions are denoted by FH_s and FT_p respectively. Wave height and wave period calculated from duration-limited growth functions are denoted by DH_s and DT_p respectively.

The key to estimating wind-wave information with these growth relations is to first determine the fetch and duration. For a finite water body with a well-defined land-water interface, the fetch and duration can be defined easily according to their definitions. However, it is difficult to directly obtain fetch and duration under inhomogeneous or unsteady wind situations, like storms. By making use of Equations (3) or (4), the equivalent fetch and equivalent duration inside the typhoon/hurricane can also be obtained from the wind-wave triplets. Based on SRA (Scanning Radar Altimeter) measurements of H_s, T_p and U_{10}, Hwang et al. [42–44] proposed three groups of typhoon/ hurricane fetch and duration models. They are:

(1) H-3Sec model

Using 60 wave spectra of hurricane Bonnie (1998), Hwang [42] gives the empirically formulas in terms of the fetch (unit: km) and duration (unit: h) for the three sectors of the storm (left, right, back). They are functions of the radial distance r (unit: km) from the hurricane center

$$x_{\eta x} = \begin{cases} -0.26r + 259.79, \; right \\ 1.25r + 58.25, \; left \\ 0.71r + 30.02, \; back \end{cases}, \quad x_{\omega x} = \begin{cases} 0.21r + 170.00, \; right \\ 2.25r + 24.85, \; left \\ 0.50r + 14.16, \; back \end{cases} \tag{5}$$

$$t_{\eta t} = \begin{cases} -0.0069r + 11.88, \; right \\ 0.066r + 3.78, left \\ 0.040r + 2.20, \; back \end{cases}, \quad t_{\omega t} = \begin{cases} 0.010r + 8.56, \; right \\ 0.110r + 2.12, \; left \\ 0.031r + 1.28, \; back \end{cases} \tag{6}$$

The subscripts ηx, ωx, ηt and ωt indicate the variables derived from the equations for $\eta_\#(x_\#)$, $\omega_\#(x_\#)$, $\eta_\#(t_\#)$, $\omega_\#(t_\#)$, respectively. This model is hereafter denoted as the H-3Sec model.

(2) H-LUT model

Hwang and Walsh [43] analyzed the full set of the SRA wave measurements collected during hurricane Bonnie (1998), which contains 233 spectra along 10 transect flights radiating from the hurricane center, and improved the ability of the H-3Sec model to simulate surface waves in the azimuthal and radial directions under storms. The fetch and duration relations are represented as

$$x_{\eta x}(r, \phi) = s_{\eta x}(\phi)r + I_{\eta x}(\phi), x_{\omega x}(r, \phi) = s_{\omega x}(\phi)r + I_{\omega x}(\phi) \tag{7}$$

$$t_{\eta t}(r, \phi) = s_{\eta t}(\phi)r + I_{\eta t}(\phi), t_{\omega t}(r, \phi) = s_{\omega t}(\phi)r + I_{\omega t}(\phi) \tag{8}$$

The slopes $s_{\eta x}$, $s_{\omega x}$, $s_{\eta t}$, $s_{\omega t}$ and intercepts $I_{\eta x}$, $I_{\omega x}$, $I_{\eta t}$, $I_{\omega t}$ in the equations are functions of azimuth angle ϕ, which is referenced to the direction of the hurricane motion. The convention is that the angles are positive counterclockwise. Similarly, an alternative set of empirical coefficients is obtained through analyzing the data from hurricane Ivan (2004). The corresponding parametric models above are denoted as the 'H-LUTB' model and 'H-LUTI' model, respectively.

(3) H-Harm model

By considering the influence of the radius of the maximum wind speed in storms in the fetch- and duration-limited simulations, the third set of models was proposed [44]. Similar to the H-LUT model, fetch and duration are also expressed with radial components r and ϕ, like Equations (7) and (8). The slope and intercept can be decomposed as Fourier series,

$$q = a_0 + 2 \sum_{n=1}^{N} (a_{n,q} \cos n\phi + b_{n,q} \sin n\phi) \tag{9}$$

where q represents the slopes $s_{\eta x}$, $s_{\omega x}$, $s_{\eta t}$, $s_{\omega t}$ and the intercepts $I_{\eta x}$, $I_{\omega x}$, $I_{\eta t}$, $I_{\omega t}$. The harmonic parameters $a_{n,q}$ and $b_{n,q}$ exhibit a systematic quasi-linear variation with the radius of maximum wind speed,

$$Y = P_{1Y} r_m + P_{2Y} \tag{10}$$

where Y represents amplitudes and $b_{n,q}$; and P_{1Y} and P_{2Y} are empirical coefficients. Given the storm wind field, $a_{n,q}$ and $b_{n,q}$ can be calculated by Equation (10), using the radius of maximum wind speed r_m. Inserting these variables into Equation (9), the slopes and intercepts in Equations (7) and (8) can be computed. Thus, with the wind speed input, we can determine the equivalent fetches and durations for any point inside the storms by (7) and (8).

Please note the value of N in the Fourier series (9) might affect the results. We denote the models with $N = 1, 2, 3$ as 'H-Harm1', 'H-Harm2', and 'H-Harm3' respectively. In addition, because of the lack of wave period data in S1A Level-2 products, we can only make comparisons with the significant wave heights for typhoon/hurricane–generated waves using wind measurements of S1A SAR wave mode data.

3. Results

3.1. Validation of H-Models by Sentinel-1A SAR Wave Mode Wind and Wave Data

3.1.1. Typhoon Krovanh

Between 14 and 21 September 2015, Sentinel-1A (S1A) tracks passed over typhoon Krovanh (2015) three times, acquiring 17 SAR wave mode images (Figure 1a). On 14 September, Krovanh proceeded to the east of the Mariana Islands, and then moved northwest. As S1A moved along its descending track on 15 September, S1A captured 6 SAR wave mode images of Krovanh. Because this acquisition occurred within a minute, the measurements of the SAR wave mode images can be regarded as instantaneous sampling relative to the typhoon time scales. The corresponding typhoon center is shown as a red cross in Figure 1a, and the radius of maximum wind speed and the maximum wind speed are 61.7 km and 21.4 m/s, respectively, at that moment. Under the typhoon reference frame (Figure 1b), it is shown that the SAR measurements occurred in the rear portions of both the left and right regions. As Krovanh continued to move northwest, the maximum wind speed continued to increase, up to 33 m/s on 16 September. Thereafter, S1A captured additional four SAR images around Krovanh. The sampling areas are located on the right side of the typhoon. The maximum wind speed reached 40.1 m/s, with radius reduced to 27 km, while the air-sea exchange remained relatively strong. The last sampling by S1A of typhoon Krovanh occurred on September 20, as the typhoon began to dissipate. At that time, the maximum wind speed decreased to 20.6 m/s and the radius increased to 74.08 km. The typhoon track, maximum wind speed and radius of maximum wind speed are provided by the Join Typhoon Warning Centre (JTWC).

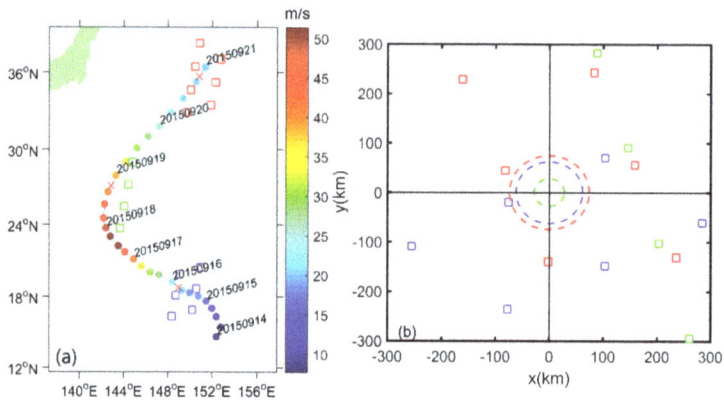

Figure 1. (**a**) The best track for typhoon Krovanh. The maximum wind speed (m/s) is denoted by the color of circles. Red crosses represent the typhoon centers as S1A passed through. The colored squares represent the SAR wave mode images observed by S1A (1st pass: blue; 2nd pass: green; 3rd pass: red). (**b**) SAR wave mode images in the typhoon reference frame (colored squares). The corresponding radius of maximum wind speed is denoted by dashed line. In (**b**), the coordinates (x, y) are rotated such that the typhoon heading is toward the top of the page. The x and y present the left-right and front-back distances with respect to the typhoon center, respectively.

By comparing H-model simulation results (FH_s and DH_s) to significant wave heights from S1A Level-2 products (Figure 2), it is found that H_s is underestimated to different extents by each model. Values from fetch models FH_s are typically slightly larger than those from duration models DH_s, and are closer to those measured by S1A. Each H-model has a better performance for the simulation of H_s less than 5 m. However, these models have large differences when simulating regions where the high waves dominate. In particular, estimates from the H-3Sec model seriously underestimate H_s in cases larger than 4 m. Secondly, the set of H-Harm models all have similar simulation performances, although in regions of higher significant wave heights, results from H-Harm2 and H-Harm3 appear to perform slightly better than the simulation by the H-Harm1 model. We conclude that the H-LUT models work best in these sets of models; and the H-LUTI model is clearly better than the H-LUTB model.

In Figure 2, the measurements by S1A are marked with three different colors (of squares), corresponding to those shown in Figure 1a (1st pass: Blue; 2nd pass: Green; 3rd pass: Red), which occurred at different stages of the typhoon development: Generation stage, full maturity stage and decay stage. The highest values of H_s occurred during the typhoon decay stage. Accordingly, the deviations in simulations for these H-models are largest during this stage. Although the maximum wind speed and *radius* of maximum wind speed during the generation stage are quite similar to those during the collapse of the typhoon, the values for H_s in this former stage are relatively smaller. Of course, this follows the processes that form the growth and development of the waves as the storm progresses through the stages of its life cycle. Thus, the performance of the H_s simulation models is good during the generation stage. The main areas with large deviations are near the radius of maximum wind speed ($r = 78.19$ km, $\phi = 104°$). Previous studies [53] have shown that wider angles between wind and wave propagation directions are always in the left regions of the typhoon, which indicates that the wave field is dominated by swell. However, H-modes are based on the wind-wave growth relationships, which perform poorly in the left regions of the typhoons.

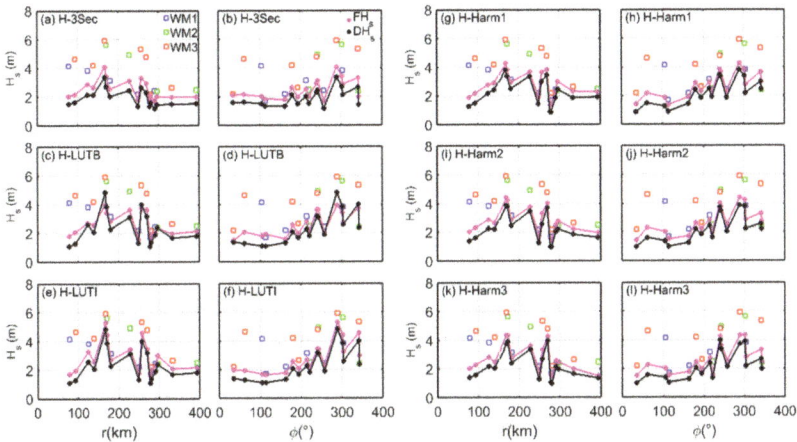

Figure 2. Radial (**a,c,e,g,i,k**) and azimuthal (**b,d,f,h,j,l**) variations of H_s provided by S1A Level-2 products (colored squares) and modeled by H-models (FH_s: magenta solid points; DH_s: black solid points). The colors of the squares denote the sampling sequence (1st pass: blue; 2nd pass: Green; 3rd pass: red).

3.1.2. 12 Pacific Typhoons

From July 2015 to December 2016, S1A acquired 86 SAR wave modes from 12 typhoons in the Northwest Pacific. Those SAR images are located less than 400 km from the typhoon center. For our selection condition, the maximum wind speed of the typhoon is relaxed to 20 m/s, including some conditions related to measurements in the generation and decay stages of the typhoon. Comparing the H-model simulation results to significant wave heights estimated by S1A, the scatter plots for wave heights (Figure 3) show that the models underestimate H_s to various extents, depending on the specific model; the deviations for results from DH_s are generally greater than those from FH_s.

The deviation of H_s from H-3Sec appears to be the most significant, especially for DH_s. The root mean square error (*rmse*) reaches to 1.86 m. The models in the H-LUT group have a much finer resolution in the azimuthal direction and therefore better represent the surface wave development inside the typhoon than the H-3Sec model. Essentially, the results from H-LUTI model using the fetch-limited function have a bias of -0.71 m and *rmse* of 1.20 m, which are better than results from H-LUTB model. The H-Harm model group doesn't perform very well in terms of the simulation error, though it is the only group of models that consider the impact of the radius of maximum wind speed. By increasing the number of components N in the Fourier series in Equation (9), the deviation and root mean square error of the associated models are reduced slightly. The model with $N = 3$, namely H-Harm3 appears to have the best performance among this group of models, corresponding to a deviation and *rmse* of -0.89 m and 1.45 m, respectively.

With respect to the correlation coefficient (R^2) for the model results, H-3Sec and the H-LUT group have similar performances, with high values for R^2 (about 0.84). They demonstrate that these models can simulate the H_s spatial distribution inside the typhoon well. The third set of models, H-Harm, has relatively poor correlation coefficients (average, 0.68). H-Harm3 is the best among this group with a correlation coefficient of 0.75 for FH_s, and 0.71 for DH_s. Therefore, based on this analysis, the H-LUTI model (with fetch) produces the most reliable spatial distribution of H_s, for a given wind speed, and has the least bias among these models.

Figure 3 compares estimates of H_s, as modeled by the H-models, with respect to measurements from S1A. The color bar represents the normalized radial distance r/r_m from the typhoon center. Higher values of r/r_m represent greater distance from the center of the typhoon. All model results show similar characteristics, namely that simulations close to the center of the typhoon have greater deviation from the H_s, compared to locations farther from the typhoon center. Taking the results from FH_s of the H-LUTI model as an example, the spatial distribution of simulation errors in waves generated by typhoons is further discussed below.

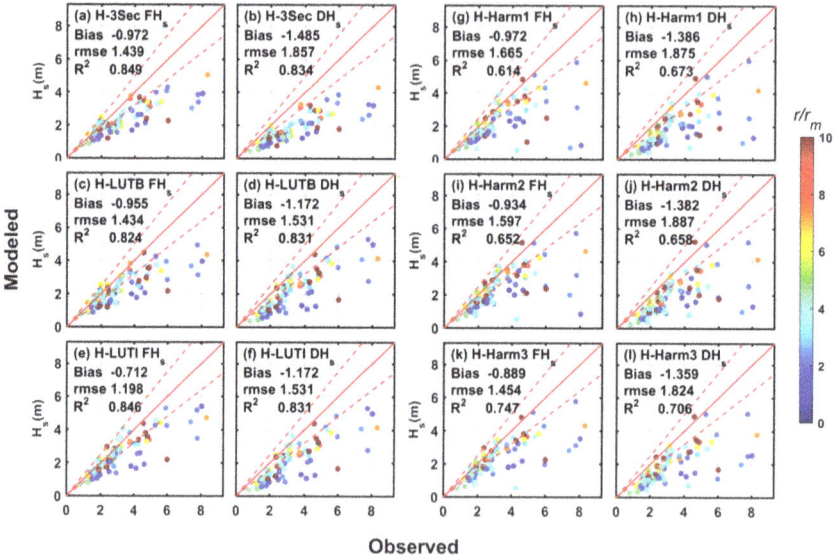

Figure 3. Comparison of H_s provided by S1A measurements and modeled by H-models. The color represents the normalized distance from the typhoon center (r/r_m). For reference, the red lines correspond to 1:1 (solid line) and 1:1.25 (1.25:1) (dashed lines).

The relative deviations of wave height $bias_r = |H_s - FH_s|/H_s$ for the H-LUTI model are shown, with the typhoon reference frame in Figure 4a. Overall, the simulated FH_s values by H-LUTI model are in good agreement with significant wave heights from S1A. However, the relative deviation of the model is larger near the typhoon center and our results are in good agreement with those of Hwang et al. [42–44]. They suggested that these deviations may result from the presence of swell contamination, which may also be complicated by the processes in the typhoon centers and not resolved by relatively simple models. The $bias_r$ values are shown as a function of normalized radial distance in Figure 4b. The H-LUTI model has maximum relative deviation (about 60%) near the radius of maximum wind speed. Moreover, $bias_r$ constantly decreases with increasing normalized radial distance, until r/r_m is more than 2.5, when the $bias_r$ is kept within about 25%.

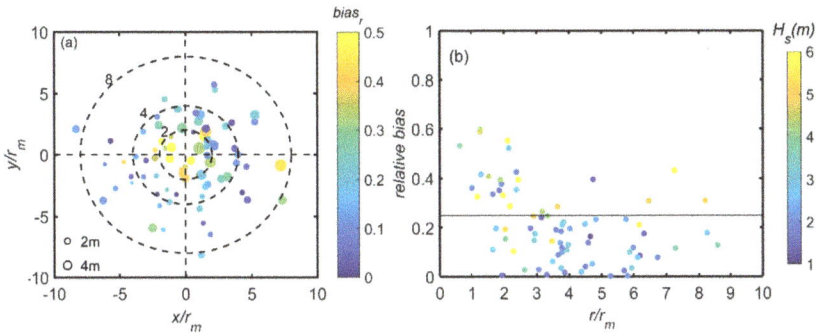

Figure 4. (**a**) The relative bias of FH_s modeled by H-LUTI shown in the typhoon reference frame, with the typhoon center located at the center of the figure and the coordinates showing the distance to the typhoon center scaled by the radius of the maximum wind speed r_m. The propagation direction of typhoon is toward the top of the page. The color and size of the solid circles correspond to $bias_r$ and H_s, respectively. The larger concentric dashed-line cycles indicate r/r_m =2, 4 and 8. (**b**) the relative bias as the function of r/r_m. The color denotes the values for H_s as measured by S1A. The solid line marks the $bias_r = 0.25$.

Without including the area near the typhoon center, where r/r_m is less than 2.5, the deviation and *rmse* for FH_s simulated by H-LUTI model are reduced to −0.43 m and 0.88 m, respectively. The corresponding correlation coefficient reaches 0.86. Moreover, other H-models also are significantly improved, as shown in Figure 5.

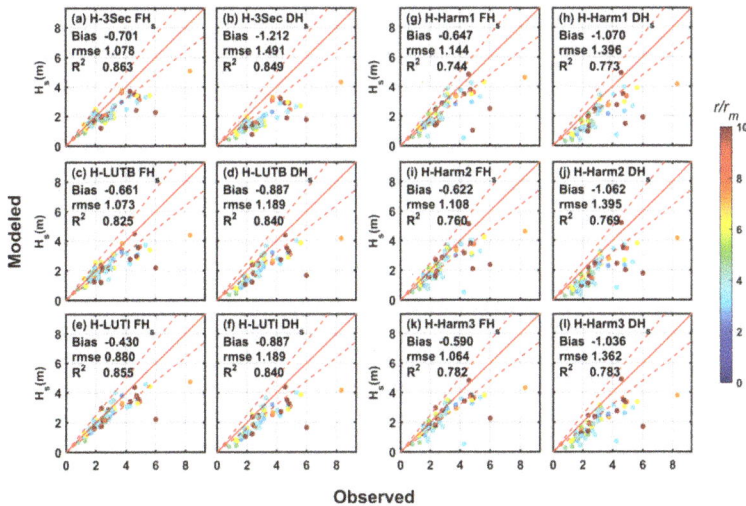

Figure 5. As Figure 4 but for comparison of H_s provided by S1A measurements to modeled estimates provided by 6 H-models except near the eye region ($r/r_m < 2.5$).

In summary, other than the area near the typhoon center, significant wave heights can be estimated well by using the H-models, driven by of SAR-derived wind speed data. Of all the models, the best one is H-LUTI, which agrees well with the wave heights obtained from S1A measurements.

3.2. Wind Waves from RADARSAT-2 ScanSAR Mode Hurricane Winds

3.2.1. Validation by Wave Buoys

We collected 6 RADARSAT-2 cross-polarization (VH) SAR images covering six hurricanes acquired during the 2007–2017, collocated with 7 National Data Buoy Center (NDBC) buoys in the Gulf of Mexico and northwest Atlantic. The 6 SAR images include the centers of these hurricanes, as shown in Figure 6. Thus, these ScanSAR mode images capture part or the entire hurricane core, not as the wave mode images discussed above which only captured a small-range measurement of the hurricane. The best track data of the hurricanes was provided by NOAA, by the Extended Best Track Dataset. For each hurricane, SAR measurements are required to be within a 30-min window. Since a hurricane continues to move and rotate during this time window, we define the location of hurricane center by interpolation of the time series. A summary of the information for these hurricanes, including the locations of the hurricane center, maximum wind speeds and their radii, is given in Table 1. By using the C-2POD model, we directly obtained the wind speeds from these images, as shown in Figure 7.

Table 1. Basic information of hurricanes, Bill 2009, Earl 2010, Igor 2010, Ingrid 2013, Arthur 2014 and Ana 2015. The u_m and r_m represent maximum wind speed and radius of maximum wind speed.

Hurricane Names	Date (yyyy-mm-dd)	Time (UTC)	Center		u_m (m/s)	r_m (km)
			Latitude	Longitude		
Bill	2009-08-23	10:40:56	41.89°N	−65.82°E	36.57	74.08
Earl	2010-08-30	09:57:38	18.36°N	−62.69°E	52.26	49.45
Igor	2010-09-19	10:11:24	29.24°N	−65.48°E	38.58	92.60
Ingrid	2013-09-15	00:20:59	21.62°N	−94.73°E	38.58	37.04
Arthur	2014-07-03	11:13:56	31.68°N	−78.84°E	40.49	39.41
Ana	2015-05-09	23:24:12	33.06°N	−78.27°E	23.15	74.08

The red circles in Figures 6 and 7 show the locations of the buoys. A total of 7 buoys collected hourly wind speed and the wave spectra. Because H-models only apply to the wind waves, the wind waves are separated from wave spectra $S(f)$ by the wave steepness method [54] developed by NDBC, in order to validate the models with buoy data. The significant wave heights of the wind sea $H_s = 4\sqrt{\int_{f_s}^{f_u} S(f)df}$ are used to validate the H-models' results, where f_u is the upper frequency limit for wave spectra measurements and f_s is the estimated separation frequency. All buoy wind speeds measured at different heights were adjusted to a reference level 10 m following [55]. To match the observation time of SAR images and buoys, the wave parameters and wind speeds are averaged over hourly intervals.

The C-2POD model is utilized to retrieve the hurricane wind fields from the SAR images, which also shows good agreement with the buoy wind measurements in this study (Figure 7g). Using the retrieved wind, the wave height and peak period can be estimated by 6 H-models with fetch- and duration- limited growth functions. Figure 8 presents the wave height comparisons between buoy measurements and retrievals from these SAR winds. The comparison with buoy wave heights supports some conclusions from the wave mode analysis. The computed wave heights using the fetch models are more accurate than those using the duration models, with smaller *rmse* and greater R^2. The duration computations contain more underestimates, causing a negative *bias* in many H-models. Moreover, the H-3Sec model and the set of H-LUT models are considerably better than H-Harm models (*rmse* values of 0.67 m to 0.98 m vs. 1.00 m to 1.21 m; correlation coefficients of 0.73 to 0.89 vs. 0.67 to 0.75). The H-LUTI model using the fetch-limited function has *rmse* of 0.86 m and R^2 of 0.77 here. It is also found that H-Harm3 with a higher value of N in the Fourier series (9) does the best simulation among the H-Harm models, whose *rmse* is 1.00 m and R^2 is 0.75 for the fetch model result.

In Figure 9, the retrieved T_p is compared with buoy data. The negative *bias* for each of the H-models implies a tendency of those models to slightly underestimate the wave period. Similar to the results for the wave heights, the fetch models have better behavior in simulating T_p than the duration models, with *rmse* of 1.06 s to 1.40 s. The simulated FT_p of H-LUTI model has the least *rmse* of 1.06 s and the highest R^2 of 0.76 among the 6 H-models, which illustrates that the H-LUTI model is effective to simulate the dominant wave periods using the fetch-limited function. The wavelengths can be estimated from the dominant wave periods (approximately from 6 s to 10 s) according to the dispersion relationship, taking the water depths of the buoys into account. Thus, wind waves corresponding to 6 s~10 s, with wavelengths less than 150 m, can be retrieved, although they cannot be imaged directly by SAR because of the limited spatial resolution [24] and because of the cutoff caused by velocity bunching [14,15], especially in high sea states typical of tropical storms (e.g., larger than 450 m) [23]. The good agreement with buoy measurements is encouraging and indicates the possibilities for H-models to calculate dominant wave periods under hurricane conditions.

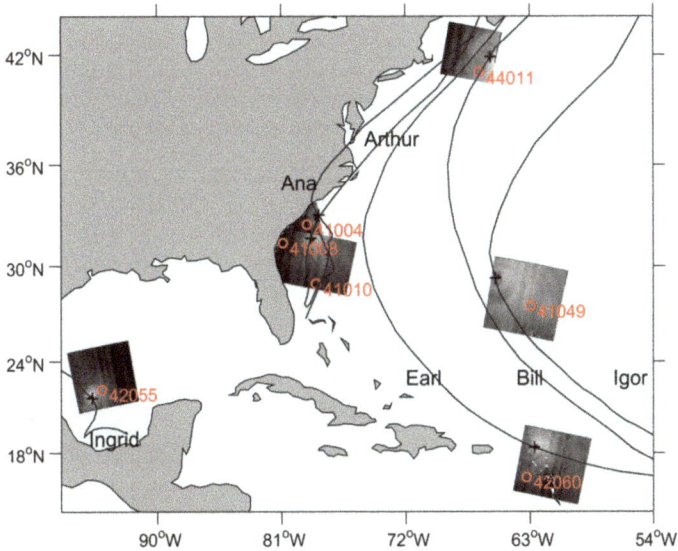

Figure 6. The RADARSAT-2 ScanSAR images for selected hurricanes with the best track data set; the 7 buoys covered by SAR images are presented as red circles.

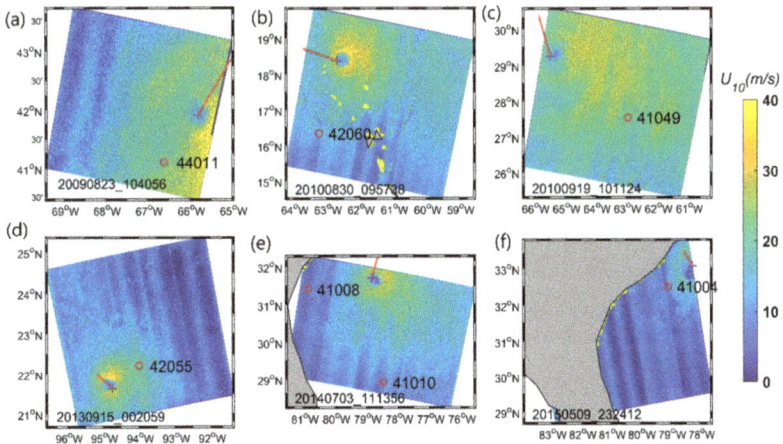

Figure 7. *Cont.*

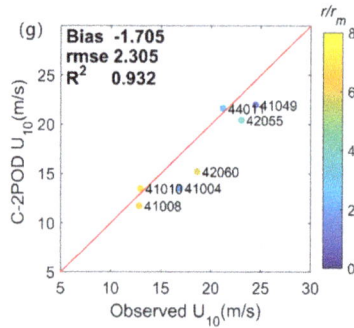

Figure 7. (**a–f**) Hurricane wind speeds retrieved by C-2POD wind retrieval model for the 6 SAR images shown in Figure 6. The superimposed arrows show the hurricane heading direction with the root of the arrow at the hurricane center. The length of the arrows represents the velocity of forward movement of hurricanes: (**a**) 55.19 km/h (**b**) 23.40 km/h (**c**) 23.40 km/h (**d**) 11.85 km/h (**e**) 18.65 km/h (**f**) 9.25 km/h. (**g**) Comparisons of retrieved wind speeds with collected buoys measurements.

Since buoys tend to have reduced observational capabilities when the wind speeds approach hurricane force conditions, there were few buoys that were able to still function and to be captured in SRA images when hurricanes pass. Although buoys used to validate the H-models are quite limited, the results for the H-models robustly agree well with the buoy measurements. Therefore, H-models can potentially be used to retrieve the wave heights and peak periods from winds retrieved from RADARSAT-2 ScanSAR images, for example with application of the C-2POD model, taking advantage of cross-polarization SAR with its good sensitivity to higher wind speeds.

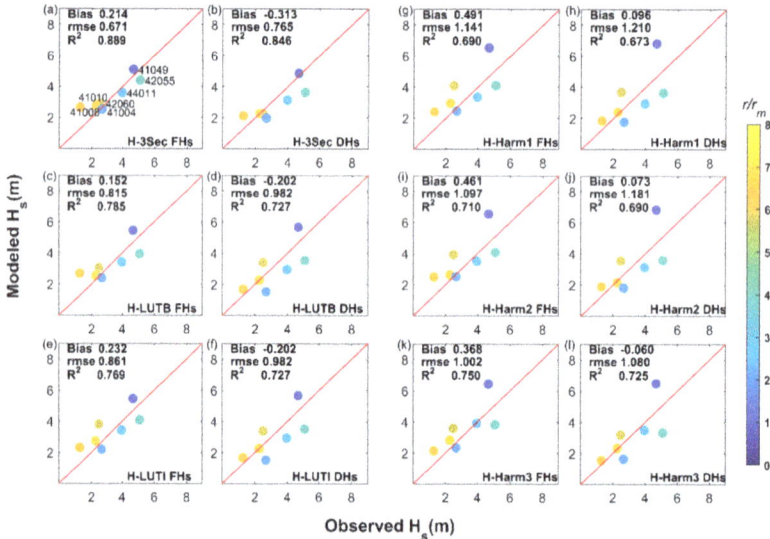

Figure 8. Comparison of H_s retrieval from SAR-derived winds, using fetch- (**a,c,e,g,i,k**) or duration-limited (**b,d,f,h,j,l**) growth models, and the buoy observations. Results for 6 models are presented: H-3Sec, H-LUTB, H-LUTI, H-Harm1, H-Harm2, and H-Harm3.

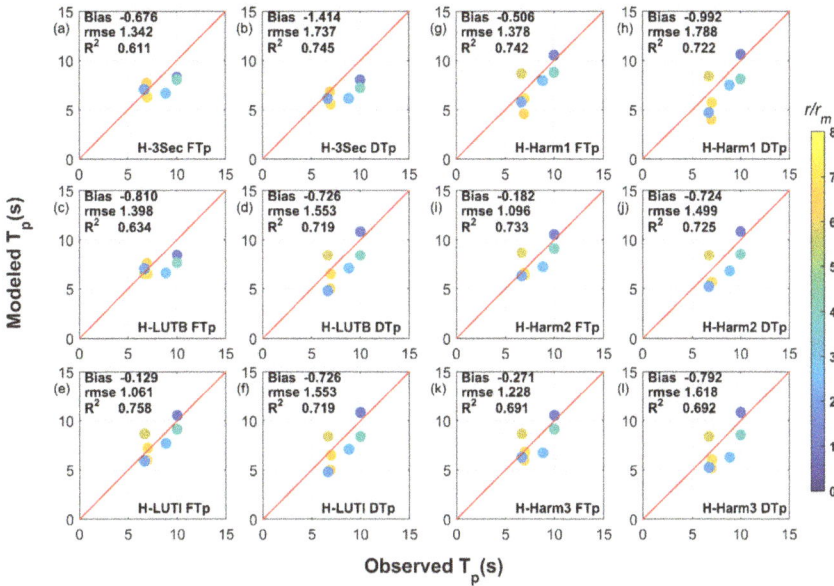

Figure 9. Comparison of T_p retrieved from SAR-derived winds, using fetch- (**a,c,e,g,i,k**) or duration-limited (**b,d,f,h,j,l**) growth models, and the buoy observations. Results for 6 models are presented: H-3Sec, H-LUTB, H-LUTI, H-Harm1, H-Harm2, and H-Harm3.

3.2.2. 2-Dimensional Application

Figure 10a shows a RADARSAT-2 SAR image (only VH-channel image) acquired in ScanSAR mode with a 500 km swath for hurricane Gustav (2008). On basis of the VH-polarization SAR image, we can generate a wind map (Figure 10b) using a newly developed wind-retrieval algorithm, Symmetric Hurricane Estimates for Wind (SHEW) model [49]. The 2D hurricane winds are plotted, with the hurricane heading direction pointing toward the top of the page (unit: km). We only show the main area controlled by the hurricane, with the wind field calculated according to the symmetry of hurricane.

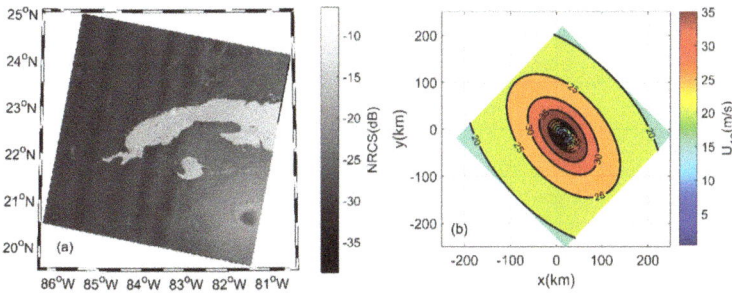

Figure 10. (**a**) RADARSAT-2 ScanSAR image acquired over hurricane Gustav at 1128UTC 30 August 2008. (**b**) SAR-retrieved wind speed.

By using the H-LUTI model with the SAR-derived hurricane wind field, the significant wave heights and wave periods can be estimated (Figure 11). The results show that the location of maximum wave heights is within the right front regions, which is consistent with the previous studies of Young [56]. Since the wind vectors tend to be approximately aligned with the direction of propagation of the hurricane, waves generated in this area tend to move forward with the hurricane and hence remain in the intense wind regions for extended period of time (extended fetch), conversely, to the left of the hurricane center. As a result, the spatial distribution of the wave field is not exactly symmetric.

The simulation of T_p is shown in Figure 11c,d. Although the simulation results from the H-LUTI model can describe the wave distribution features well for the longer wave periods on the right front side of the hurricane, the accuracy of the results still needs to be verified in additional studies with more buoy measurements.

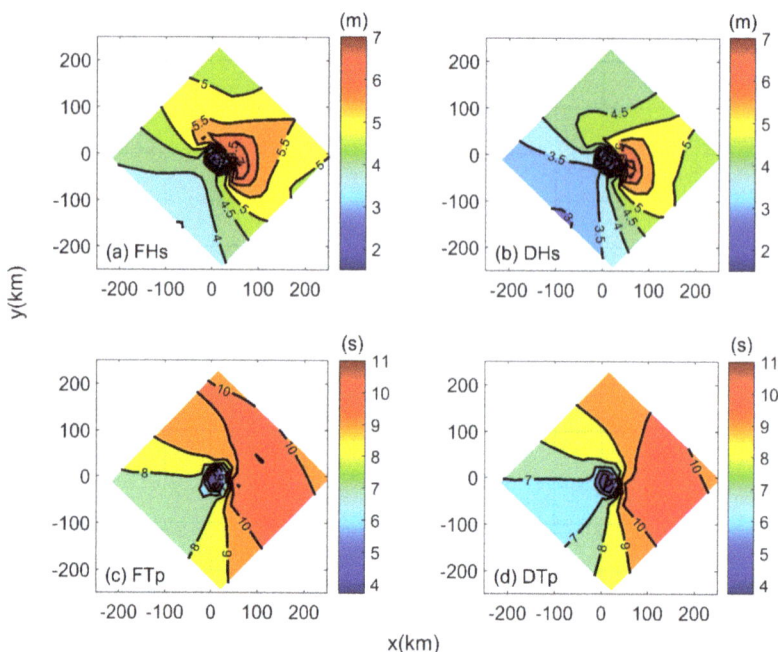

Figure 11. (a) Wave height from fetch-limited growth function FH_s, (b) wave height from duration-limited growth function DH_s, (c) wave period from fetch-limited growth function FT_p and (d) wave period from duration-limited growth function DT_p modeled by H-LUTI.

4. Discussion

All of the models are based on the implicit initial assumption regarding the essential role of nonlinear wave-wave interactions in maintaining the wave spectrum similarity. Moreover, many studies have shown that most of the spectra are monomodal under extreme conditions, similar to the spectra generated under fetch-limited, steady wind conditions. However, bi-modal spectra are also found in both measurements and model results under intense cyclone conditions [11,29,31], in which case the accuracy of the parametric models used in this study can be degraded. For instance, this is the case in the cyclone's left forward quadrant where the direction of wind deviates considerably from the wave direction [31]. As shown in a previous study [11], the analysis of directional spectra observed by Extreme Air-Sea Interaction buoys shows that a variety of spectral geometries can exist close to the eyes of typhoons. Thus, the effectiveness of the simple fetch-limited parametric models should be further discussed, in terms of the cyclone quadrant under consideration, and the rate of development or change the intensity of the storm.

In this context, we suggest that the H-LUTI model is the best among the three H-models. Regarding the original studies that developed the H-models, it is not difficult to infer the reasons for this result. *Firstly*, the H-LUT models simulate wind wave parameters along comparatively more transects radiating from the storm center, improving on the H-3Sec model's ability to simulate the development of the azimuthal and radial variations of the surface waves. Moreover, as shown in Figure 1 of [44], the dataset from Ivan (2014) used to develop the H-LUTI model contains more observational transects inside the storm coverage region than the dataset from Bonnie (1998) used in development of the H-LUTB model, thus providing a better azimuthal resolution for fitting the empirical model. Many previous studies [39–41,56] clearly demonstrate that the equivalent fetch and duration of storm is associated with the relative position of the storm, the velocity of forward movement of the storm, maximum wind speed and the radius of maximum wind speed. However, for the third set of models, the H-Harm models, a systematic quasi-linear variation of the harmonic parameters $a_{n,q}$ and $b_{n,q}$ with the radius of maximum wind r_m (Equation (10)) was established based on only 4 storm datasets. These 4 datasets have different values for r_m, which have only a limited coverage range, leading to their relatively poor performance. Therefore, in future studies, it is particularly important to collect a large number of simultaneous wind and wave measurements under conditions appropriate for these storms in order to optimize the wave model.

5. Conclusions

Making use of the fetch- or duration-limited H-models, the basic typhoon/hurricane wind wave parameters can be estimated by only using the SAR-derived wind field data. This approach provides a new method for detecting typhoon/hurricane wind waves from SAR measurements. We show that the H-models can effectively calculate the significant wave heights inside the typhoon based on wind observations from Sentinel-1A (S1A) SAR images, except in the area near the typhoon center. Comparing the results with wave heights measured by S1A, we show that the wave heights calculated from the fetch-limited function (FH_s) are always larger than those calculated by the duration-limited function (DH_s), and in good agreement with the S1A wave height estimates. Among the results of these three set of H-models, the best one is the H-LUTI model using the fetch-limited function, which has a root mean square error of 0.88 m, and correlation coefficient of 0.86. Operating in ScanSAR mode, the H-models also have the potential to reliably simulate H_s and T_p for wind waves inside hurricanes from RADARSAT-2 ScanSAR mode observations, based on similar statistical properties derived from verifications by buoy data. The H-LUTI model is especially notable with results using the fetch function that are good, with *rmse* of 0.86 m and R^2 of 0.77 for H_s, and *rmse* of 1.06 s and R^2 of 0.76 for T_p. Furthermore, this model works well to describe the high values of significant wave heights and dominant wave periods in the right frontal regions of the typhoons/hurricanes.

Author Contributions: G.L. contributed to the idea of this study and suggested for the topic; L.Z. collected and analyzed the data; L.Z. wrote the original draft; G.Z. helped a lot to prepare the wind finds; G.L., W.P. and Y.H. assisted in manuscript preparation and revision.

Funding: This research was funded by the National Key Research and Development Program of China 2016YFC1401407; National Natural Science Foundation of China under Grant 41506028; Natural Science Foundation of Jiangsu Province under Grant BK20150913; the Startup Foundation for Introducing Talent of NUIST; the International cooperation project of National Natural Science Foundation of China under Grant 41620104003; National Program on Global Change and Air-Sea Interaction under Grant GASI-IPOVAI-04; the Office of Energy Research and Development (OERD) project 1B00.003C; the Canadian Space Agency Data Utilization and Applications Program (DUAP) project 14SURM006; MEOPAR (Marine Environmental Observation Prediction and Response Network) project 1P1.2 and in part by National Natural Science Youth Foundation of China under Grant 41706193.

Acknowledgments: We thank Paul Hwang for sharing the fetch- and duration-limited parametric models' codes at https://www.researchgate.net/publication/315772258_HurricaneFetchDurPackage.

Conflicts of Interest: The authors declare no conflict of interest.

References

1. Craig, P.D.; Banner, M.L. Modeling wave-enhanced turbulence in the ocean surface layer. *J. Phys. Oceanogr.* **1994**, *24*, 2546–2559. [CrossRef]
2. Craig, P.D. Velocity profiles and surface roughness under breaking waves. *J. Geophys. Res.* **1996**, *101*, 1265–1277. [CrossRef]
3. Toffoli, A.; McConochie, J.; Ghantous, M.; Loffredo, L.; Babanin, A.V. The effect of wave-induced turbulence on the ocean mixed layer during tropical cyclones: Field observations on the Australian North-West Shelf. *J. Geophys. Res.* **2012**, *117*, 1–8. [CrossRef]
4. Reichl, B.G.; Wang, D.; Hara, T.; Ginis, I.; Kukulka, T. Langmuir turbulence parameterization in tropical cyclone conditions. *J. Phys. Oceanogr.* **2016**, *46*, 863–886. [CrossRef]
5. Perrie, W.; Toulany, B.; Roland, A.; Dutour-Sikiric, M.; Chen, C.; Beardsley, R.C.; Chen, C.; Beardsley, R.C.; Qi, J.; Hu, Y.; et al. Modeling North Atlantic Nor'easters with modern wave forecast models. *J. Geophys. Res.* **2018**, *123*, 533–557. [CrossRef]
6. Cardone, V.J.; Jensen, R.E.; Resio, D.T.; Swail, V.R.; Cox, A.T. Evaluation of contemporary ocean wave models in rare extreme events: The Halloween Storm of October 1991 and the Stormof the Century of March 1993. *J. Atmos. Ocean. Technol.* **1996**, *13*, 198–230. [CrossRef]
7. Beal, R.C.; Gerling, T.W.; Irvine, D.E.; Monaldo, F.M. Spatial variations of ocean wave directional spectra from the Seasat synthetic aperture radar. *J. Geophys. Res.* **1986**, *91*, 2433–2449. [CrossRef]
8. Wright, C.W.; Walsh, E.J.; Vandemark, D.; Krabill, W.B.; Garcia, A.W.; Houston, S.H.; Powell, M.D.; Black, P.G.; Marks, F.D. Hurricane directional wave spectrum spatial variation in the open ocean. *J. Phys. Oceanogr.* **2001**, *31*, 2472–2488. [CrossRef]
9. Walsh, E.J.; Wright, C.W.; Vandemark, D.; Krabill, W.B.; Garcia, A.W.; Houston, S.H.; Murillo, S.T.; Powell, M.D.; Black, P.G.; Marks, F.D., Jr. Hurricane directional wave spectrum spatial variation at landfall. *J. Phys. Oceanogr.* **2002**, *32*, 1667–1684. [CrossRef]
10. Forristall, G.Z.; Ward, E.G.; Cardone, V.J.; Borgmann, L.E. The directional spectra and kinematics of surface gravity waves in tropical storm Delia. *J. Phys. Oceanogr.* **1978**, *8*, 888–909. [CrossRef]
11. Collins, C.O.; Potter, H.; Lund, B.; Tamura, H.; Graber, H.C. Directional wave spectra observed during intense tropical cyclones. *J. Geophys. Res.* **2018**, *123*, 773–793. [CrossRef]
12. Collins, C.O. Typhoon Generated Surface Gravity Waves Measured by NOMAD-Type Buoys. Ph.D. Thesis, University of Miami, Coral Gables, FL, USA, 2014.
13. Xu, Y.; He, H.; Song, J.; Hou, Y.; Li, F. Observations and Modeling of Typhoon Waves in the South China Sea. *J. Phys. Oceanogr.* **2017**, *47*, 1307–1324. [CrossRef]
14. Hasselmann, K.; Hasselmann, S. On the nonlinear mapping of an ocean wave spectrum into a synthetic aperture radar image spectrum and its inversion. *J. Geophys. Res.* **1991**, *96*, 10713–10729. [CrossRef]
15. Mastenbroek, C.D.; Valk, C.D. A semiparametric algorithm to retrieve ocean wave spectra from synthetic aperture radar. *J. Geophys. Res.* **2000**, *105*, 3497–3516. [CrossRef]
16. Sun, J.; Guan, C.L. Parameterized first-guess spectrum method for retrieving directional spectrum of swell-dominated waves and huge waves from SAR images. *Chin. J. Oceanol. Limnol.* **2006**, *24*, 12–20.
17. Sun, J.; Kawamura, H. Retrieval of surface wave parameters from SAR images and their validation in the coastal seas around Japan. *J. Oceanogr.* **2009**, *65*, 567–577. [CrossRef]
18. Shao, J.; Li, X.; Sun, J. Ocean wave parameters retrieval from TerraSAR-X images validated against buoy measurements and model results. *Remote Sens.* **2015**, *7*, 12815–12828. [CrossRef]
19. Schulz-Stellenfleth, J.; Lehner, S.; Hoja, D. A parametric scheme for the retrieval of two-dimensional ocean wave spectra from synthetic aperture radar look cross spectra. *J. Geophys. Res.* **2005**, *110*, 297–314. [CrossRef]
20. Hasselmann, K.; Raney, R.K.; Plant, W.J.; Alpers, W.; Shuchman, R.A.; Lyzenga, D.R.; Rufenach, C.L.; Tucker, M.J. Theory of synthetic aperture radar ocean imaging: A MARSEN view. *J. Geophys. Res.* **2005**, *90*, 4659–4686. [CrossRef]
21. Schulz-Stellenfleth, J.; König, T.; Lehner, S. An empirical approach for the retrieval of integral ocean wave parameters from synthetic aperture radar data. *J. Geophys. Res.* **2007**, *112*, 1–14. [CrossRef]
22. Li, X.; Lehner, S.; Bruns, T. Ocean wave integral parameter measurements using Envisat ASAR wave mode data. *IEEE Trans. Geosci. Remote Sens.* **2011**, *49*, 155–174. [CrossRef]

23. Stopa, J.E.; Mouche, A. Significant wave heights from Sentinel-1 SAR: Validation and applications. *J. Geophys. Res.* **2017**, *122*, 1827–1848. [CrossRef]
24. Romeiser, R.; Graber, H.C.; Caruso, M.J.; Jensen, R.E.; Walker, D.T.; Cox, A.T. A new approach to ocean wave parameter estimates from C-band ScanSAR images. *IEEE Trans. Geosci. Remote Sens.* **2015**, *53*, 1320–1345. [CrossRef]
25. Zhang, B.; Li, X.; Perrie, W.; He, Y. Synergistic measurements of ocean winds and waves from SAR. *J. Geophys. Res.* **2015**, *120*, 6164–6184. [CrossRef]
26. Zhang, B.; Perrie, W.; He, Y. Validation of RADARSAT-2 fully polarimetric SAR measurements of ocean surface waves. *J. Geophys. Res.* **2010**, *115*, 1–11. [CrossRef]
27. Xie, T.; Perrie, W.; He, Y.; Li, H.; Fang, H.; Zhao, S.; Yu, W. Ocean surface wave measurements from fully polarimetric SAR imagery. *Sci. China Earth Sci.* **2015**, *58*, 1849–1861. [CrossRef]
28. Young, I.R. Observations of the spectra of hurricane generated waves. *Ocean Eng.* **1998**, *25*, 361–376. [CrossRef]
29. Young, I.R. Directional spectra of hurricane wind waves. *J. Geophys. Res.* **2006**, *111*, 1–14. [CrossRef]
30. Ochi, M.K. *Hurricane-Generated Seas*; Elsevier: Oxford, UK, 2003; pp. 25–53.
31. Hu, K.; Chen, Q. Directional spectra of hurricane-generated waves in the Gulf of Mexico. *Geophys. Res. Lett.* **2011**, *38*, 1–7. [CrossRef]
32. Kudryavtsev, V.; Golubkin, P.; Chapron, B. A simplified wave enhancement criterion for moving extreme events. *J. Geophys. Res-Oceans* **2016**, *120*, 7538–7558. [CrossRef]
33. Badulin, S.I.; Pushkarev, A.N.; Resio, D.; Zakharov, V.E. Self-similarity of wind-driven seas. *Nonlinear Proc. Geoph.* **2005**, *12*, 891–945. [CrossRef]
34. Zakharov, V.E. Theoretical interpretation of fetch limited wind-drivensea observations. *Nonlinear Proc. Geoph.* **2005**, *12*, 1011–1020. [CrossRef]
35. Resio, D.; Perrie, W. A numerical study of nonlinear energy fluxes due to wave-wave interactions Part 1. Methodology and basic results. *J. Fluid Mech.* **1991**, *223*, 603–629. [CrossRef]
36. Resio, D.; Long, C.; Perrie, W. The effect of nonlinear fluxes on spectral shape and energy source-sink balances in wave generation. *J. Phys. Oceanogr.* **2011**, *41*, 781–801. [CrossRef]
37. Banner, M.L.; Young, I.R. Modeling spectral dissipation in the evolution of wind waves. Part 1. Assessment of existing model performance. *J. Phys. Oceanogr.* **1994**, *24*, 1550–1671. [CrossRef]
38. The SWAMP Group. *Sea Wave Modelling Project (SWAMP). An Intercomparison Study of Wind Wave Prediction Models. Part 1: Principal Results and Conclusions*; Ocean Wave Modeling; Plenum Press: New York, NY, USA, 1985.
39. Young, I.R. Parametric hurricane wave prediction model. *J. Waterw. Port Coast. Ocean Eng.* **1988**, *114*, 637–652. [CrossRef]
40. Young, I.R.; Burchell, G.P. Hurricane generated waves as observed by satellite. *Ocean Eng.* **1996**, *23*, 761–776. [CrossRef]
41. Young, I.R.; Vinoth, J. An "extended fetch" model for the spatial distribution of tropical cyclone wind–waves as observed by altimeter. *Ocean Eng.* **2013**, *70*, 14–24. [CrossRef]
42. Hwang, P. Fetch-and duration-limited nature of surface wave growth inside tropical cyclones: With applications to air–sea exchange and remote sensing. *J. Phys. Oceanogr.* **2016**, *46*, 41–56. [CrossRef]
43. Hwang, P.; Walsh, E.J. Azimuthal and radial variation of wind-generated surface waves inside tropical cyclones. *J. Phys. Oceanogr.* **2016**, *46*, 2605–2621. [CrossRef]
44. Hwang, P.; Fan, Y. Effective fetch and duration of tropical cyclone wind fields estimated from simultaneous wind and wave measurements: Surface wave and air–sea exchange computation. *J. Phys. Oceanogr.* **2017**, *47*, 447–470. [CrossRef]
45. Zhang, B.; Perrie, W. Cross-polarized synthetic aperture radar: A new potential measurement technique for hurricanes. *Bull. Am. Meteorol. Soc.* **2012**, *93*, 531–541. [CrossRef]
46. Zhang, B.; Perrie, W.; Zhang, J.A.; Uhlhorn, E.W.; He, Y. High-resolution hurricane vector winds from C-band dual-polarization SAR observations. *J. Atmos. Ocean. Technol.* **2014**, *31*, 272–286. [CrossRef]
47. Horstmann, J.; Wackerman, C.; Falchetti, S.; Maresca, S. Tropical cyclone winds retrieved from synthetic aperture radar. *Oceanography* **2013**, *26*, 46–57. [CrossRef]

48. Van Zadelhoff, G.-J.; Stoffelen, A.; Vachon, P.W.; Wolfe, J.; Horstmann, J.; Belmonte Rivas, M. Retrieving hurricane wind speeds using cross-polarization C-band measurements. *Atmos. Meas. Tech.* **2014**, *7*, 437–449. [CrossRef]

49. Zhang, G.; Perrie, W.; Li, X.; Zhang, J.A. A hurricane morphology and sea surface wind vector estimation model based on C-band cross-polarization SAR imagery. *IEEE Trans. Geosci. Remote Sens.* **2017**, *55*, 1743–1751. [CrossRef]

50. Zhang, G.; Li, X.; Perrie, W.; Hwang, P.A.; Zhang, B.; Yang, X. A Hurricane wind speed retrieval model for C-band RADARSAT-2 cross-polarization ScanSAR images. *IEEE Trans. Geosci. Remote Sens.* **2017**, *55*, 4766–4774. [CrossRef]

51. Mouche, A.A.; Chapron, B.; Zhang, B.; Husson, R. Combined co- and cross-polarized SAR measurements under extreme wind conditions. *IEEE Trans. Geosci. Remote Sens.* **2017**, *55*, 6746–6755. [CrossRef]

52. Engen, G.; Johnsen, H. SAR-ocean wave inversion using image cross spectra. *IEEE Trans. Geosci. Remote Sens.* **1995**, *33*, 1047–1056. [CrossRef]

53. Hwang, P.; Fan, Y.; Ocampo-Torres, F.J.; García-Nava, H. Ocean Surface Wave Spectra inside Tropical Cyclones. *J. Phys. Oceanogr.* **2017**, *47*, 2393–2417. [CrossRef]

54. Gilhousen, D.B.; Hervey, R. Improved estimates of swell from moored buoys. *Ocean Wave Meas. Anal.* **2001**, *2002*, 387–393.

55. Atlas, R.; Hoffman, R.N.; Ardizzone, J.; Leidner, S.M.; Jusem, J.C.; Smith, D.K.; Gombos, D. A cross-calibrated, multiplatform ocean surface wind velocity product for meteorological and oceanographic applications. *Bull. Am. Meteorol. Soc.* **2011**, *92*, 157–174. [CrossRef]

56. Young, I.R. A Review of Parametric Descriptions of Tropical Cyclone Wind-Wave Generation. *Atmosphere* **2017**, *8*, 194. [CrossRef]

remote sensing

MDPI

Article

Developing a Quality Index Associated with Rain for Hurricane Winds from SAR

Hui Shen [1,2,*], Chana Seitz [3], William Perrie [2], Yijun He [1] and Mark Powell [3]

[1] School of Marine Sciences, Nanjing University of Information Science and Technology, Nanjing 210044, China; yjhe@nuist.edu.cn

[2] Bedford Institute of Oceanography, Fisheries and Oceans, Dartmouth, NS B2Y4A2, Canada; William.Perrie@dfo-mpo.gc.ca

[3] Risk Management Solutions, Tallahassee, FL 32304, USA; Chana.Seitz@rms.com (C.S.); Mark.Powell@rms.com (M.P.)

* Correspondence: Hui.Shen@dfo-mpo.gc.ca; Tel.: +1-902-426-3147

Received: 20 August 2018; Accepted: 7 November 2018; Published: 10 November 2018

Abstract: Differences in synthetic aperture radar (SAR)-retrieved hurricane wind speeds from co-polarization and cross-polarization measurements are found to be correlated with rain rate. A quality index is proposed for the SAR-retrieved wind speed product to recognize heavy rain-affected areas by taking account of the different imaging mechanisms of the radar backscattering from the ocean surface via cross-polarization and co-polarization observations. A procedure is proposed to rectify wind retrievals in the rain-contaminated areas within the hurricane core, based on the theoretical physical profile for hurricanes. The effectiveness of the proposed methodology for heavy rain area recognition and wind speed reconstruction in the rain-affected areas is validated against step frequency microwave radiometer measurements from hurricane reconnaissance missions and the hurricane surface wind analysis product (HWIND). The quality flags provide confidence levels of hurricane surface winds from SAR, which together with the proposed method to correct wind retrievals in rain-contaminated areas, can contribute to improved operational applications of SAR-derived winds under hurricane conditions.

Keywords: synthetic aperture radar (SAR); hurricane; rain; wind; dual-polarization

---

1. Introduction

From the very first spaceborne synthetic aperture radar (SAR)-SEASAT (seafaring satellite), storm signatures have been seen, mainly in the unique structure of storm-related rain footprints over the ocean surface [1,2]. Unique mesoscale features such as rain bands [3], eye morphology [4], and vortices [5] have also been observed by follow-on SAR missions. Significant efforts for the retrieval of quantitative information from SAR have been conducted for ocean surface wind. After several decades of development, the algorithms for SAR wind retrieval have reached operational applications for routine sea conditions; for example, the Alaskan coastal SAR program [6] and the Canadian National SAR Wind program [7]. These developments have especially benefited from large datasets of associated microwave frequency band scatterometer wind products. Based on the continuous improvements in wind speed retrieval under conventional wind conditions, it is appropriate to test the potential capability of high wind speed retrieval from SAR, for hurricanes. The motivation is highlighted by the critical demands for enhanced wind observations during hurricanes, especially over the ocean, before storms make landfall.

Modern SAR instruments have a capability for multiple polarization measurements. Among these, co-polarization (hereafter: co-pol), i.e., HH and/or VV polarization (H: Horizontal, V: vertical, with the first letter standing for radar transmission polarization and the second letter for receiving polarization), is mostly used in ocean studies, especially in early satellite SAR missions e.g., Reference [6]. The wind vector retrieval methods based on these co-pol SAR measurements have been successful, based on a series of geophysical model functions (GMFs), such as CMOD5.N [8], for C-band SAR, which are leading to the operational monitoring of wind fields under low to moderate wind conditions (i.e., 0–35 m/s). For example, ocean wind products are operationally provided in the Sentinel SAR L2 product data by the European Space Agency. Attempts to retrieve hurricane-force wind from co-pol SAR have also been made, as shown in References [9,10]. Mathematically, it is straightforward to apply the SAR wind algorithm derived for conventional wind conditions to SAR hurricane images, however, challenges remain. For example, studies show that co-pol SAR signals suffer signal saturation under high wind speeds e.g., Reference [8], which result in decreased sensitivity of radar backscattered signals with increased wind speed. This is thought to be mainly induced by suppressed Bragg waves under high wind conditions, due to sea spray [11] or changes in the atmospheric surface layer [12]; however, the detailed mechanism is still unclear due to limited in situ observations. Under low radar incidence angles (<30°), the normalized radar cross section (NRCS) even appears to decrease as observed by laboratory [13] and aircraft measurements [14], leading to speed ambiguity in SAR wind retrievals for high wind speeds [15]. The CMOD5.N GMF captures this natural saturation effect; thus, an ambiguity removal scheme needs to be applied so that hurricane-force wind speed can be obtained from co-pol radar returns [16].

Meanwhile, the capability of cross-polarization (hereafter: cross-pol) measurements for high wind speed retrieval has been revealed, benefiting from the high-quality radiometric calibration performance of C-band RADARSAT-2 SAR. Compared to higher frequency microwave bands, for example X-band, C-band radar is less influenced by rain, and thus more widely used for hurricane wind retrieval. Studies show that cross-pol has increased sensitivity under high wind speeds, which makes it especially suitable for high wind speed monitoring. Data analysis has revealed that the cross-pol radar NRCS monotonically increases with wind speed, with no dependence on wind direction and no or little dependence on radar incidence angle. Thus, by collocating observations of radar NRCS with wind data from buoys, dropsondes, models etc., empirical GMFs have been developed [17–19]. For example, the authors of study [18] proposed a linear model between radar NRCS and wind speed for quad-polarization SAR measurements; and the authors of study [19] proposed a piecewise linear model with a noise suppression procedure for ScanSAR mode RADARSAT-2 SAR data, which is the most widely used mode for hurricane observations [incidence angles, 20–49°]. Since wind direction is not needed in these models, wind speed can be directly retrieved from radar measurements without additional information regarding wind direction as required for wind speed retrieval from co-pol radar signals. These empirical GMFs have been successful in quantifying the relationship between radar backscattered signals and wind speed, which can be difficult to establish for a theoretical model, due to the complicated state of the ocean surface under high wind speeds, involving processes such as wave breaking induced sea spray and foam.

Although empirical geophysical model functions are able to quantify the relationship between the radar NRCS and wind vectors, and accommodate different wind-induced dynamical processes implicitly, there are non-wind-induced external processes in the ocean which can contribute additional radar backscattered signals [20] Rain is one of these processes. Naturally, heavy precipitation is present in tropical storms and heavy rainfall-induced flooding is a major threat to human society. As mentioned previously, the signatures of heavy rainfall have been observed from the very beginning of SAR satellite remote sensing, e.g., Reference [2]. Many studies have focused mainly on the morphology of rain signatures in SAR images. In 1994, Atlas [2] first explained these hurricane footprints in SAR imagery as a result of rain downdrafts. A recent 2016 study by Alpers et al. [21] suggested a C-band SAR imaging mechanism for rain under low to moderate wind conditions. The rain morphology [3] apparent from SAR images indicates possible contamination to the retrieved wind speeds in affected areas. Attempts to build a scattering model due to rain effects on the rough sea surface have also been pursued e.g., References [22,23]. However, most of these studies rely on rain rate measurements from other sources. A method to retrieve wind and rain simultaneously from ERS scatterometer data has been tested and is possible [24] under moderate wind conditions. Thus far, no GMF explicitly includes rain estimation for wind retrieval under hurricane conditions. For hurricane-force wind, rain can contribute up to 100% error ([25] Figure 2) for wind retrieval from airborne stepped frequency microwave radiometer (SFMR) observations (incidence angles −40°–+40°). Compared to SFMR, wind retrieval from SAR suffers similar apparent effects from heavy rain on the ocean surface, because the presence of rain not only changes the brightness temperature which is measured by SFMR, but also surface roughness which is captured by SAR. Thus it is necessary to consider the rain effect in hurricane wind retrieval from SAR.

Finding a method to recognize rain-affected wind cells in SAR images is not a simple task. Because of its antenna design, the scatterometer can assign a quality flag for each retrieved wind vector based on the consistency of multiple measurements from different incidence angles [24,26]. SAR has only one incidence angle for each observation, and thus it is difficult to make a similar quality assessment based on its measurements, although the removal of directional ambiguity has been possible based on the concentric wind structure that is generally followed for hurricanes [10].

However, similar to the multi-looking directional measurements from scatterometers, modern SAR instruments have special multiple-polarization capabilities, such as RADARSAT-2 SAR, which is able to conduct dual-polarization (dual-pol hereafter) and quad-polarization measurements. Dual-pol mode can be operated at ScanSAR swath, with swath widths as much as 500 km, which is suitable for observing hurricanes. The two polarization measurements of dual-pol mode are cross-pol (VH or HV) and co-pol (VV or HH), which have different imaging mechanisms. For cross-pol, volume scattering dominates, whereas for co-pol Bragg scattering is more important [19]. Study [21] indicated that non-Bragg scattering may be a dominant scattering mechanism for rain cells. Therefore, the combination of both measurements has been important for wind vector retrieval, in rain-free areas, owing to the advantage of wind speed retrieval from cross-pol measurements and the wind direction sensitivity of co-pol measurements [27]. This combination has also been shown to be useful for the detection of various targets, for example, oil [28], wind turbines [29], macroalgae bloom patches [30] etc. Under hurricane conditions, a recent study showed that both co- and cross-polarized SAR measurements can be used for extreme wind retrieval. Using their combined geophysical model functions, they obtained wind speeds up to 60 m/s [31].

In this paper, we extend the application of multiple polarization measurements for rain under hurricane conditions. This is achieved by introducing the principle of rain recognition from SAR described in Section 2, based on SAR measurements and co-located SFMR measurements. In Section 3, results are given for SAR-retrieved hurricane wind speeds with a rain flag as a quality index. A methodology to correct the rain-contaminated areas is introduced in Section 4 and the rain-corrected wind field is validated by HWIND data from Risk Management Solutions (RMS; www.rms.com/models/hwind), which are post-analysis winds based on objective analysis of best available observations during the storm. Discussions of uncertainties and future plans are presented in Section 5, followed by conclusions in Section 6.

2. Principle of Rain Recognition from SAR

Although microwave radar is able to penetrate through cloud and light rain, because of its long wavelength as compared to the size of rain drops, heavy rainfall will modulate the radar signals through various mechanisms [21,22]. Study [22] summarized the effect of rain into two categories: In the atmosphere, raindrops induce volumetric scattering and attenuation of radar waves, whereas on the ocean surface, rain alters the roughness of the ocean surface by the competitive functions of rain damped ocean surface waves, and rain splashing enhanced ring waves. Although both negative and positive effects exist in the atmosphere and in ocean surface components of radar backscattered signals, the negative effects are dominant for high wind speeds [25]. A detailed quantification of each component is complicated. Nevertheless, all these effects will eventually be combined together and represented by the radar NRCS. In contrast to the stronger NRCS caused by higher winds under hurricane conditions, rain makes a relatively smaller contribution to the NRCS in high winds as compared to that made under moderate wind conditions. Figure 1 shows an example of rain features in Hurricane Patricia (2015), for moderate wind speeds (15 m/s), as compared to high wind speeds (35 m/s), where rain features are more prominent in (a,b) than in (c,d).

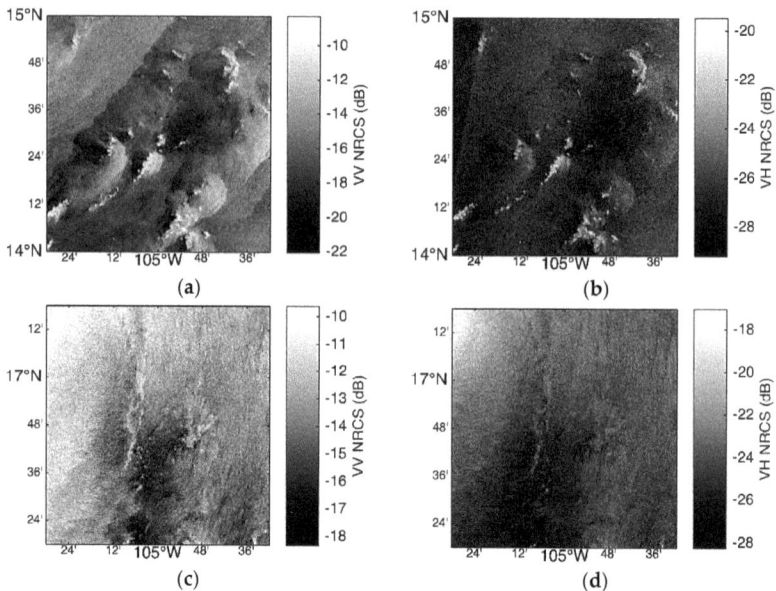

Figure 1. Rain cells in the outer range areas of synthetic aperture radar (SAR) images of Hurricane Patricia. Upper panels (**a,b**) are 125 km from the eye where the wind speed is 15 m/s. Lower panels (**c,d**) are 40 km from the eye where wind speed is 35 m/s. Left panels (**a,c**) show VV normalized radar cross section (NRCS). Right panels (**b,d**) show VH NRCS. Note the differences in the grayscale bars.

For solely wind-induced radar roughness, both the co-pol and the cross-pol radar NRCSs are constrained by the geophysical model functions; radar measurements from either polarization or combined can lead to wind speed retrieval. However, for radar roughness generated by other processes, for example rain, it is hypothesized that the same rain rate causes different radar modulations for co-pol radar as compared to cross-pol radar measurements. This is shown in Figure 1, and it results from the different imaging mechanisms for cross-pol and co-pol SAR measurements, which will show up as different wind speed retrievals in the rain-affected areas. These differences can potentially lead to detection of non-wind-induced features in the SAR images themselves. This principle will be explained in more detail in the following sections.

3. Rain Recognition from Dual-Pol SAR Imagery

In order to develop a methodology to recognize the rain contaminated ocean surface areas under hurricane conditions, we studied RADARSAT-2 SAR imagery of the most intense winds observed in Hurricane Patricia (2015), together with SFMR wind speeds and rain rate measurements from hurricane reconnaissance missions. During 20–24 October 2015, Hurricane Patricia intensified from a tropical storm to a Category 5 hurricane and made landfall on the Pacific coast of Mexico (Figure 2). According to aircraft measurements conducted by the United States National Hurricane Center (NHC) of NOAA (National Ocean and Atmosphere Administration), the lowest pressure was 879 mbar and maximum sustained winds were 200 mph (~89 m/s) at 05:33PM UTC 23 October 2015, which confirms Hurricane

Patricia as the most intense cyclone recorded in the western hemisphere in terms of barometric pressure, and the most intense globally in terms of measured maximum sustained winds [32].

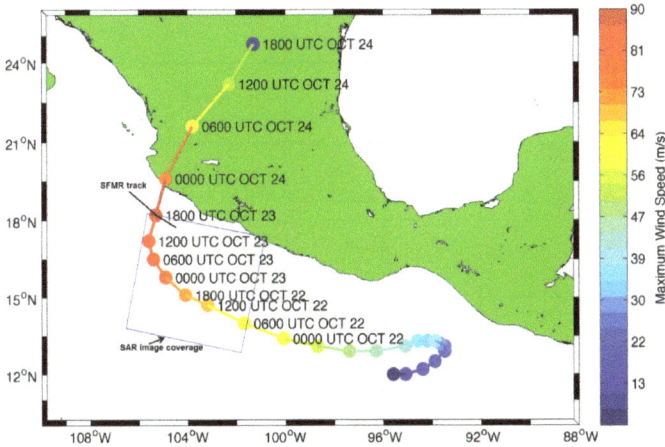

Figure 2. Track of Hurricane Patricia (2015).

RADARSAT-2 SAR measurements during the most intense phase of Hurricane Patricia took place at 12:45PM UTC on 23 October 2015. RADARSAT-2 SAR has multiple imaging modes and polarizations. Almost all the hurricane images taken by RADARSAT-2 SAR are in ScanSAR mode with dual-polarizations (VV and VH). For ScanSAR, only signal intensity is recorded. For other modes, phase information can be processed by request. The SAR measurements for Hurricane Patricia (2015) were taken in dual-pol mode, and two images with co-pol VV polarization and cross-pol VH polarization were captured. These two RADARSAT-2 SAR images are also notable for capturing the most intense wind speed that has thus far been recorded by SAR measurements over the ocean surface.

Hurricane track data are from the NHC (www.nhc.noaa.gov/). The color of the track as indicated with dots shows the strength of the maximum wind speeds. The square box shows the coverage of the RADARSAT-2 SAR image at 12:45 UTC on 23 October 2015, and the black line shows the NOAA 43 SFMR track in the hurricane core area.

The flowchart of the proposed procedure to recognize rain areas from SAR is shown in Figure 3. Firstly, RADARSAT-2 SAR dual-pol measurements are processed to obtain the radiometric calibrated NRCS (dB). Hurricane wind speeds are then retrieved from the cross-pol mode image, using a VH dual-pol GMF [19]. Similar to other cross-pol GMFs, this GMF [19] was developed by collocating SAR measurements with data from in situ buoys. However, this model function targets VH dual-pol data only instead of also including quad-pol data which has different radiometric accuracy. Since most, if not all, of the cross-pol RADARSAT-2 SAR hurricane images are taken in the VH dual-pol mode, this approach is expected to better fit the objectives of this study. In previous work [19], we also introduced a noise removal scheme which significantly removed the apparent "seams" between different beams of the ScanSAR mode image. The VH dual-pol GMF was further evaluated in Reference [33] by comparing SAR winds with the state-of-art hurricane wind analysis product, HWND, from NOAA (now provided by RMS). HWIND uses expertly standardized and quality controlled wind observations from multiple platforms (aircraft, surface-based stations, buoys, remote sensing, etc.) in a storm-relative framework to map a tropical cyclone's wind field [34]. Validation results show that wind patterns retrieved from the VH dual-pol GMF are consistent with HWIND, with SAR speed underestimates of -1.3 m/s over the whole hurricane area, and -7.05 m/s in the hurricane core area (within 100 km of the hurricane center). The root mean square difference is 4.5 m/s. Heavy rain may contribute to a larger bias in the core area, which is a focus of the present paper.

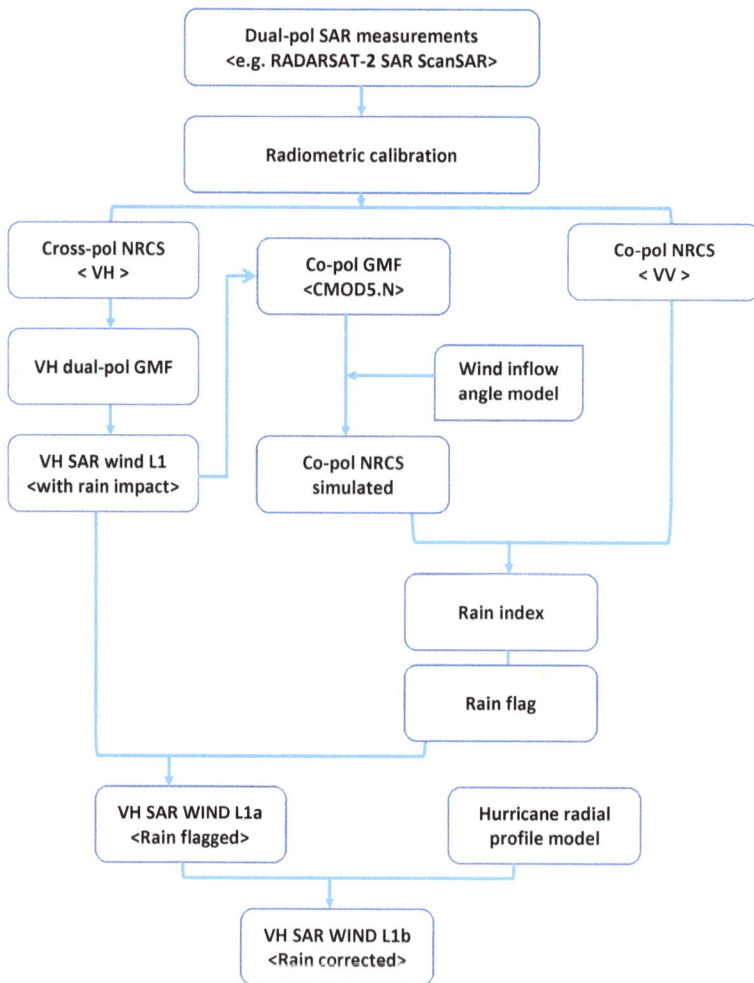

Figure 3. Flowchart of the hurricane wind quality index from SAR with proposed rain correction procedure.

We simulated the co-pol radar NRCS based on wind speed retrieved from VH SAR measurements (hereafter, VH wind speed), using the co-pol GMF CMOD5.N [8], which would be used for comparison with the measured co-pol data. For the co-pol GMF, wind direction is required to be known a priori. We adopted the hurricane sea surface inflow angle parameterization model from Reference [35] to obtain the hurricane wind direction. To decrease any possible bias induced from uncertainty in the wind direction away from the hurricane center, we focus on the wind field within 100 km of the hurricane eye.

We then compared the simulated co-pol radar NRCS to the NRCS from observations. The difference between the simulated and measured co-pol NRCS was taken as a quality index as shown in Equation (1). To avoid possible errors induced by speed ambiguity from co-pol wind retrieval under low incidence angles, we did not directly compare the wind retrievals from the two polarization modes; instead, we used an alternate approach to quantify the rain-induced differences in the dual-pol measurements. Note these attenuations are a natural phenomenon in NRCS for high winds measurements by co-pol SAR. We simulated the co-pol radar NRCS based on the wind speed retrieved from the VH wind speed, using the co-pol GMF CMOD5.N.

When there was an apparent difference between the simulated co-pol radar NCRS and the measurements, we labeled the corresponding areas as rain contaminated. Ideally, there should be no difference between these areas and rain-free areas. However, there is always random noise in SAR measurements. RADARSAT-2 SAR has a radiometric accuracy of 0.3 dB [36]; a SAR wind White Paper [37] suggested that the radiometric error should be less than 0.5 dB in order to obtain high resolution wind retrieval from SAR. By accommodating these factors, we set the threshold value for the difference to be 0.5 dB. The low wind area (here <20 m/s) within the hurricane eye was not labeled for quality assessment, based on two considerations: firstly, that the hurricane eye area may have complex wind dynamics, for example mesovortices [38], thus the wind direction parametric model adopted for co-pol simulation may not apply, as it was developed mostly for areas outside the eyewall [35]; and secondly, the eye area is usually cloud-free as observed from optical imagery, thus is not directly affected by rain. For Hurricane Patricia (2015), this low wind eye area is within a 7 km radius of the hurricane center.

$$Index = \left| \sigma_{cmod5.N(U,\varphi,\theta)} - \sigma_{SAR_{obs}} \right| \tag{1}$$

where, σ is NRCS in dB, U is VH wind speed, ϕ is radar incidence angle, θ is wind direction.

SAR-retrieved wind speeds were compared with SFMR measurements from the NOAA hurricane hunter mission. A storm-relative coordinate system was adopted by firstly adjusting the SAR hurricane center to the SFMR observed hurricane center, and thus we retrieved SAR observations along the SFMR track. The hurricane center from SAR is determined by fitting the position of the hurricane maximum wind speed (eyewall) as retrieved from cross-pol image to an ellipse as in study [4]. The center of the ellipse is taken as the center of hurricane. We note that Hurricane Patricia remained relatively steady from 12:00PM to 06:00PM as indicated by the hurricane best track analysis by the National Hurricane Center. The lowest surface pressure was estimated to be 872 mbar at around 12:00PM and 879 at 05:39PM, according to 43 SFMR measurements [30].

Figure 4 shows a comparison of both measurements along the SFMR track. The SAR-retrieved wind speed from the cross-polarization data reaches 85 m/s, and the SFMR peak wind speed indicates about 93 m/s, with simultaneous rain measurements showing strong precipitation in the hurricane eyewall area, located to the right of the peak wind as indicated by SFMR observations (Figure 4a). By adopting the hurricane inflow angle model [35] to CMOD5.N GMF, wind speed is also obtained from co-pol SAR measurements; the peak wind speed obtained by co-pol SAR observations is over 100 m/s (Figure 4b). Generally, for the rain-free area, both estimates for SAR winds are consistent with SFMR measurements. However, in the heavy rain area (>30 mm/h), the winds derived from co-pol SAR data underestimate the wind speed by about 20–30 m/s, whereas the cross-pol observations seem less affected. Figure 4 shows the underestimation of the SAR-derived winds in the heavy precipitation area, which once again confirms the overall attenuation effect of rain under hurricane-force wind conditions. This has previously been shown in the analysis of QuikSCAT measurements [24] and in our data in Figure 1.

Figure 5a shows the performance of the co-pol backscattered signals along the SFMR tracks. In the heavy precipitation area, the co-pol NRCS measurements decrease by 2–4 dB, compared to the nearby rain-free observations. Comparing the two lines in Figure 4, it is found that the co-pol data are more heavily affected by rain than the cross-pol data. Consequently, the rain leads to greater underestimation of the wind speed retrieval from co-pol SAR measurements than from cross-pol SAR measurements, with cross-pol showing no apparent underestimation of wind speed, and co-pol showing an underestimation of wind speed of about ~20 m/s.

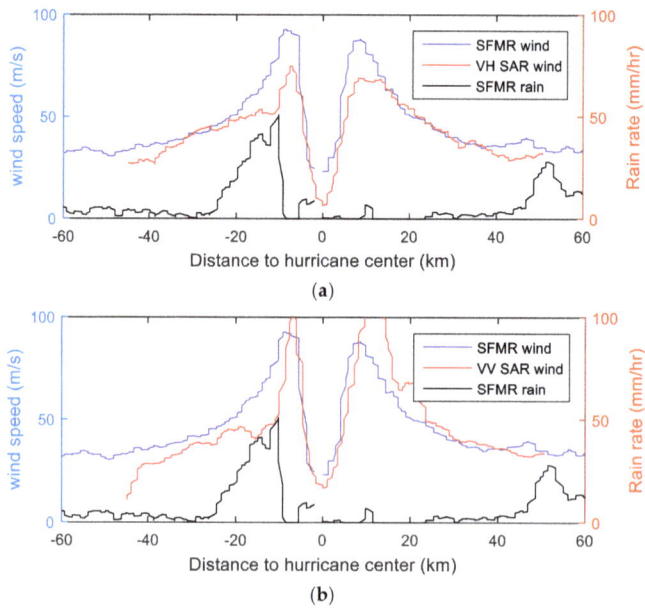

Figure 4. SAR wind speed retrieved from (**a**) cross-polarization (cross-pol), and (**b**) co-polarization (co-pol) profiles, overlaid with the rain rate along the airborne track of the NOAA43 stepped frequency microwave radiometer (SFMR) measurements.

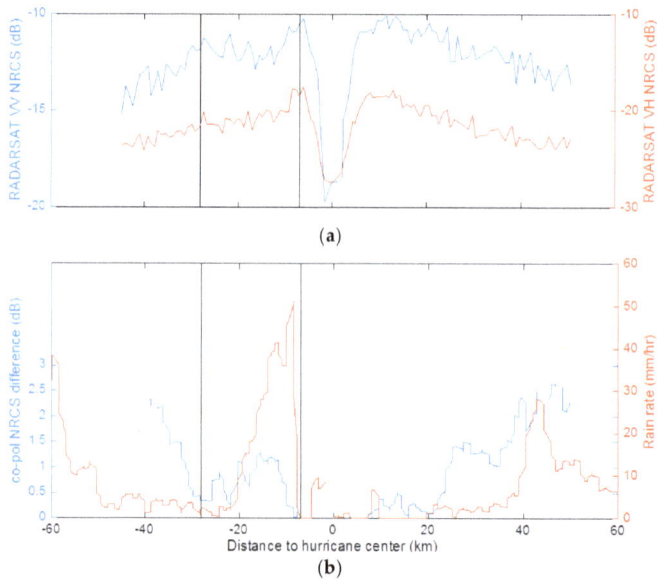

Figure 5. NRCS of (**a**) SAR co-pol, and (**b**) difference between SAR and simulated co-pol profiles, overlaid with the rain rate along the airborne track of the NOAA43 SFMR measurement. The area in the box corresponds to heavy rainfall in the eyewall. The lower panel also shows rain-induced co-pol NRCS changes as recognized by SAR observations and model simulations.

For areas with rain rates exceeding 30 mm/h, the wind difference between the two GMFs output is larger than 10 m/s (Figure 4), which is more than three times the standard deviation of SAR-retrieved winds in the 3 × 3 image pixels.

By applying the proposed rain flag methodology as expressed in Equation (1), we label the rain-affected areas in the SAR image and compare them with the SFMR measured rain rate. Overall, the retrieved rain area is consistent with the high rate domain as suggested by SFMR measurements (Figure 5b). The two peak index values at ~−15 km and 40 km correspond to high rain rates (>20 mm/h) in these areas. However, the index value does not monotonically increase with higher rain rates. Outside the eyewall area, a larger index value is achieved, benefiting from the higher sensitivity to rain in lower wind speed areas as shown previously in Figure 1. The relatively lower rain rate areas (3–5 mm/h) at −40 km and 35 km also have high index values of 1.5–2 dB. The index values in these areas are even higher than in the very high rain rate areas around the eyewall at around −10–−20 km, where the rain rate reaches 50 mm/h. Figure 4 suggests that the wind speed in these corresponding high rain rate areas is around >60 m/s, whereas the wind speed in the relatively low rain rate areas is around <40 m/s.

Comparing this quality index with SFMR measurements suggests that the 0.5 dB threshold corresponds to a 5 mm/h rain rate in hurricane eyewall areas and 2–3 mm/h in moderate wind speed areas. In the full 2 Dimension spatial domain of the SAR image, Figure 6 presents results of the high precipitation areas of Hurricane Patricia (2015). With the proposed index as a third dimension, this figure shows the 3 Dimension view of possible rain contamination in the SAR image. A clearly evident rain band stands out. This rain band is shaped like the number '6', with a larger index value on the left side of the hurricane eyewall. This is consistent with the radar reflectivity measurements obtained from hurricane reconnaissance mission at 05:35PM 23 October 2015 (Figure 7a) which shows a similar rain band pattern as revealed in Figure 6. The radar reflectivity shown in Figure 7a is from the lower fuselage radar mounted on the NOAA WP-3D aircraft which operated at the same C-band as RADARSAT-2 SAR (http://www.aoml.noaa.gov/hrd/about_hrd/HRD-P3_radar.html). The WP-3D radar recognizes areas of strong precipitation by measuring radar reflectivity when flying through hurricanes. In Figure 7a, the strong precipitation area is located at the hurricane eyewall; with another strong precipitation area ~100 km away from the hurricane center and outside of the SAR image coverage. The strong precipitation close to the eyewall is in a spatial pattern of the number '6'.

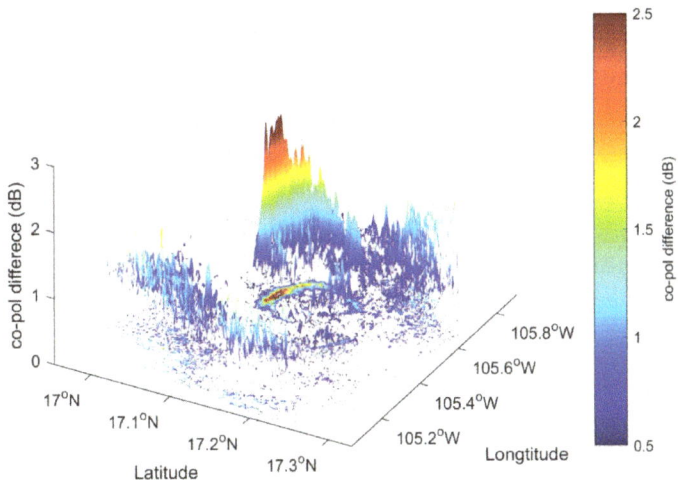

Figure 6. The rain index for Hurricane Patricia (2015) as derived from the proposed quality index based on the difference of co-pol NRCSs between SAR observations and model simulation.

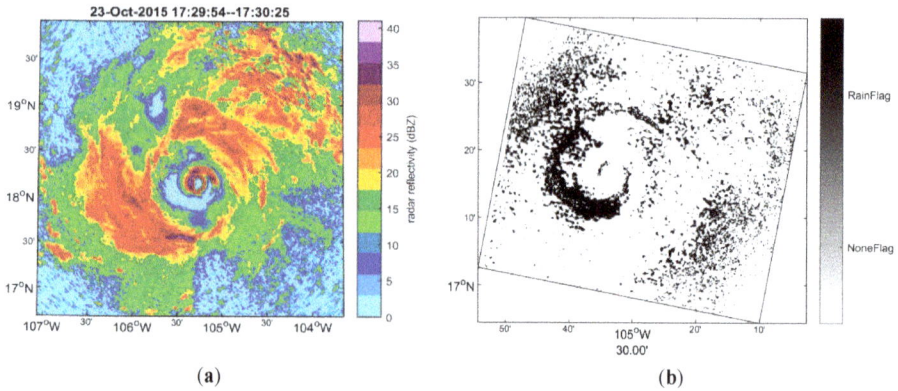

Figure 7. Radar reflectivity measured on board NOAA 43 flight at 05:30PM 23 October 2015: (**a,b**) rain flag of Hurricane Patricia at 12:45 UTC on 23 October 2015 as retrieved from SAR. Both figures show a heavy precipitation pattern in the shape of the number "6" in Hurricane Patricia. The observed spatial pattern in (**a**) is consistent with that of the rain index from SAR in figure (**b**).

We label the retrieved rain area by taking the absolute value of Equation (1) as an index of the quality of the SAR-retrieved wind speed. The recognized rain area is given a quality flag of 1, which indicates the **presence** of strong precipitation, and other areas are given a quality flag of 0, indicating the **absence** of strong precipitation. Figure 7b shows the results of the quality flag corresponding to Hurricane Patricia as shown in Figure 6. Figure 7b is a 2D view of Figure 6, where spatial locations of possible rain areas are presented. Figure 7b may be directly compared to Figure 7a, as both figures show the horizontal 2D structure of rain. Despite the time differences of the two measurements (~5 h), recall that Hurricane Patricia (2015) maintains very strong intensity with relatively steady state during 12:00 and 06:00PM [39]. Moreover, the two datasets show the consistent spatial patterns of rain in the shape of the number '6'. Figure 7b misses the high radar reflectivity area to the north of the hurricane eye; this bias suggests areas for potential improvement for the proposed method, as discussed in the following section. The quality flag presents quality assessments for each grid cell in a SAR image indicating possible contamination by rain. This flag can be used as a reference when SAR-retrieved wind is used in operational analyses.

4. Correction for Rain-Contaminated Wind Cells

The rain flag provides a valuable quality assessment for SAR-retrieved hurricane wind speed. Since rain causes contamination for these flagged wind vector cells, caution is required when using SAR-derived wind products in rain conditions. One way to resolve the problem is to mask all the rain-flagged areas. However, the mask leaves a number of areas that are devoid of any wind information, which may be filled by assimilating SAR winds into comprehensive numerical prediction models (NWP). Sometimes it may be desirable to obtain quick estimations of the full hurricane wind field, for example, in order to facilitate rapid decision making and response. Under these circumstances, the following simple methodology may be used to quickly fill these missing values in a quality-flagged wind field.

4.1. Hurricane Wind Radial Profile Model

A hurricane is a strong mesoscale atmospheric low pressure system, defined by a low pressure center with low wind speeds and an eyewall with very high wind speeds, and maintains a unique wind profile along the radial direction. There exists a strong physical relationship between such a radial profile and the strength of the hurricane, which can be represented by a radial profile model, for example [40,41]. Radial profile models have been widely used to represent and reconstruct hurricane wind fields from hurricane best tracks. As shown in Reference [40], a Rankine combined vortex model shown in Equation (2) below can be used to simulate the hurricane wind profile along each radial transect. More sophisticated models involving more parameters, such as pressure, and wind measurements along the radial profile have been proposed by the authors of [41]. For this study, we use Equation (2), which requires no additional information for input. Thus, we demonstrate the methodology for wind correction for rain-contaminated wind cells. Wind speed is given by:

$$v = \begin{cases} v_m (r/r_{v_m}) & r < r_{v_m} \\ v_m (r_{v_m}/r)^{0.5} & r \geq r_{v_m} \end{cases} \tag{2}$$

where v is the wind speed, r_m is radial distance of maximum wind speed v_m from the hurricane eye, and r is the radial distance. Both r_m and v_m are parameters to be retrieved based on the VH wind speed with the rain flag of 0, in each radial direction.

Figure 8 gives examples of the model (2) applied to the wind profile of Hurricane Patricia's wind field derived from the VH SAR image. The parameters in model (2) are obtained by applying a least squares method to fit the model to the VH SAR wind data along each radial direction. The model represents the pattern of the hurricane wind profile in the radial direction very well. Therefore, for wind data with partly missing values, the model can be used to estimate and reconstruct the missing values. Note that the model does not account for the high frequency variations shown in Figure 7b.

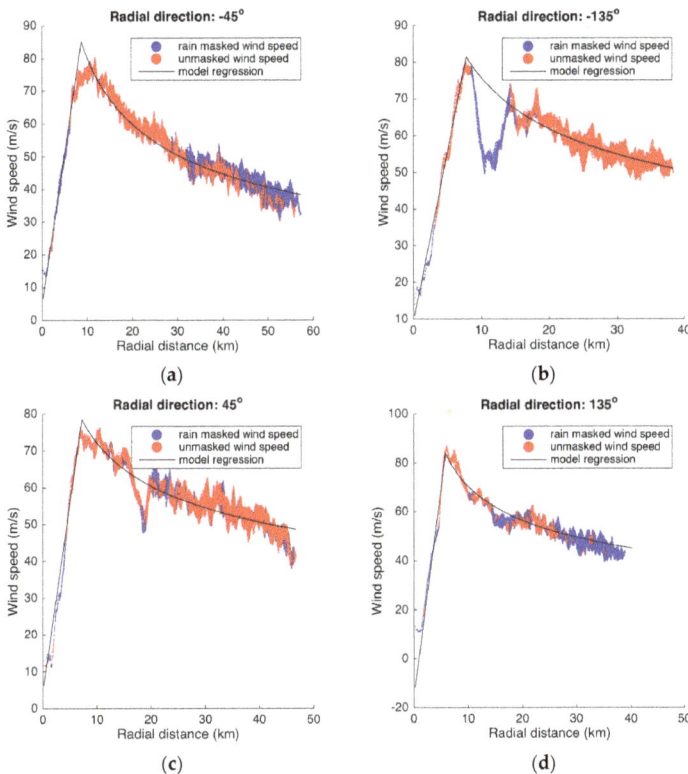

Figure 8. Wind speed in four radial profiles overlaid by Holland's (1980) regression model in four directions: (**a**) −45°; (**b**) −135°; (**c**) 45°; and (**d**) 135°. The red curve is the VH SAR wind with rain-flagged values removed; the blue curve is with rain; black is the regression model.

4.2. Wind Correction for Rain-Contaminated Cells

In Section 4.1, we have shown that the hurricane wind profile along the radial direction can be used to rebuild the missing values due to the rain flag. In this section, we adopt this method to correct the rain-contaminated wind retrievals from SAR. Based on the proposed quality index methodology, the heavy rain-contaminated wind cells are flagged. We hereafter treat these rain- flagged areas as pixels with missing values. Thus, we use this radial profile simulation method to rebuild the entire wind profile as a best estimate for the wind field values in the heavy rain areas.

For Hurricane Patricia, Figure 9 presents the corrected wind field, as well as the original wind retrievals from SAR with rain, and the results for rain-flagged winds. The strong underestimation bias for wind speeds in the left and lower portions of the eyewall (Figure 9a) are recognized as heavy rain-contaminated areas (Figure 9b). The results show a more consistent circular wind pattern for Hurricane Patricia. The wind reconstruction method (Figure 3) takes advantage of all the quality-controlled wind data, both outside the eyewall and within the eyewall. Therefore, the wind radial profile can be reconstructed, and further used to correct underestimated wind retrievals due to heavy rainfall contamination. Note that we take all the qualified SAR winds along the radial direction to obtain estimates of the physical wind profile, rather than using the maximum wind speed along the corresponding radial direction only. This is different from when the Holland model was originally developed, as this was targeting limited hurricane parameters from forecasts, such as the maximum wind radius etc. In fact, benefiting from the dense data points from high resolution SAR measurements, the model has the ability to retrieve the maximum wind speed v_m and the associated radius r_m along each radial direction. Therefore, the method does not assume symmetric hurricane wind structure, which in principle is useful for all hurricane wind configurations.

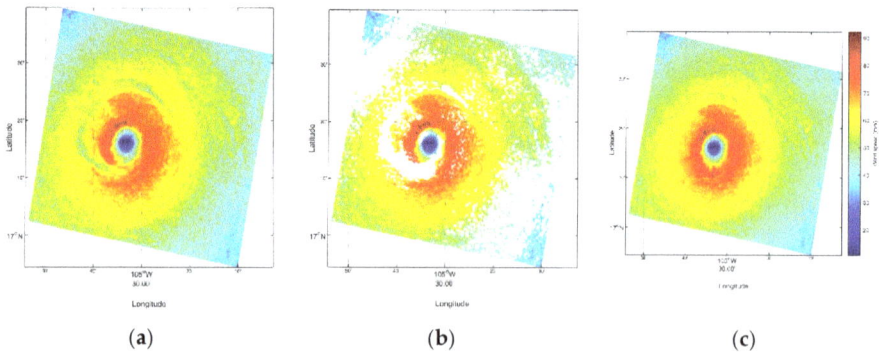

(a)   (b)   (c)

Figure 9. Wind field of Hurricane Patricia: (**a**) with rain; (**b**) with rain flagged; (**c**) after rain correction.

The rain flag and the rain correction method were evaluated against SFMR measurements along the flight track (Figure 1). Figure 10 shows both original wind retrieval where rain is neglected, and rain labeled/corrected wind results, as compared to SMFR measured wind speeds. As shown previously, the quality index was able to flag the rain-contaminated areas as in Figure 10a,b. The reconstructed wind field is able to somewhat correct the wind bias induced by the rain. Since the wind correction model adopted all SAR winds along each of the wind profiles, the maximum wind speed also shows somewhat better performance as compared to SFMR measurements. Figure 10c shows that most of the corrections bring the scattered data closer to the line of equality. The root mean square difference (rms) is reduced from 5.86 m/s to 3.78 m/s (Figure 10c).

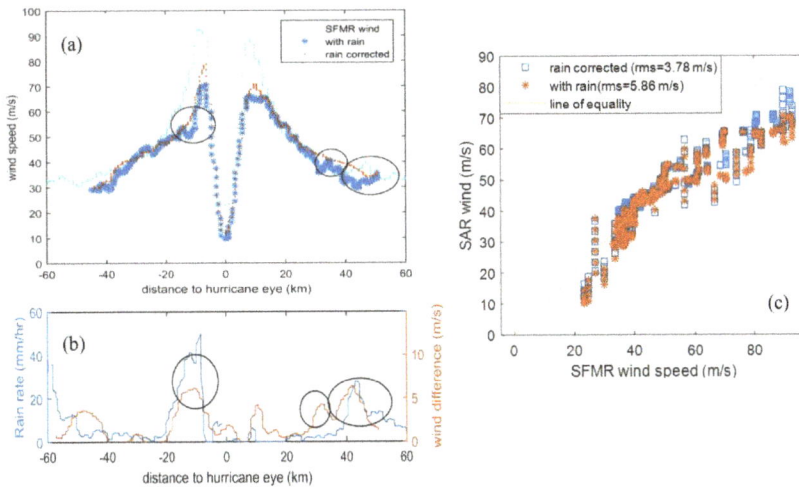

Figure 10. Comparisons of wind speed before and after rain flag along the NOAA flight. (**a**) Wind speed along SFMR track, (**b**) rain rate from SFMR measurements and associated speed difference from SAR, (**c**) scatterplot of wind speed before and after rain correction as compared to SFMR wind measurements. Note: the capability of the proposed method to recognize the strong precipitation areas is highlighted in the circled areas of (**a,b**).

Similar analysis was conducted for additional examples of hurricane measurements. Eight more hurricane cases (Table 1) were studied. Although the limited number of SAR images is far from a thorough validation, the objective here was to further demonstrate the effectiveness of the method. The winds retrieved from cross-pol mode SAR images of these hurricanes have been studied previously [33] without taking into account the possible rain contamination in the wind speed retrieval algorithm. We used the rain flag and wind correction method introduced in this study for these examples. Comparisons against hurricane surface wind analysis data HWIND [34] are summarized in Table 1. Compared to previous studies when rain is not removed, the rain correction methodology proposed in this study provides improved wind analysis, which confirms the influence of heavy rain on wind retrieval from SAR.

Table 1. Statistics for VH SAR and HWIND wind speed before and after wind correction due to heavy rain contamination: 'a' columns show results without rain correction, 'b' columns show results with rain correction. The hurricane category is indicated following the Saffir-Simpson scale at the time of the SAR observation.

ID	Hurricane	SAR Time	Hurricane Category	Bias		RMS Error		R	
				A	b	a	b	a	b
1	Gustav	11:27, 30 August 2008	3	−6.86	−3.67	4.05	4.01	0.85	0.91
2	Ike	23:54, 10 September 2008	2	−9.18	−5.66	4.73	3.25	0.42	0.56
3	Ike	23:56, 10 September 2008	2	−7.73	−3.39	4.49	2.46	0.40	0.71
4	Bill	22:27, 22 August 2009	1	−3.31	−2.57	3.73	1.99	0.53	0.67
5	Bill	10:40, 23 August 2009	1	−0.09	−2.33	2.56	2.27	0.81	0.86
6	Earl	22:59, 02 September 2010	2	−12.3	−2.09	4.97	2.38	0.85	0.89
7	Igor	10:11, 19 September 2010	1	−8.11	−6.13	3.97	1.92	0.79	0.88
8	Rina	11:30, 26 October 2011	2	−8.12	−4.28	3.85	2.48	0.61	0.89

5. Discussion

Reliable forecasts of hurricanes depend on accurate understanding of the physical processes related to their formation and development and should be based on accurate observations. Although the hurricane modeling community claims that the basic physics are well understood [42,43], estimating and forecasting hurricane winds is still challenging [44] due to the difficulty in obtaining accurate hurricane observations, especially surface winds. SAR has been proven to be capable of obtaining hurricane-force winds over the ocean surface from co-pol and cross-pol satellite measurements.

Despite the progress made on the methodology, and algorithms based on improved understanding of hurricane imaging theory and advanced SAR instruments in the past decades, the operational application of SAR-retrieved hurricane winds has still not been achieved. Many factors are at play in this situation. For example, due to the high power requirements, SAR only acquires data when ordered. Thus, temporal continuity and intervals separating data are irregular. This is different from other conventional wind observation systems, such as scatterometer, with regular revisiting periods. Another approach is Sentinel SAR, which continuously acquires images in the globally pre-defined areas. Numerical models could benefit by assimilating these high resolution SAR time series. However, hurricanes don't necessarily occur in these pre-defined areas, and particular data analysis and assimilation schemes are needed to enable these temporally irregular SAR observations to help hurricane forecasting. Meanwhile, SAR's capability to provide detailed wind structure within the hurricane core makes it unique for capturing high frequency wind variations, such as strong wind shear, hurricane-related atmospheric boundary layer rolls etc. [5]. Abrupt variations of wind speed can happen on small spatial scales due to turbulence etc., which are important for hurricane-related disaster assessment, risk management, insurance industries etc. Considering these advantages and potential applications of SAR hurricane winds, it is of urgency to introduce these data into operational applications. Such methodology has been shown to be useful [45] to improve the track and intensity forecasting of Hurricane Isabel (2003) when SAR-retrieved hurricane wind vectors were assimilated into a numerical weather prediction model.

In this paper, we developed a methodology to assess the quality of SAR-retrieved wind speed based on two different modes of SAR observations. We demonstrated a capability to label poor-quality wind retrievals where the two observation modes are inconsistent. The capability of SAR to recognize strong precipitation areas in hurricanes is important, since it will not only lessen the dependence on external measurements to obtain the corresponding information, but also presents a useful tool for assessment of the quality of SAR-retrieved winds, in each wind cell. These differences in the rain effect on SAR-retrieved wind speeds under co-pol and cross-pol measurements are due to the different contributions of rain on the SAR Bragg-scattering and volume scattering mechanisms. Although cross-pol is relatively weakly affected by rain, as compared to co-pol, the rain bias on wind speed is still visible (Figure 9a). Thus, the difference in wind speeds retrieved from co-pol and cross-pol can be linked to non-wind contributions. For hurricanes, this non-wind contribution is mostly due to the contamination of heavy precipitation. Therefore, the methodology provides assessments of the validity of SAR-retrieved wind speed, and can be used as a quality index for the SAR wind product. To the best of our knowledge, this is the first time that a quality flag has been applied to SAR-retrieved wind speed, which we hope will be helpful for the operational application of SAR wind products. The proposed methodology is motivated by the quality flag for scatterometer wind products, where the product quality flag is based on various measurements from different antenna [24], comparable to the different polarizations of SAR dual-pol measurements that are applied here. Scatterometer data flagged as poor-quality are often eliminated from wind estimation. For the specific processes considered here, i.e., hurricanes, a dynamically consistent wind profile along the radial direction is adopted to rebuild the wind field for the rain-contaminated area, providing a tool for wind correction for poor quality SAR wind data. This methodology takes advantage of the particular wind structure of hurricanes, which is a robust physical mechanism for hurricanes. Such methodology has been successfully adopted for deriving hurricane wind direction information from co-pol SAR [10], for removing the speed ambiguity problem for high wind retrieval from co-pol SAR [16], and for rebuilding the full wind field of SAR images that only partly cover hurricanes [46].

Optimal performance of the proposed quality index depends on the accuracy of GMFs for wind retrieval from co-pol and cross-pol SAR measurements. With continued advancement in our understanding of the air-sea boundary layer physics underlying hurricanes, and more comprehensive collocated datasets for SAR measurements and wind vectors, the related SAR wind GMFs are expected to continuously improve. Thus, the quality of the proposed SAR winds index will be improved accordingly. The hurricane wind profile model along the radial direction adopted here considers a single eyewall hurricane only. For double or triple eyewall hurricanes, a different model would be developed and applied to represent the radial wind profile. SAR may be used to discriminate the unique multiple-eyewall structure based on its high-resolution imaging capability as shown in Reference [4]. Our proposed method will benefit from future improvement of parametric hurricane models for improved wind correction of rain-contaminated areas. The present study assumes consistent radar returns from co-pol and cross-pol SAR measurements induced by the wind to recognize non-wind-induced features by examining the inconsistent radar returns in the SAR measurements of the two different polarizations. Even though the concept introduced here does not rely on the GMF itself, the methodology and performance of the outcome would be affected by the accuracy of GMFs. Therefore, future advancement in the development of GMFs will also be beneficial. By including hurricane images with various spatial structures and at different stages, it will be possible to build a generalized dataset, which can be used to optimize the concept and methodology into broad hurricane conditions, and therefore provide improved, more robust accuracy assessments.

The threshold value of 0.5 dB was chosen based on comparisons between wind differences from two measurements of SAR and the SFMR rain rate. Our study shows this threshold value corresponds to rain rates higher than 5 mm/h in high wind areas, and 2–3 mm/h in moderately high wind speed conditions. The index has increased sensitivity at lower wind speeds. This confirms the decreased contamination of rain on radar backscattered signals as revealed by Figure 1. The threshold value is also based on current state of radiometric calibration accuracy of this SAR instrument, and applies to most of the current SAR instruments. Future SAR sensors might be able to achieve higher performance, thus with more accurate GMFs for wind retrieval, a smaller threshold value might be achieved. For hurricanes, the 0.5 dB threshold value should remain valid to detect rain areas, since rain is usually heavy in hurricanes. Figures 5 and 6 present the relationship of the rain-induced NRCS difference in regard to the rain rates; therefore, a formula for rain rate retrieval based on SAR imagery seems possible in the future. In order to achieve this goal, a comprehensive dataset needs to be built with SAR measurements and collocated simultaneous rain information. By recognizing the complicated processes related to rain modulation of the ocean surface roughness, and the fact that the sensitivity of the proposed quality index changes under different wind speeds, this may lead to a multi-factor non-linear relationship.

Despite decades of efforts in improved monitoring of hurricane core structures, methods for high resolution wind measurements are limited. The hurricane hunter radar reconnaissance missions conducted by NHC NOAA provide valuable but limited data in a temporally and spatially changing coordinate. Synthetic aperture radar, which is suitable for conducting high spatial resolution hurricane monitoring, is capable of filling this gap. By conducting simultaneous wind measurements over hurricanes, SAR captures fine-scale wind features within hurricanes. The comparison of SAR wind to in situ measurements of high wind speeds is challenging. The time difference of the measurements could be critical for highly dynamical hurricane processes. Future studies may consider developing an enhanced dataset by pairing multiple SAR images with the collocated flight/buoy measurements at SAR observation times. With more SAR platforms going into orbit, SAR-derived winds are expected to play a key role in future hurricane prediction systems.

6. Conclusions

Synthetic aperture radar provides unique wind observations over the ocean surface under hurricane conditions, with very high spatial resolution. Heavy precipitation can modify radar backscattered signals and contaminate SAR wind retrievals, making it difficult to adopt SAR winds in operational applications for hurricane forecasting. A hurricane wind quality index was developed

Remote Sens. **2018**, *10*, 1783

to evaluate SAR wind retrievals from dual-pol SAR measurements. This index is used as a flag for rain-contaminated wind cells, thus providing a reference for SAR wind data quality control.

A methodology for wind correction under heavy rain-contaminated cells was also developed, based on the unique radial profile structure of hurricane winds. Therefore, the rain-contaminated wind retrieval can be rebuilt, providing a reliable estimate of hurricane wind analysis from SAR. The proposed methodologies are solely based on dual-polarization radar measurements, and do not rely on any external dataset, which makes it feasible for them to be adopted into operational applications.

Looking to the future, the launch of future SAR missions, such as the RADARSAT Constellation Mission (RCM), plus the combination of multiple satellites, will significantly improve the possibility of SAR images that can observe hurricanes. With reliable quality control, the hurricane wind data from SAR are expected to play a more important role in hurricane forecasting and related decision making processes.

Author Contributions: Initiation of idea, H.S. and W.P.; SAR processing and development of algorithms, H.S.; data processing for HWIND, C.S.; writing-review and editing, all authors contributed; supervision, W.P., M.P. and Y.H.; funding acquisition, W.P. and Y.H.

Funding: This research was funded by the International Cooperation Project of the National Natural Science Foundation of China, Grant 41620104003; the National Programme on Global Change and Air-Sea Interaction, Grant GASI-IPOVAI-04; the Canadian Space Agency DUAP program; and the Office of Energy Research and Development (OERD).

Acknowledgments: The authors would like to thank the academic editor and the five anonymous reviewers for their detailed and insightful comments which have greatly improved our manuscript. We thank the Canadian Space Agency for the RADARSAT-2 SAR images through the Hurricane Watch program, the Hurricane Research Division of NOAA for providing aircraft-based data measurements, and Risk Management Solutions for HWIND re-analysis data.

Conflicts of Interest: The authors declare no conflicts of interest.

References

1. Fu, L.L.; Holt, B. *Seasat Views Oceans and Sea Ice with Synthetic Aperture Radar*; Jet Propulsion Laboratory: Pasadena, CA, USA, 1982; pp. 81–120.
2. Atlas, D. Origin of storm footprints on the sea seen by synthetic aperture radar. *Science* **1994**, *266*, 1364–1366. [CrossRef] [PubMed]
3. Katsaros, K.B.; Vachon, P.W.; Liu, W.T.; Black, P.G. Microwave remote sensing of tropical cyclones from space. *J. Oceanogr.* **2002**, *58*, 137–151. [CrossRef]
4. Li, X.; Zhang, J.A.; Yang, X.; Pichel, W.G.; DeMaria, M.; Long, D.; Li, Z. Tropical cyclone morphology from spaceborne Synthetic Aperture Radar. *Bull. Amer. Meteor. Soc.* **2013**, *94*, 215–230. [CrossRef]
5. Foster, R. Signature of large aspect ratio roll vortices in synthetic aperture radar images of tropical cyclones. *Oceanography* **2013**, *26*, 58–67. [CrossRef]
6. Monaldo, F.; Jackson, C.R.; Pichel, W.G.; Li, X. A weather eye on coastal winds. *Eos Trans. Am. Geophys. Union* **2015**, *96*, 16–19. [CrossRef]
7. Khurshid, S.; Bradley, D.; Manore, M. *National SAR Wind Product—User Requirements Document*; Meteorological Service of Canada, Environment Canada: Montreal, QC, Canada, 2012.
8. Hersbach, H. Comparison of C-Band scatterometer CMOD5.N equivalent neutral winds with ECMWF. *J. Atmos. Oceanic Technol.* **2010**, *27*, 721–736. [CrossRef]
9. Horstmann, J.; Thompson, D.R.; Monaldo, F.; Iris, S.; Graber, H.C. Can synthetic aperture radars be used to estimate hurricane force winds? *Geophys. Res. Lett.* **2005**, *32*, L22801. [CrossRef]
10. Shen, H.; Perrie, W.; He, Y. A new hurricane wind retrieval algorithm for SAR images. *Geophys. Res. Lett.* **2006**, *33*, L21812. [CrossRef]
11. Andreas, E.L. Spray stress revisited. *J. Phys. Oceanogr.* **2004**, *34*, 1429–1440. [CrossRef]
12. Smith, R.K.; Montgomery, M.T. On the existence of the logarithmic surface layer in the inner core of hurricanes. *Q.J.R. Meteorol. Soc.* **2014**, *140*, 72–81. [CrossRef]

13. Donelan, M.A.; Haus, B.K.; Reul, N.; Plant, W.J.; Stiassnie, M.; Graber, H.C.; Brown, O.B.; Saltzman, E.S. On the limiting aerodynamic roughness of the ocean in very strong winds. *Geophys. Res. Lett.* **2004**, *31*, L18306. [CrossRef]

14. Fernandez, D.E.; Carswell, J.R.; Frasier, S.; Chang, P.S.; Black, P.G.; Marks, F.D. Dual-polarized C- and Ku-band ocean backscatter response to hurricane-force winds. *J. Geophys. Res.* **2006**, *111*, C08013. [CrossRef]

15. Shen, H.; Perrie, W.; He, Y. On SAR wind speed ambiguities and related geophysical model functions. *Can. J. Remote Sens.* **2009**, *35*, 310–319. [CrossRef]

16. Shen, H.; He, Y.; Perrie, W. Speed ambiguity in hurricane wind retrieval from SAR imagery. *Int. J. Remote Sens.* **2009**, *30*, 2827–2836. [CrossRef]

17. Vachon, P.W.; Wolfe, J. C-band cross-polarization wind speed retrieval. *IEEE Geosci. Remote Sens. Lett.* **2011**, *8*, 456–459. [CrossRef]

18. Zhang, B.; Perrie, W. Cross-polarized Synthetic Aperture Radar: A new potential measurement technique for hurricanes. *Bull. Am. Meteorol. Soc.* **2012**, *93*, 531–541. [CrossRef]

19. Shen, H.; Perrie, W.; He, Y.; Liu, G. Wind speed retrieval from VH dual-polarization radarsat-2 SAR images. *IEEE Trans. Geosci. Remote Sens.* **2014**, *52*, 5820–5826. [CrossRef]

20. Pugliese Carratelli, E.; Dentale, F.; Reale, F. Numerical PSEUDO—Random simulation of SAR sea and wind response. In *Advances in SAR Oceanography from ENVISAT and ERS Missions, Proceedings of the SEASAR 2006 (ESA SP-613), Frascati, Italy, 23–26 January 2006*; Lacoste, H., Ed.; ESA Publications Division: Noordwijk, The Netherlands, 2006.

21. Alpers, W.; Zhang, B.; Mouche, A.; Zeng, K.; Chan, P.W. Rain footprints on C-band synthetic aperture radar images of the ocean—Revisited. *Remote Sens. Environ.* **2016**, *187*, 169–185. [CrossRef]

22. Zhang, G.; Li, X.; Perrie, W.; Zhang, B.; Wang, L. Rain effects on the hurricane observations over the ocean by C-band Synthetic Aperture Radar. *J. Geophys. Res. Oceans* **2016**, *121*, 14–26. [CrossRef]

23. Xu, F.; Li, X.; Wang, P.; Yang, J.; Pichel, W.G.; Jin, Y.Q. A backscattering model of rainfall over rough sea surface for Synthetic Aperture Radar. *IEEE Trans. Geosci. Remote Sens.* **2015**, *53*, 3042–3054. [CrossRef]

24. Nie, C.; Long, D.G. A C-Band scatterometer simultaneous wind/rain retrieval method. *IEEE Trans. Geosci. Remote Sens.* **2008**, *46*, 3618–3631. [CrossRef]

25. Klotz, B.; Uhlhorn, E.W. Improved stepped frequency microwave radiometer tropical cyclone surface winds in heavy precipitation. *J. Atmos. Ocean. Technol.* **2014**, *31*, 2392–2408. [CrossRef]

26. Mears, C.A.; Smith, D.; Wentz, F.J. Detecting rain with QuikScat. In Proceedings of the IEEE International Geoscience and Remote Sensing Symposium, (Cat. No.00CH37120), Honolulu, HI, USA, 24–28 July 2000; Volume 3, pp. 1235–1237. [CrossRef]

27. Zhang, B.; Perrie, W.; Vachon, P.W.; Li, X.; Pichel, W.G.; Guo, J.; He, Y. Ocean vector winds retrieval from C-band fully polarimetric SAR measurements. *IEEE Trans. Geosci. Remote Sens.* **2012**, *50*, 4252–4261. [CrossRef]

28. Zhang, B.; Perrie, W.; Li, X.; Pichel, W.G. Mapping sea surface oil slicks using RADARSAT-2 quad-polarization SAR image. *Geophys. Res. Lett.* **2011**, *38*. [CrossRef]

29. Li, H.; Perrie, W.; He, Y.; Lehner, S.; Brusch, S. Target detection on the ocean with the relative phase of compact polarimetry SAR. *IEEE Trans. Geosci. Remote Sens.* **2013**, *51*, 3299–3305. [CrossRef]

30. Shen, H.; Perrie, W.; Liu, Q.; He, Y. Detection of macroalgae blooms by complex SAR imagery. *Mar. Pollut. Bull.* **2014**, *78*, 190–195. [CrossRef] [PubMed]

31. Mouche, A.; Chapron, B.; Zhang, B.; Husson, R. Combined co- and cross-polarized SAR measurements under extreme wind conditions. *IEEE Trans. Geosci. Remote Sens.* **2017**. [CrossRef]

32. Kimberlain, T.B.; Blake, E.S.; Cangialosi, J.P. Hurricane Patricia—National Hurricane Center—NOAA. National Hurricane Center Tropical Cyclone Report. EP202015; 2016. Available online: http://www.nhc.noaa.gov/data/tcr/EP202015_Patricia.pdf (accessed on 23 May 2018).

33. Shen, H.; Perrie, W.; He, Y. Evaluation of hurricane wind speed retrieval from cross-dual-pol SAR. *Int. J. Remote Sens.* **2016**, *37*, 599–614. [CrossRef]

34. Powell, M.D.; Houston, S.H.; Amat, L.R.; Morisseau-Leroy, N. The HRD real-time hurricane wind analysis system. *J. Wind Eng. Ind. Aerodyn.* **1998**, *77–78*, 53–64. [CrossRef]

35. Zhang, J.A.; Uhlhorn, E.W. Hurricane sea surface inflow angle and an observation-based parametric model. *Mon. Wea. Rev.* **2012**, *140*, 3587–3605. [CrossRef]

36. Luscombe, A.P. RADARSAT-2 SAR image quality and calibration operations. *Can. J. Remote Sens.* **2014**, *30*, 345–354. [CrossRef]

37. Dagestad, K.-F.; Horstmann, J.; Mouche, A.; Perrie, W.; Shen, H.; Zhang, B.; Li, X.; Monaldo, F.; Pichel, W.; Lehner, S.; et al. Wind retrieval from Synthetic Aperture Radar—An overview, SAR Wind Whitepaper. In Proceedings of the SEASAR 2012 Advances in SAR Oceanography, ESA SP-709, Tromso, Norway, 18–22 June 2012.

38. Kossin, J.; Schubert, W. Mesovortices in hurricane ISABEL. *Bull. Am. Meteorol. Soc.* **2004**, *85*, 151–153. [CrossRef]

39. Rogers, R.F.; Aberson, S.; Bell, M.M.; Cecil, D.J.; Doyle, J.D.; Kimberlain, T.B.; Morgerman, J.; Shay, L.K.; Velden, C. Rewriting the Tropical Record Books: The Extraordinary Intensification of Hurricane Patricia (2015). *Bull. Amer. Meteor. Soc.* **2017**, *98*, 2091–2112. [CrossRef]

40. Holland, G.J. An analytic model of the wind and pressure profiles in hurricanes. *Mon. Weather Rev.* **1980**, *108*, 1212–1218. [CrossRef]

41. Holland, G.J.; Belanger, J.I.; Fritz, A. A revised model for radial profiles of hurricane winds. *Mon. Weather Rev.* **2010**, *138*, 4393–4401. [CrossRef]

42. Emanuel, K. Tropical cyclones. *Annu. Rev. Earth Planet Sci.* **2003**, *31*, 75–104. [CrossRef]

43. Chan, J.C.L. The physics of tropical cyclone motion. *Annu. Rev. Fluid Mech.* **2005**, *37*, 99–128. [CrossRef]

44. Rappaport, E.N.; Franklin, J.L.; Avila, L.A.; Baig, S.R.; Beven, J.L., II; Blake, E.S.; Burr, C.A.; Jiing, J.-G.; Juckins, C.A.; Knabb, R.D.; et al. Advances and challenges at the National Hurricane Center. *Weather Forecast.* **2009**, *24*, 395–419. [CrossRef]

45. Perrie, W.; Zhang, W.; Bourassa, M.; Shen, H.; Vachon, P.W. Impact of satellite winds on marine wind simulations. *Weather Forecast.* **2008**, *23*, 290–303. [CrossRef]

46. Zhang, G.; Perrie, W.; Li, X.; Zhang, J.A. A hurricane morphology and sea surface wind vector estimation model based on C-Band cross-polarization SAR imagery. *IEEE Trans. Geosci. Remote Sens.* **2017**, *55*, 1743–1751. [CrossRef]

remote sensing

MDPI

Article

A Wind Speed Retrieval Model for Sentinel-1A EW Mode Cross-Polarization Images

Yuan Gao [1], Changlong Guan [1], Jian Sun [1,*] and Lian Xie [2]

[1]   Physical Oceanography Laboratory, Ocean University of China, Qingdao 266100, China;
      ygao24@ncsu.edu (Y.G.); clguan@ouc.edu.cn (C.G.)
[2]   Department of Marine, Earth and Atmospheric Sciences, North Carolina State University, Raleigh, NC 27607,
      USA; xie@ncsu.edu
*   Correspondence: sunjian77@ouc.edu.cn; Tel.: +86-532-66786228

Received: 30 December 2018; Accepted: 14 January 2019; Published: 15 January 2019

Abstract: In contrast to co-polarization (VV or HH) synthetic aperture radar (SAR) images, cross-polarization (CP for VH or HV) SAR images can be used to retrieve sea surface wind speeds larger than 20 m/s without knowing the wind directions. In this paper, a new wind speed retrieval model is proposed for European Space Agency (ESA) Sentinel-1A (S-1A) Extra-Wide swath (EW) mode VH-polarized images. Nineteen S-1A images under tropical cyclone condition observed in the 2016 hurricane season and the matching data from the Soil Moisture Active Passive (SMAP) radiometer are collected and divided into two datasets. The relationships between normalized radar cross-section (NRCS), sea surface wind speed, wind direction and radar incidence angle are analyzed for each sub-band, and an empirical retrieval model is presented. To correct the large biases at the center and at the boundaries of each sub-band, a corrected model with an incidence angle factor is proposed. The new model is validated by comparing the wind speeds retrieved from S-1A images with the wind speeds measured by SMAP. The results suggest that the proposed model can be used to retrieve wind speeds up to 35 m/s for sub-bands 1 to 4 and 25 m/s for sub-band 5.

Keywords: Sentinel-1; cross-polarization; wind retrieval; SMAP

---

1. Introduction

A large number of geophysical model functions (GMF) have been presented to retrieve wind speeds from co-polarization (VV or HH) SAR images. According to many C-band VV-polarized GMF models, the normalized radar cross section (NRCS) is dependent upon the wind speed at 10-m height, wind direction and radar incidence angle. However, wind speed retrieval from co-polarization SAR images is known to have a number of limitations. First, due to the saturation of the backscattering signal under strong wind condition, the retrieval results may have large error for wind speed higher than 20 m/s [1,2]. Second, the difficulty to obtain a collocated high-resolution wind direction field often leads to a decrease in the accuracy of wind speed retrieval [3–6]. Third, the co-polarization NRCS is dampened at certain incidence angles, leading to a wind speed ambiguity problem [7].

The backscattering signals of both co-polarization and cross-polarization (CP for VH or HV) are induced by the Bragg scattering from sea surface [8–10]. However, at moderate to high wind conditions, the CP backscattering signal could trace the surface wave breaking efficiently, which causes the non-Bragg contribution [11,12]. The NRCS of CP SAR image is barely dependent upon wind direction and radar incidence angle. The CP signal remains sensitive to sea surface wind speed with high signal-to-noise ratio under more extreme conditions [12–15]. Moreover, the CP NRCS in decibels linearly increases with wind speed, indicating that it could potentially be used to retrieve tropical cyclone winds. Comparing with co-polarization SAR images, the CP SAR images are more suitable for high winds (>20 m/s) retrieval [2,12,16–18].

With the development of the SAR technology, more and more wind retrieval models are proposed for CP SAR images, promoting the progress of ocean wind retrieval by SAR. In some models, wind speed is the only factor [15,16,18,19]. Based on Radarsat-2 (R-2) fine quad-polarization mode SAR images and wind speed observations from National Data Buoy Center (NDBC), the C-band Cross-Polarization Ocean model (C-2PO) is proposed as a linear relationship between VH-polarized NRCS and wind speeds ranging from 8 to 26 m/s [18]. Compared with wind speeds from the H*Wind data, the retrieved wind speeds by C-2PO have a bias of about −0.88 m/s and a root mean square error (RMSE) of approximately 4.47 m/s. Monaldo et al. retrieved the wind speed field from a S-1A image of Typhoon Lionrock utilizing the C-2PO model [2]. They found that the retrieval results in the near-range beam (sub-band 1) seem to be higher than those in the other beams (sub-bands 2–5). In 2011, an empirical model similar to the C-2PO model is proposed by Vachon et al., utilizing R-2 fine quad-polarization mode images and wind measurements from operational weather buoys [15]. The highest wind speed in their dataset is 22.5 m/s. In 2014, Zhang et al. presented a new linear wind speed retrieval model (C-2POD) for R-2 dual polarization images, expanding the wind speed retrieval range up to 39.7 m/s [16]. Compared with the measurements from Quikscat, the retrieved wind speeds by C-2POD have a bias of −1.21 m/s and a centered RMSE of 2.75 m/s. In 2014, van Zadelhoff et al. proposed a wind speed retrieval model for strong-to-severe wind conditions (20–45 m/s) [19]. They found that the relationship between VH-polarized NRCS and wind speeds has distinct characteristics in low-to-strong (<20 m/s) and strong-to-severe (>20 m/s) wind regimes.

Some VH GMF models are considered to be functions of two parameters: wind speed and incidence angle, e.g., H14, MS1A, and C-3POD [11,12,20]. In 2015, Hwang et al. presented a wind speed retrieval model (H14) according to R-2 dual-polarization data and massive wind speed data from buoys, the NOAA/Hurricane Research Division's (HRD) Stepped-Frequency Microwave Radiometer (SFMR), H*Wind and European Centre for Medium-Range Weather Forecasts (ECMWF) [11]. H14 is a power law function relating VH-polarized NRCS in linear units to wind speeds (up to 56 m/s) and radar incidence angle. In 2017, Mouche et al. presented the MS1A wind speed retrieval model, based on the Soil Moisture Active Passive (SMAP) brightness temperature data and Sentinel-1A (S-1A) extra-wide swath (EW) mode images for several hurricanes [12]. The MS1A model is a power law function similar to the H14 model and works well for wind speeds higher than 25 m/s. Compared with the SMAP measurements, the wind speeds retrieved by MS1A have a bias of 3.35 m/s and a standard deviation (Std) of 4.85 m/s. Based on the Radarsat-2 data and the SFMR wind speeds, Zhang et al. proposed the C-3PO wind speed retrieval model, which is an empirical function of VH-polarized NRCS, wind speed and incidence angle [20]. It can be used to retrieve wind speeds up to 40 m/s. A validation was made by comparing the retrieval results and SFMR observations, showing a RMSE less than 3 m/s.

In 2017, Huang et al. made a technical evaluation on Sentinel-1 Interferometric Wide swath (IW) mode CP images and proposed an empirical retrieval model with three factors: wind speed, wind direction and incidence angle [21]. Their model can be applied to retrieving wind speeds under 15 m/s. Validating against the wind speed observations from ASCAT, the wind speeds retrieved by their model have a bias of 0.42 m/s and a RMSE of 1.26 m/s.

The aim of this study is to develop a new wind speed retrieval model for S-1A EW mode VH-polarized images according to the relationships between noise-free NRCS, sea surface wind speed and radar incidence angle. In this paper, 19 S-1A EW mode VH-polarized images under tropical cyclone conditions are studied. The SAR-collocated wind speed data are collected from SMAP radiometer for model construction and validation. The samples cover low-to-severe wind regimes (2–35 m/s). For each sub-band of the S-1A image, a basic retrieval model is proposed with VH NRCS and wind speed. Based on incidence angle, a new correction methodology is proposed to improve the accuracy of the basic model. The effect of incidence angle on VH NRCS under different wind conditions is then simulated by proposing a modified wind speed and incidence angle coupled model. Due to the ambiguous relationship between VH NRCS and wind direction, the wind direction parameter is not included in the proposed model. Finally, the proposed model is validated against dataset 2 to evaluate the retrieval accuracy.

The remaining sections of this paper are organized as follows. Section 2 describes the S-1A images and SMAP data. In Section 3, the relationships between VH-polarized NRCS, wind speed, wind direction and radar incidence angle are analyzed. In Section 4, the basic wind retrieval model and the corrected wind retrieval model are proposed. In Section 5, the two models are validated, compared and discussed. Conclusions are summarized in Section 6.

2. Dataset

In this study, 19 Sentinel-1A VH-polarized EW mode images under tropical cyclone conditions are collected. The matching SMAP radiometer wind speeds are collected for comparison and model validation. The data are divided into two datasets. Dataset 1 is used for analyzing the relationships between NRCS, wind vector and incidence angle and proposing model. Dataset 2 is used for validation and comparison.

2.1. Sentinel-1A Data

The Sentinel-1A (S-1A) satellite is designed by the European Space Agency (ESA). The C-SAR boarded on the S-1A satellite can provide single-polarization (HH or VV) and dual-polarization (VV, VH or HH, HV) data with 4 sensor modes: the Stripmap (SM) mode, the Interferometric Wide swath (IW) mode, the Extra-Wide swath (EW) mode and the Wave (WV) mode [21]. The Level-1 products can be one of two product types, either Single Look Complex (SLC) or Ground Range Detected (GRD).

The SAR data analyzed in this study are the S-1A EW mode VH-polarized GRD products. The EW mode image covers incidence angles from about 18.9 ° to 47.0 ° and is up to 410-km wide with a spatial resolution of 93 m × 87 m (range by azimuth) and a pixel spacing of 40 m × 40 m. Each EW mode image has five sub-bands in range direction. In this paper, the sub-bands are named sub-band 1, sub-band 2, sub-band 3, sub-band 4, and sub-band 5 with increasing distance from the sub-satellite point. Compared with VV-polarized signal, the VH-polarized signal does not saturate for wind speeds as strong as 55 m/s and is insensitive to wind direction [20,22,23]. The GRD products consist of focused SAR data that has been detected, multi-looked and projected to ground range.

The S-1A products are openly available from ESA. During the 2016 hurricane season, the Satellite Hurricane Observation Campaign (SHOC) was designed by the ESA Sentinel-1 mission planning team to gather hurricane images [12]. The S-1A data used in this paper are collected from the SHOC. Data information is shown in Table 1.

Table 1. S-1A data information.

Tropical Cyclone Name	Sensing Time (UTC)	Number of Matching Points	Dataset
Lester	2016-08-26 13:39	241	1
Lester	2016-08-30 14:45	202	1
Lester	2016-08-31 03:15	184	1
Gaston	2016-08-27 09:22	257	2
Gaston	2016-08-29 21:41	112	2
Gaston	2016-08-29 21:42	125	2
Gaston	2016-09-01 20:29	279	2
Lionrock	2016-08-27 20:52	225	1
Lionrock	2016-08-29 20:34	264	2
Lionrock	2016-08-29 20:35	263	1
Namtheum	2016-09-04 09:20	253	2
Hermine	2016-09-04 22:31	279	1
Hermine	2016-09-04 22:32	282	2

Table 1. *Cont.*

Tropical Cyclone Name	Sensing Time (UTC)	Number of Matching Points	Dataset
Karl	2016-09-23 22:22	176	1
Karl	2016-09-23 22:23	171	1
Karl	2016-09-24 10:25	154	1
Karl	2016-09-24 10:26	166	1
Karl	2016-09-24 10:27	183	1
Megi	2016-09-26 09:34	232	1

Figure 1 shows the Noise Equivalent Sigma Zero (NESZ) of the S-1A EW mode data in range direction and the incidence angle ranges in different sub-bands. The distribution of NESZ in each sub-band is different, showing a low level in the middle of each sub-band and a high level at the inter-band boundaries, which may cause a discontinuity of the image [24]. In this study, Sentinel Application Platform (SNAP) 4.0 is used for radiometric calibration. After radiometric calibration, all measurement samples have higher decibel values than the NESZ values.

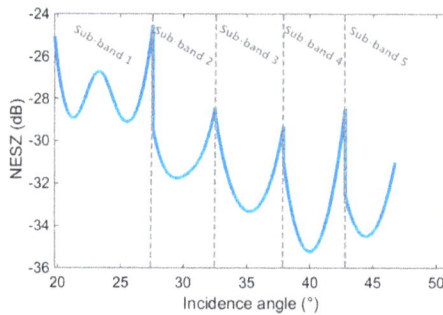

Figure 1. The distributions of NESZ and incidence angle in sub-bands 1 to 5.

Due to the difference in spatial resolution between the SAR data and the SMAP data, the NRCS is averaged within each SMAP cell (27 km × 27 km) for data matching. However, the different number of pixels for averaging (calculation resolution) might lead to homogeneity variation of SAR data in a calculation cell. Based on dataset 1 and dataset 2, the Std variation of NRCS in a calculation cell with calculation resolutions between 8 × 8 and 1048 × 1048 pixels is shown in Figure 2. The Std increases from 1.23 dB to 1.51 dB within a SMAP cell, indicating that the homogeneity decreases with calculation resolution. In this paper, to ensure the quality of the matching data, a calculation resolution of 16 × 16 pixels is utilized for averaging.

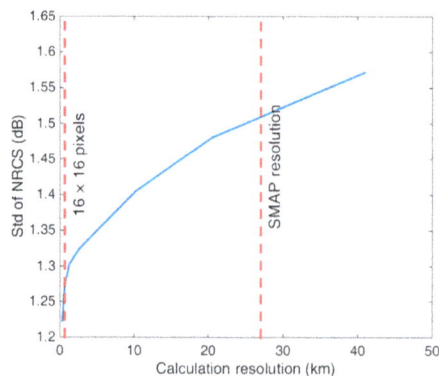

Figure 2. The variation of NRCS Std with different calculation resolutions.

2.2. SMAP Data

In this study, the Soil Moisture Active Passive (SMAP) Level-2 wind measurements are downloaded from Remote Sensing Systems (RSS) as references for the wind vectors. The National Aeronautics and Space Administration (NASA) SMAP winds are retrieved from brightness temperatures measured by L-band passive radiometer, which are largely unaffected by rain [25]. The SMAP can provide excellent sensitivity to wind speed even in very high winds [12,25,26]. The SMAP Level-2 wind dataset has a spatial resolution of 0.25° × 0.25° (about 27 km × 27 km) and a swath width of 1000 km. The difference between SMAP and WindSat wind speeds yields a global RMS of about 1.5 m/s for rain-free ocean scenes [25]. In this study, to ensure the accuracy of the matching data, the sensing time differences between SMAP and S-1A are controlled within one hour.

The S-1A images and SMAP references are divided into two datasets. Figure 3 shows the numbers of matching points in different wind ranges and different sub-bands. Both datasets cover wind speeds ranging from 5 to 35 m/s. There is a total of 4048 matching samples: 2476 in dataset 1 and 1572 in dataset 2. Note that, the width of sub-band 1 is larger than the widths of sub-bands 2–5 in range direction. For some images, there are no matching points in sub-bands 4 and 5. Therefore, the number of matching samples decreases from sub-bands 1 to 5.

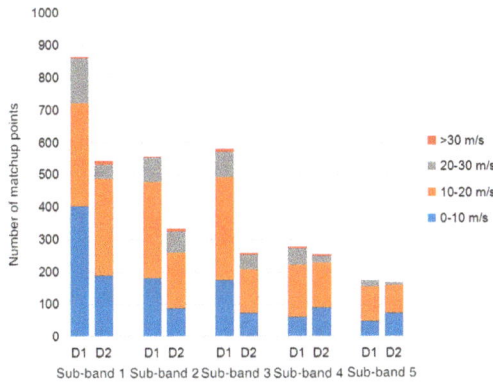

Figure 3. The numbers of matching points in different wind ranges and different sub-bands. D1 is an abbreviation of dataset 1 and D2 is an abbreviation of dataset 2.

3. Data Analyses

As mentioned above, the NRCS of VH-polarized signal is mainly dependent on wind speed and is barely dependent on wind direction and incidence angle, which makes VH-polarized images suitable for high wind retrieval. In this section, based on dataset 1, the relationships between VH NRCS, wind speed, wind direction, and incidence angle will be analyzed.

Figure 4 shows the relationships between VH NRCS and SMAP wind speed observations in different sub-bands. The wind ranges are 2–32 m/s, 2–35 m/s, 2–31 m/s, 7–32 m/s, and 7–24 m/s for sub-bands 1–5, respectively. The NRCS samples with different incidence angles cover the whole wind range in each sub-band.

As shown in Figure 4, the NRCS increases with wind speed in all sub-bands. For sub-bands 1–3, the NRCS increases linearly. For sub-bands 4 and 5, the slopes decrease in the entire wind ranges. Compared with sub-band 1, sub-bands 2–5 have lower NRCS levels under the same wind speed. The correlation coefficients (r) between NRCS and wind speed are 0.86, 0.91, 0.82, 0.76, and 0.78 for sub-bands 1–5, respectively. Based on the strong dependence of NRCS on wind speed, wind speed retrieval model will be presented in Section 4 for each sub-band.

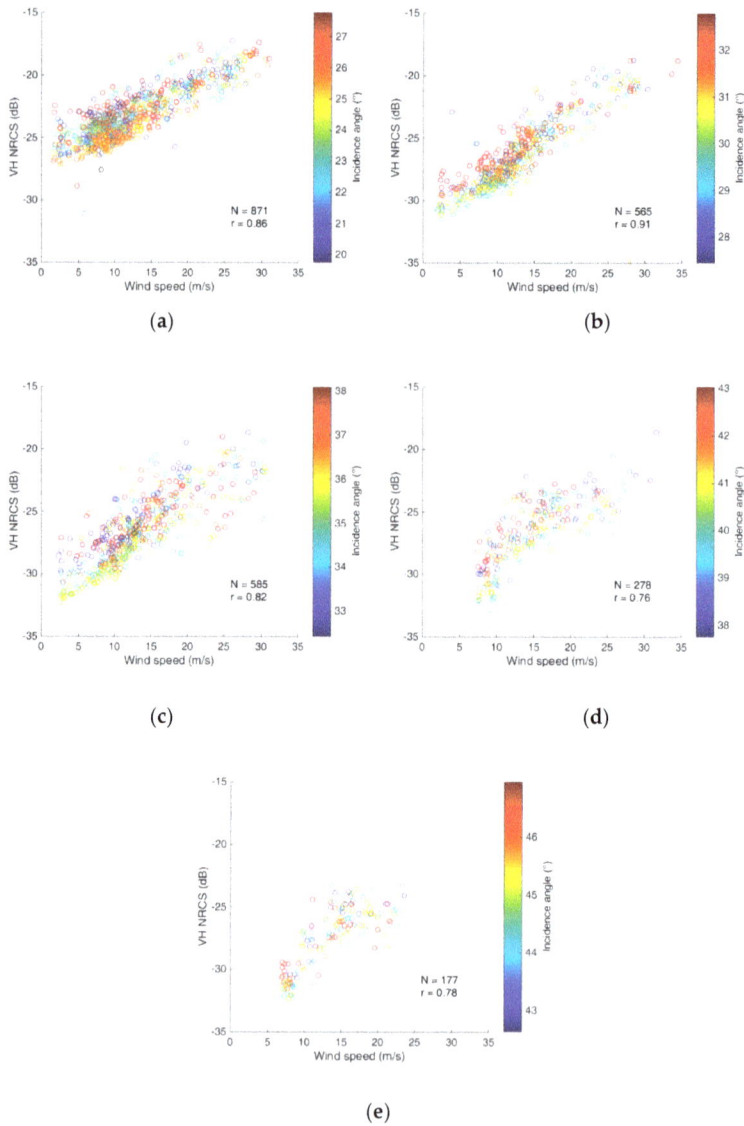

Figure 4. Relationships between VH NRCS and SMAP wind speed for (**a**) sub-band 1, (**b**) sub-band 2, (**c**) sub-band 3, (**d**) sub-band 4, and (**e**) sub-band 5. N is the number of matching points and r stands for the correlation coefficient.

The relationships between the VH-polarized NRCS and the incidence angle under different wind speeds are shown in Figure 5. For S-1A EW mode data, the incidence angles are about 19.75–27.55°, 27.55–32.55°, 32.55–37.95°, 37.95–42.85°, and 42.85–46.95° for sub-bands 1–5, respectively.

The features of NESZ mentioned above can also be found in Figure 5. For sub-band 1, the NRCS under the same wind speed has three peaks: one in the middle of the band and two at the boundaries. For sub-bands 2-5, the NRCS has a low level in the middle of the band and a high level at the inter-band boundaries. As is shown in Figure 5, the incidence angle has a strong influence on NRCS under low wind speed (<10 m/s). In addition, under the same wind speed level, the fluctuation of NRCS is up to

5 dB, which may influence the precision of the NRCS simulation and wind retrieval. According to the role of incidence angle in backscattering, the corrected functions of NRCS will be proposed in Section 4.

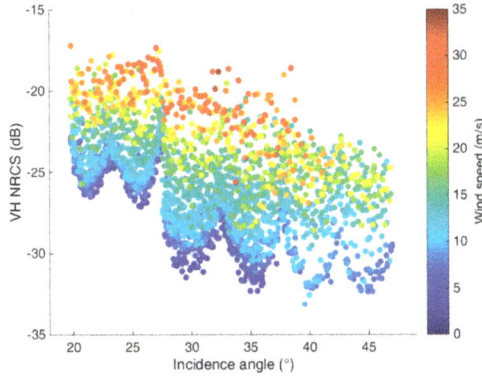

Figure 5. The relationship between VH NRCS and incidence angle under different wind speeds.

In GMF, the wind direction is the radar relative wind direction, which is the angle between the sea surface wind direction and radar azimuth look direction. Based on dataset 1, the scatterplots in Figure 6a–e show the distributions of NRCS for wind speeds at 5, 10, 15, 20, and 25 m/s with a range of ± 2.5 m/s in each sub-band. Then, the NRCS samples are averaged for wind speeds at 5, 10, 15, 20, and 25 m/s with a range of ± 2.5 m/s. The relationships between NRCS and wind direction under different wind speeds are shown in Figure 6f. The average NRCS values are calculated at different wind directions within a range of 15°.

As shown in Figure 6, the NRCS increases with wind speed and has an irregular fluctuation with the change of wind direction. The fluctuations under wind speeds 5 and 10 m/s are stronger than the fluctuations under wind speeds 15, 20, and 25 m/s. Since the incidence angle has a stronger influence on NRCS under low wind speeds, as shown in Figure 5. These phenomena indicate that the dependence of NRCS on incidence angle is stronger than on wind direction. Note that for the whole wind direction range (0–360°), the amount of matching data in dataset 1 is not enough to indicate the correlations between NRCS and wind direction under every incidence angles. Therefore, the dependence of NRCS on wind direction is assumed to be weak. In this paper, the wind direction factor is not considered in the construction of the model.

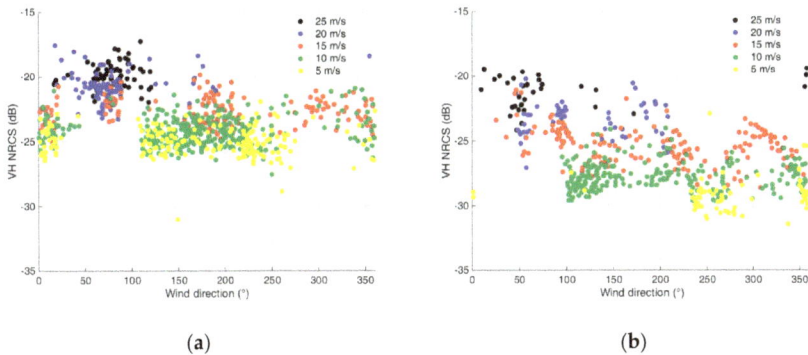

(**a**)

(**b**)

Figure 6. *Cont.*

(c)

(d)

(e)

(f)

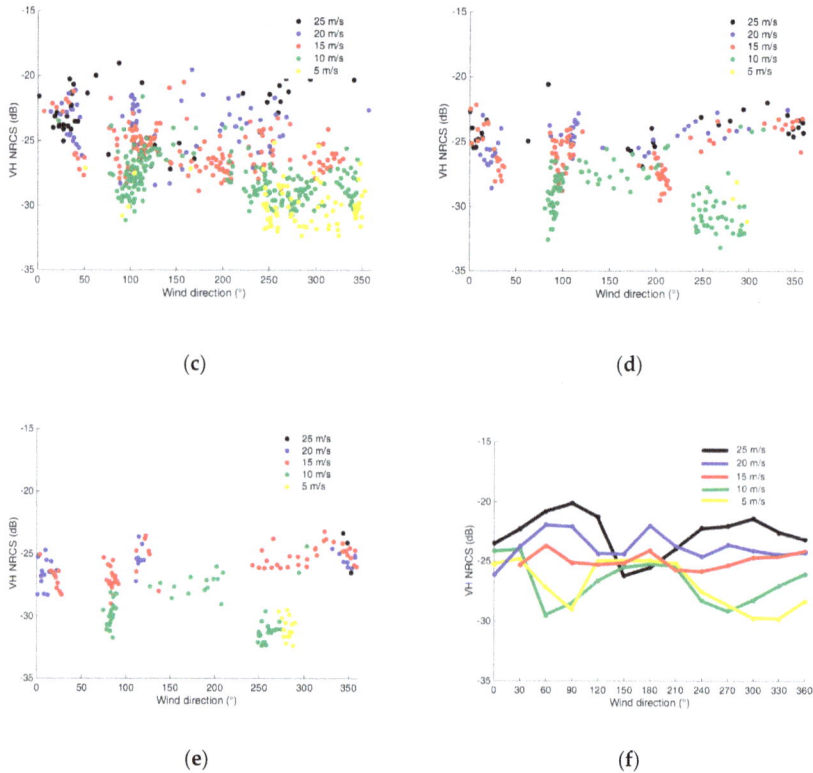

Figure 6. The relationships between VH NRCS and wind direction under different wind speeds for (**a**) sub-band 1, (**b**) sub-band 2, (**c**) sub-band 3, (**d**) sub-band 4, and (**e**) sub-band 5. (**f**) The variation of average VH NRCS with wind direction under different wind speeds.

4. Wind Retrieval Model

4.1. Basic Model

According to the distribution of the data samples in Figure 4 and the strong correlation between VH-polarized NRCS and wind speed, linear function and power law function are used to fit the points for each sub-band. The fitting functions are linear functions for sub-bands 1–3 and power law functions for sub-bands 4 and 5. The fitting results are shown in Figure 7 (red curves). These basic empirical functions are proposed as:

$$f_0(U_{10}) = \begin{cases} 0.26U_{10} - 26.58, & \text{sub-band 1} \\ 0.37U_{10} - 31.07, & \text{sub-band 2} \\ 0.39U_{10} - 31.80, & \text{sub-band 3} \\ -50.74U_{10}^{-0.25}, & \text{sub-band 4} \\ -49.38U_{10}^{-0.23}, & \text{sub-band 5} \end{cases} \tag{1}$$

where f_0 is the VH-polarized NRCS, U_{10} represents the sea surface wind speed in 10-m height. The units of f_0 and U_{10} are decibels and meters per second, respectively.

Based on the SMAP wind speeds in dataset 1, the NRCS values are simulated by the basic model to make a comparison with the observed NRCS. The comparisons between the observed and the simulated NRCS for each sub-band are shown in Figure 8 and Table 2. The correlation coefficients between the observed and the simulated NRCS are 0.83, 0.90, 0.82, 0.80, and 0.83 for sub-bands 1–5, respectively. The biases between the observed and the simulated NRCS are 0.06, −0.03, −0.06, −0.07, and 0.03 dB for sub-bands 1–5, respectively. The standard deviations (Std) between the observed and the simulated NRCS are 1.19, 1.19, 1.63, 1.62, and 1.38 dB for sub-bands 1–5, respectively. Through curve fitting, Equation (1) ensures that the bias of the simulation is minimized.

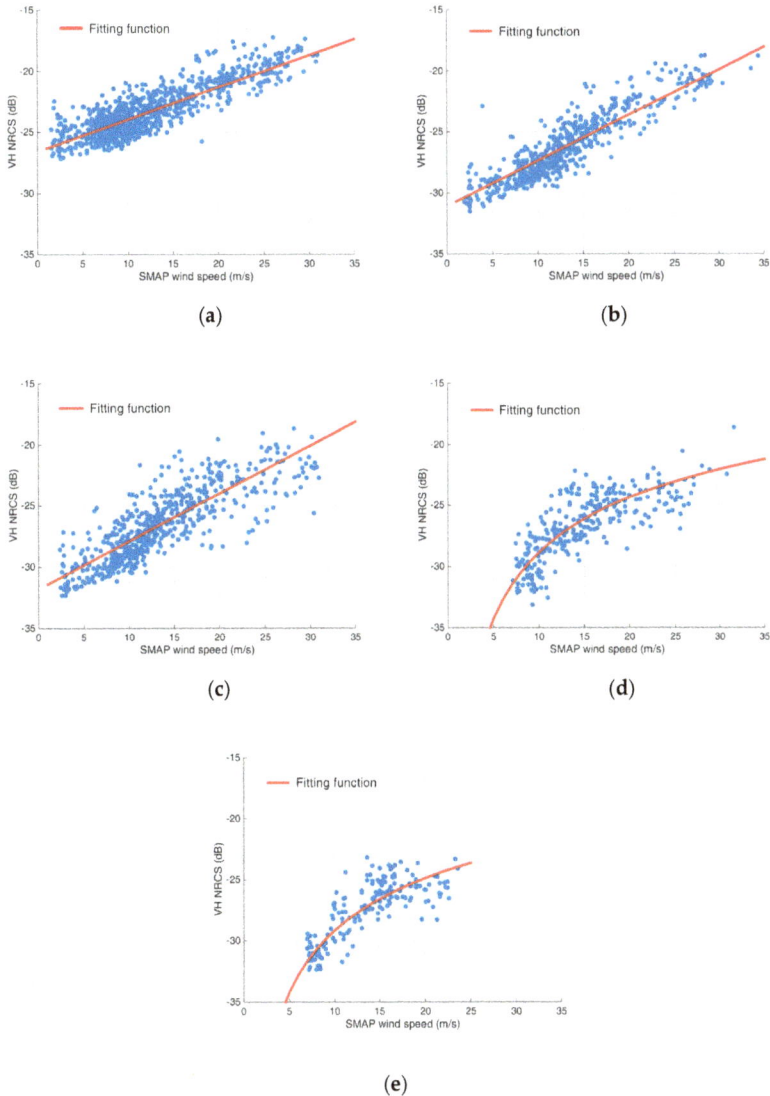

Figure 7. Fitting functions (red curves) between VH NRCS and SMAP wind speeds for (**a**) sub-band 1, (**b**) sub-band 2, (**c**) sub-band 3, (**d**) sub-band 4, and (**e**) sub-band 5.

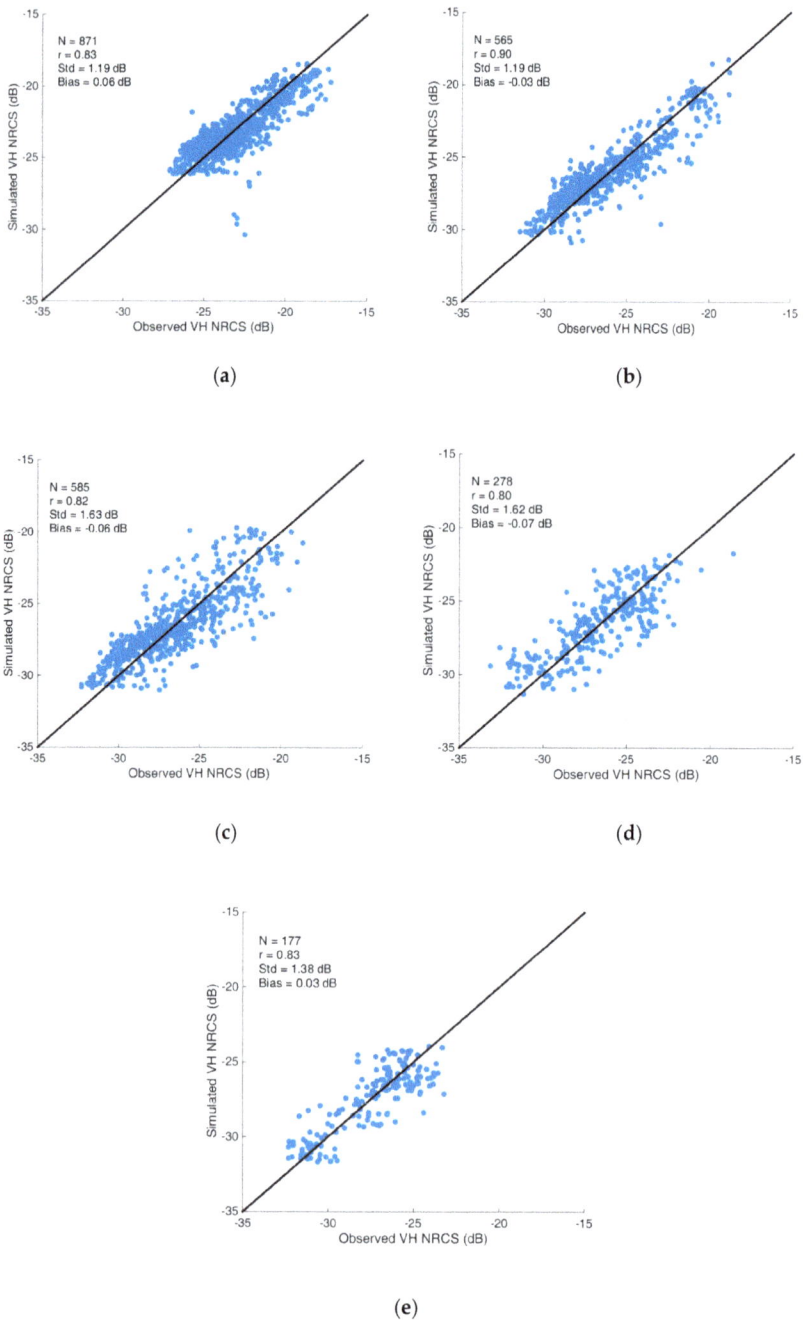

Figure 8. Comparisons between simulated VH NRCS and observed VH NRCS for (**a**) sub-band 1, (**b**) sub-band 2, (**c**) sub-band 3, (**d**) sub-band 4, and (**e**) sub-band 5. N, r, Std, Bias represents the number of matching points, correlation coefficient, standard deviation, and bias between observed NRCS and simulated NRCS with the proposed basic model.

Table 2. Correlation coefficient, Std and bias between the observed NRCS and the simulated NRCS with basic model.

Sub-Band	r	Std (dB)	Bias (dB)
1	0.83	1.19	0.06
2	0.90	1.19	−0.03
3	0.82	1.63	−0.06
4	0.80	1.62	−0.07
5	0.83	1.38	0.03

4.2. Corrected Model

Based on the dependence of VH NRCS on radar incidence angle, the basic model is corrected in this section. Trigonometric function and quadratic function are used to fit the variations of NRCS with incidence angle for sub-band 1 and sub-bands 2–5, respectively. In this paper, the samples with wind speeds higher than 20 m/s are only 14.7% of all samples, thus, the curve fitting is only carried out for samples with wind speeds lower than 20 m/s. The fitting results are proposed as follows:

$$f_1(\theta) = \begin{cases} 24.05\sin(3.36\theta + 548.22) + 0.95\sin(84.56\theta - 802.73), & \text{sub-band 1} \\ 0.24\theta^2 - 14.11\theta + 183.70, & \text{sub-band 2} \\ 0.32\theta^2 - 22.57\theta + 370.50, & \text{sub-band 3} \\ 0.32\theta^2 - 25.55\theta + 486.70, & \text{sub-band 4} \\ 0.21\theta^2 - 19.23\theta + 403.40, & \text{sub-band 5} \end{cases} \tag{2}$$

where f_1 is the VH-polarized NRCS, and θ represents the incidence angle. The units of f_1 and θ are decibels and degrees, respectively. The proposed fitting functions are shown in Figure 9.

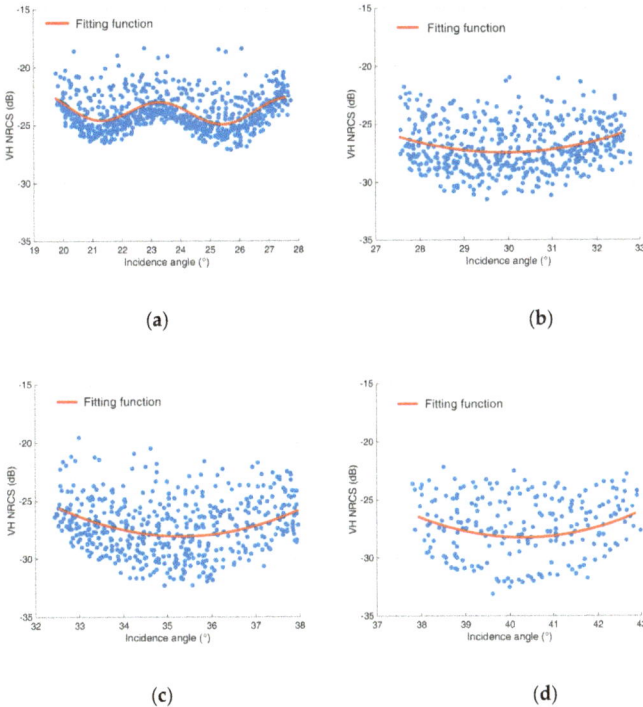

(a)

(b)

(c)

(d)

Figure 9. *Cont.*

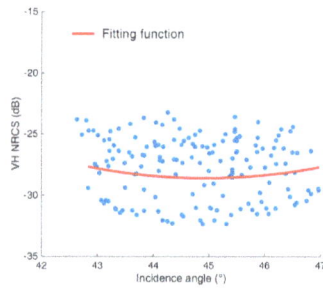

(e)

Figure 9. Fitting functions (red curves) between VH NRCS and incidence angle for (**a**) sub-band 1, (**b**) sub-band 2, (**c**) sub-band 3, (**d**) sub-band 4, and (**e**) sub-band 5.

In this paper, the proposed basic model is based on the average distribution of the matching data. Due to the fluctuation of NRCS with incidence angle, the retrieved wind speed from the basic model is too high at the peak of NRCS and too low at the trough of NRCS, leading to a high Std of wind retrieval. To minimize the Std, Equation (2) is used for making the fluctuation of NRCS as smooth as possible:

$$Std = \sqrt{\frac{\sum_{i=1}^{N}\left(\sigma_{Obs\ i}^{0} - f_1(\theta_i)f_2\right)^2}{N}} \tag{3}$$

where N is the number of matching points for each sub-band in dataset 1, f_2 is the correction factor, $\sigma_{Obs\ i}^{0}$ and θ_i are the observed VH NRCS and incidence angles of the data samples. The units of $\sigma_{Obs\ i}^{0}$ and θ_i are decibels and degrees, respectively. During the Std minimization, the f_2 values are calculated at 4, 6, 8, 10, 12, 14, 16 and 18 m/s bounded by ± 1 m/s interval. Based on linear fitting, the empirical function f_2 is:

$$f_2(U_{10}) = a_1 U_{10} + a_2 \tag{4}$$

Figure 10 shows the f_2 functions for each sub-band under different wind speeds. The correction factor decreases linearly with U_{10}. The parameters a_1, a_2 are reported in Table 3.

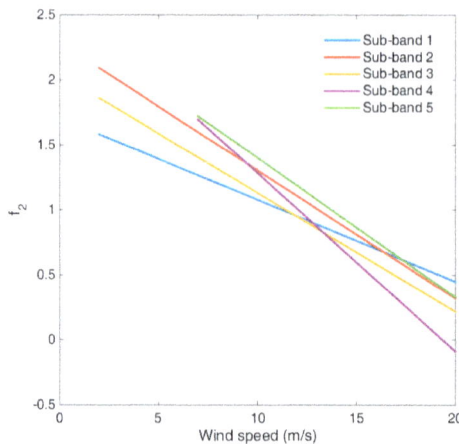

Figure 10. The empirical functions f_2 for each sub-band under different wind speeds.

Table 3. Parameters for Equation (4).

Sub-Band	a_1	a_2
1	−0.06	1.71
2	−0.10	2.29
3	−0.09	2.04
4	−0.14	2.66
5	−0.11	2.47

Based on Equations (1)–(4), Equation (5) is established to eliminate the overflow of NRCS in the process of Std minimization and decrease bias:

$$\text{Bias} = \frac{\sum_{i=1}^{N}\left(\sigma_{Obs\,i}^0 - f_1(\theta_i)f_2(U_{10\,i}) - f_3\right)}{N} \tag{5}$$

$$f_3(U_{10}) = b_1 U_{10} + b_2 \tag{6}$$

where $U_{10\,i}$ is the SMAP wind speed in dataset 1. f_3 is the correction factor which is a function of wind speed. Linear functions are used to fit f_3 for each sub-band. The fitting parameters b_1, b_2 are shown in Table 4.

Table 4. Parameters for Equation (6).

Sub-Band	b_1	b_2
1	−1.55	41.35
2	−2.63	61.28
3	−2.44	55.43
4	−3.70	72.42
5	−3.02	69.70

Finally, a Std-minimized and bias-corrected wind retrieval model is proposed:

$$\sigma_{VH}^0(U_{10}, \theta) = f_0(U_{10}) + f_1(\theta)f_2(U_{10}) + f_3(U_{10}) \tag{7}$$

which is referred to as the corrected model. This model can be used for simulating NRCS of S-1A VH-polarized EW mode images or retrieving sea surface wind speeds up to 20 m/s from S-1A VH-polarized EW mode images. Figure 11 is an example of comparison between the basic model and the corrected model at 10 m/s wind speed.

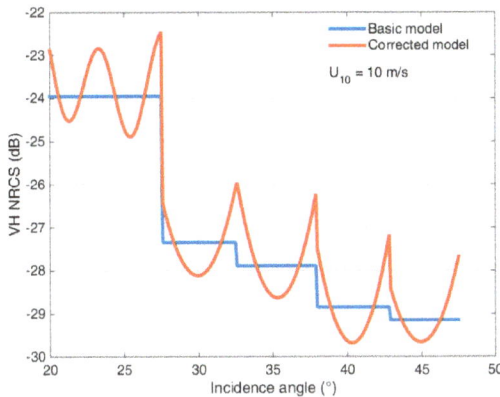

Figure 11. The comparison between basic model and corrected model at 10 m/s wind speed.

5. Validation and Discussion

As mentioned previously, the basic wind retrieval model is a function of VH-polarized NRCS and sea surface wind speed. The corrected model is a function of VH-polarized NRCS, sea surface wind speed, and radar incidence angle. Based on dataset 2, the proposed basic model and corrected model are validated and discussed in this section.

5.1. Comparison of Basic Model and Corrected Model

Experiments are carried out to compare the retrieval performance of the basic model and the corrected model for wind speeds lower than 20 m/s. The results of each sub-band are illustrated in Figure 12 and Table 5. There are 489, 260, 209, 230, and 161 samples for sub-bands 1, 2, 3, 4, and 5, respectively.

The blue points in Figure 12 illustrate the comparison of wind speeds retrieved by basic model and wind speeds from SMAP. For sub-bands 1–5, the correlation coefficients are 0.68, 0.81, 0.87, 0.81, and 0.81, the Std are 4.17, 3.89, 3.75, 3.39, and 3.20 m/s, and the biases are −0.04, −0.49, −0.39, −0.47 and −0.35 m/s, respectively.

The comparison of retrieved wind speeds by the corrected model and the wind speeds from SMAP is illustrated by the red points in Figure 12. For sub-bands 1–5, the correlation coefficients are 0.79, 0.83, 0.89, 0.81, and 0.82, the Std are 3.50, 3.50, 3.18, 3.17, and 3.11 m/s, and the biases are 0.55, −0.81, −0.31, −0.10, and −0.46 m/s, respectively.

According to the retrieval results, the results of the basic model have smaller biases. However, the wind speeds retrieved by the corrected model have larger correlation coefficients and smaller Std. Due to the weaker dependence of NRCS on incidence angle in sub-bands 4 and 5, the decrease of Std is smaller in sub-bands 4 and 5 than in sub-bands 1–3.

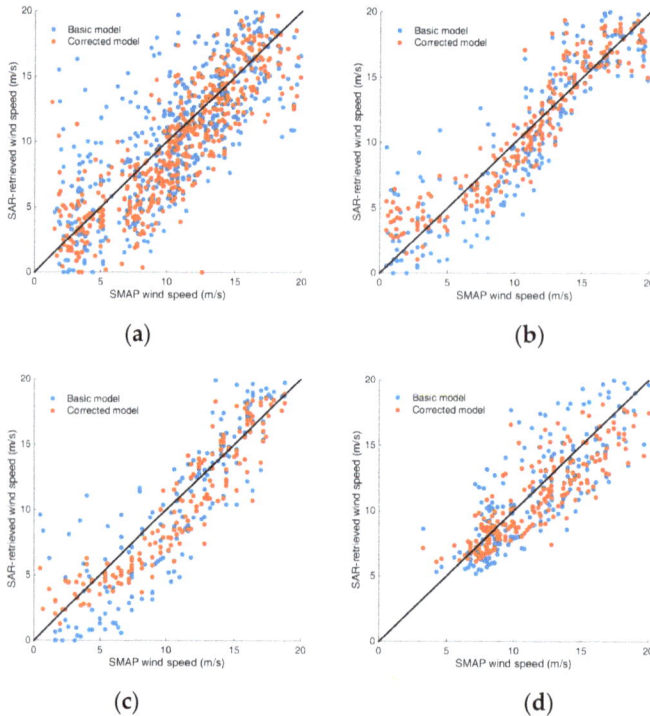

Figure 12. *Cont.*

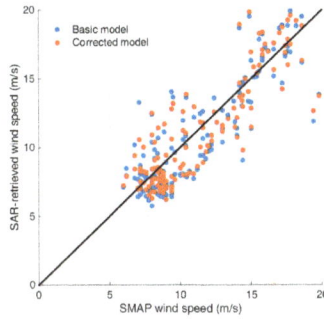

(e)

Figure 12. SAR-retrieved wind speeds with basic model and corrected model vs SMAP wind speeds in (a) sub-band 1, (b) sub-band 2, (c) sub-band 3, (d) sub-band 4, and (e) sub-band 5.

Table 5. Correlation coefficient, Std, and bias between wind speed from SMAP and retrieved wind speed with basic model and corrected model.

Sub-Band	r		Std (m/s)		Bias (m/s)	
	Basic Model	Corrected Model	Basic Model	Corrected Model	Basic Model	Corrected Model
1	0.68	0.79	4.17	3.50	−0.04	0.55
2	0.81	0.83	3.89	3.50	−0.49	−0.81
3	0.87	0.89	3.75	3.18	−0.39	−0.31
4	0.81	0.81	3.39	3.17	−0.47	−0.10
5	0.81	0.82	3.20	3.11	−0.35	−0.46

A case study is carried out by retrieving wind speeds from the S-1A VH-polarized EW mode image of Tropical Storm Lester on 26 August 2016. The retrieved wind speed fields using the basic model and the corrected model are shown in Figure 13a,b. In Figure 13b, the wind speeds lower than 20 m/s are corrected with incidence angles. The collocated SMAP wind observation is shown in Figure 13c.

In Figure 13a, wind speeds are high at the boundaries of each sub-band and in the middle of sub-band 1. In Figure 13b, for wind speeds lower than 20 m/s, such phenomena are not as obvious as in Figure 13a, indicating the Std-minimization ability of the corrected model. In this case, the maximum wind speed retrieved by the basic model is 38.7 m/s. According to the National Hurricane Center (NHC)'s report, the maximum wind speed of Tropical Storm Lester was about 55–60 knots (28.3–30.9 m/s) at the SAR sensing time. The maximum retrieved wind speed is much higher than the NHC report. Therefore, the basic model is not recommended for retrieving wind speeds higher than 30 m/s. More samples are needed to explore the wind speed retrieval model under severe wind conditions in the future. In addition, the scalloping burstwise variation is maintained in the process of wind retrieval, showing some periodic streaks in sub-band 1 in Figure 13a,b.

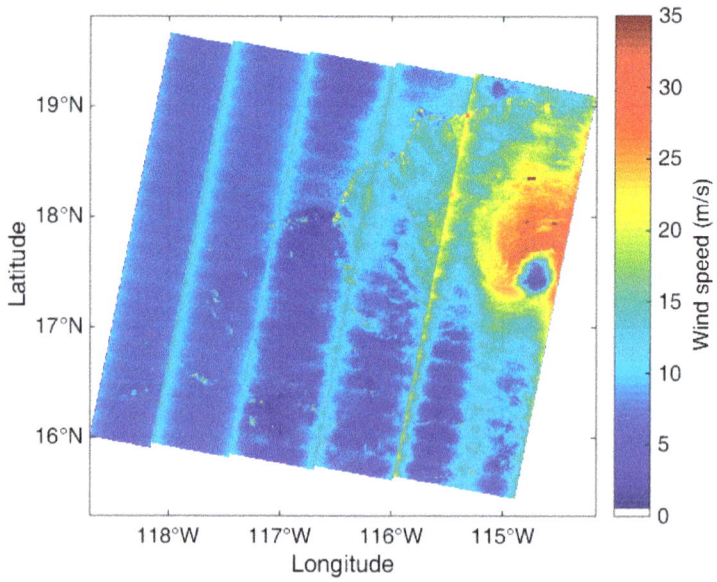

(a)

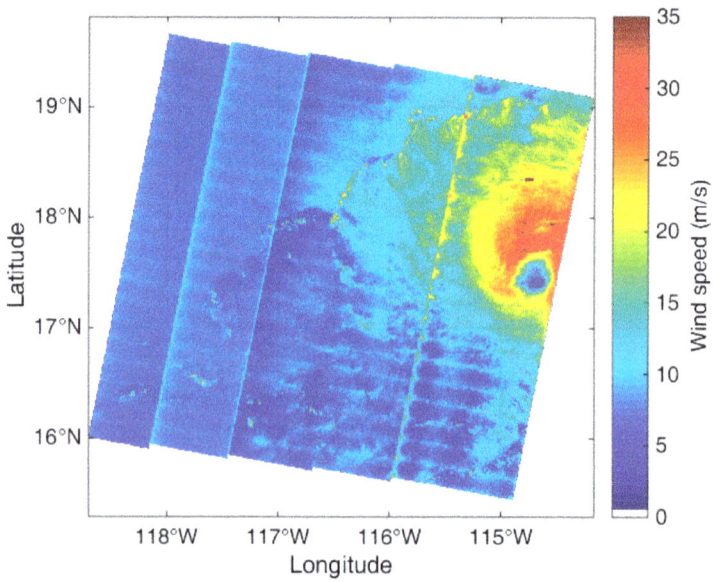

(b)

Figure 13. *Cont.*

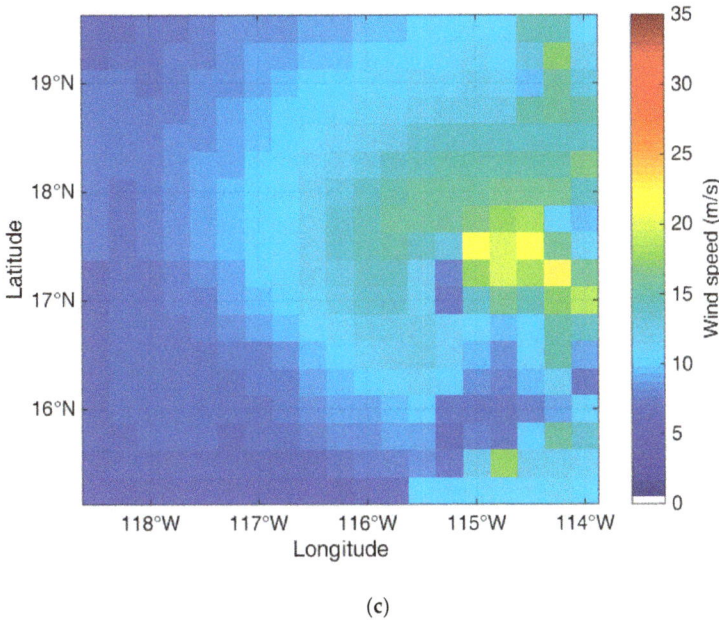

Figure 13. Retrieved sea surface wind speed of Tropical Storm Lester using (**a**) basic model, (**b**) basic model and corrected model, and (**c**) SMAP wind observation.

5.2. Model Validation

In this section, the proposed model is compared with the MS1A model proposed by Mouche et al. [12]. The MS1A model is established with Sentinel-1A VH-polarized data and collocated wind speeds from SMAP:

$$\sigma_0^{VH}(\theta, |U_{10}|) = A_{n-1}(\theta) U_{t_{n-1}}^{a_{n-1}}(\theta) \tag{8}$$

$$A_n(\theta) = A_{n-1} U_{t_n}^{a_{n-1}-a_n}, \text{ if } n > 1 \tag{9}$$

MS1A model is a power law function. σ_0^{VH} stands for the NRCS in linear scale. U_{t_n} represents the 10-m height ocean surface wind speed corresponding to the transitions in the NRCS regime. A_n and a_n are dimensionless coefficients. The correlation coefficients, Std, and biases between the SMAP winds and the wind retrievals utilizing MS1A and the model proposed in this study are calculated for each sub-band. The comparison results are illustrated in Figure 14 and Table 6, showing that the retrieved wind speeds by the model proposed this study have higher correlation coefficients and lower Std and biases in most sub-bands. The large difference of retrieval results of the two models is mainly caused by the quality of the SMAP data and the SAR data used in the two studies. On one hand, Mouche et al. used SMAP brightness temperature data to compute the wind speeds. In this paper, SMAP Level-2 data are downloaded and then used directly. On the other hand, the NRCS values they used seem to be higher than ours. In [12], there are many NRCS observations below the NESZ values, leading to higher retrieval results by MS1A as measured by the SMAP Level-2 dataset.

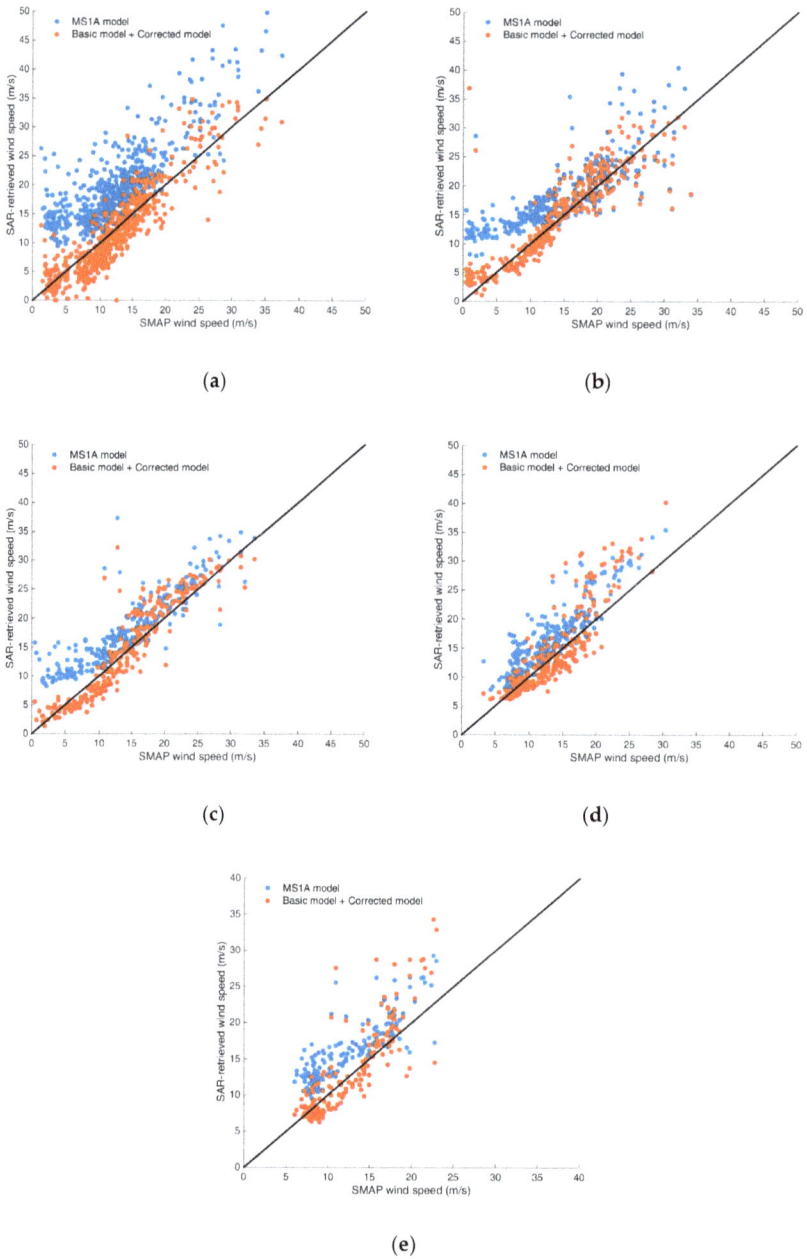

Figure 14. SAR-retrieved wind speeds by our model (red points) and MS1A model (blue points) vs SMAP wind speeds in (**a**) sub-band 1, (**b**) sub-band 2, (**c**) sub-band 3, (**d**) sub-band 4, and (**e**) sub-band 5.

Table 6. Correlation coefficient, Std and bias between wind speeds from SMAP and retrieved wind speeds by our model and MS1A model.

Sub-Band	r		Std (m/s)		Bias (m/s)	
	Our Model	MS1A Model	Our Model	MS1A Model	Our Model	MS1A Model
1	0.87	0.75	3.66	4.84	0.45	−7.38
2	0.87	0.70	3.79	5.35	−0.47	−3.85
3	0.92	0.87	3.10	3.41	−0.36	−3.47
4	0.88	0.88	3.56	2.67	−0.56	−3.52
5	0.85	0.83	3.43	2.45	−0.69	−3.73

5.3. Error Analyses

Under tropical cyclone conditions, low spatial resolution will lead to a smoothed wind field, potentially missing small regions with high wind speeds. Due to the resolution difference between the S-1A data and the SMAP data, the pixel number of S-1A image used for averaging might influence the retrieval results. In order to evaluate the performance of the proposed model for datasets with different pixel numbers, wind speeds are retrieved from dataset 2 with an averaging of 8×8, 16×16, 32×32, 64×64, 128×128, 256×256, and 512×512 pixels in one cell, respectively. Correlation coefficient, Std, and bias between the retrieved wind speeds and the SMAP winds are illustrated in Table 7, showing the stability of the proposed model. In addition, the number of matching data might influence the experiment results, especially under strong-to-severe wind conditions in this study. The proposed model can be improved when more observations with higher spatial resolution (for example SFMR or H*Wind) become available in the future.

Table 7. Performance of the model for dataset with different pixel number.

Pixel Number	r	Std (m/s)	Bias (m/s)
8×8	0.86	3.42	−0.31
16×16	0.88	3.51	−0.3
32×32	0.90	3.31	−0.38
64×64	0.89	3.20	−0.27
128×128	0.87	3.40	−0.30
256×256	0.88	3.59	0.12
512×512	0.92	3.63	0.09

In this paper, the methodology and the accuracy of data could influence the parameters of the proposed model and the validation results. On one hand, the methodology of noise removal could lead to an error of NRCS. The S-1A VH-polarized EW mode data have noise variation in the azimuth direction, called azimuth scalloping [24]. The areas near the burst edges are brighter than those in the burst center because of their higher noise power of azimuth scalloping. The azimuth scalloping attenuates from sub-bands 1 to 5. In sub-band 1, the azimuth scalloping can lead to an error of NRCS up to 1.5 dB. In this study, a large number of S-1A images are collected to minimize the azimuth scalloping error.

On the other hand, tropical cyclones are always accompanied with rainfall which can strongly dampen the NRCS, leading to significant underestimates in wind speeds [18,27]. In this study, there is no matching data for precipitation. As mentioned in Section 3, the proposed model has a low slope under strong-to-severe wind speeds for sub-bands 4 and 5. According to the proposed model, a NRCS error of 1 dB might cause a wind retrieval error up to 5 m/s. In addition, the SMAP wind speeds and WindSat observations have a global RMS of 1.5 m/s, which might influence the precision of the proposed model.

Finally, the collocation time difference is very important for modeling and validation. Requiring a smaller time difference may lead to a reduced, and insufficient quantity of data samples. In this paper, if the time difference is restricted to 30 min, nearly half of the samples will be lost. It will be difficult to propose and validate the retrieval model, especially for high wind speeds. However, if the time difference is increased to more than one hour or even two hours, the motion of tropical cyclones and the variation of wind fields will influence the accuracy of wind retrieval. If more hurricane SAR images could be acquired in the future, the time difference could be reduced. In addition, experiments could also be made to test to what extent the collocation time difference influences modeling.

6. Conclusions

In this paper, a new model is developed for retrieving sea surface wind speed from S-1A EW mode VH-polarized images. 19 noise-free S-1A images and matching data from SMAP radiometer under tropical cyclone conditions are collected and analyzed. According to 12 S-1A images and matching data, the VH NRCS has a strong correlation with wind the speeds in each sub-band of the S-1A images. With the change of incidence angle, the VH NRCS has a high level at the boundaries of each sub-band and in the middle of sub-band 1.

Based on the relationship between VH NRCS and wind speed, a basic model is proposed to construct a wind retrieval model. In addition, a corrected model is proposed to improve the accuracy of the basic model, according to the relationship between NRCS and incidence angle.

In order to validate the validity of the wind retrieval model, the wind speeds retrieved by the corrected model are compared with the wind speeds retrieved by the basic model and the MS1A model in 7 S-1A images. A case study is also carried out by retrieving the wind speed field from the S-1A image of Tropical Storm Lester. Validating against the winds from SMAP, the wind speeds retrieved by the corrected model are more accurate than the basic model for wind speeds lower than 20 m/s, especially in the middle of the sub-band and at the inter-band boundaries.

For sub-bands 1–5, the correlation coefficients, Std, and biases between the retrieved winds and the SMAP winds are 0.68–0.89, 3.11–4.17 m/s, and −0.81–0.55 m/s, respectively. The retrieval results are fairly accurate, indicating that the proposed wind speed retrieval model is reliable. Finally, error sources of the proposed model and our experiments are analyzed with respect to the proposed methodology and the matching data.

Author Contributions: Initiation of the idea: Y.G. and J.S.; data processing and model proposing: Y.G.; writing and editing: all authors contributed; supervision: L.X.; funding acquisition: J.S. and C.G.

Funding: This research was funded by National Key Research and Development Program of China under grant number 2016YFC1401405 and the National Science Foundation of China under grant number 41376010.

Acknowledgments: The authors would like to thank the European Space Agency for making Sentinel-1A data publicly available. We thank the National Aeronautics and Space Administration for the SMAP data, and the National Hurricane Center for its hurricane report.

Conflicts of Interest: The authors declare no conflict of interest.

References

1. Hwang, P.A.; Fois, F. Surface roughness and breaking wave properties retrieved from polarimetric microwave radar backscattering. *J. Geophys. Res. Oceans* **2015**, *120*, 3640–3657. [CrossRef]
2. Monaldo, F.M.; Jackson, C.; Li, X. On the use of Sentinel-1 cross-polarization imagery for wind speed retrieval. In Proceedings of the 2017 IEEE International Geoscience and Remote Sensing Symposium, Fort Worth, TX, USA, 23–28 July 2017; pp. 392–395.
3. Gerling, T. Structure of the surface wind field from the Seasat SAR. *J. Geophys. Res. Oceans* **1986**, *91*, 2308–2320. [CrossRef]
4. Alpers, W.; Brümmer, B. Atmospheric boundary layer rolls observed by the synthetic aperture radar aboard the ERS-1 satellite. *J. Geophys. Res. Oceans* **1994**, *99*, 12613–12621. [CrossRef]
5. Du, Y.; Vachon, P.W.; Wolfe, J. Wind direction estimation from SAR images of the ocean using wavelet analysis. *Can. J. Remote Sens.* **2002**, *28*, 498–509. [CrossRef]
6. Gao, Y.; Guan, C.; Sun, J.; Xie, L. A New Hurricane Wind Direction Retrieval Method for SAR Images without Hurricane Eye. *J. Atmos. Ocean. Technol.* **2018**, *35*, 2229–2239. [CrossRef]

7. Shen, H.; He, Y.; Perrie, W. Speed ambiguity in hurricane wind retrieval from SAR imagery. *Int. J. Remote Sens.* **2009**, *30*, 2827–2836. [CrossRef]
8. Valenzuela, G. Depolarization of EM waves by slightly rough surfaces. *IEEE Trans. Antennas Propag.* **1967**, *15*, 552–557. [CrossRef]
9. Voronovich, A.G.; Zavorotny, V.U. Depolarization of microwave backscattering from a rough sea surface: Modeling with small-slope approximation. In Proceedings of the 2011 IEEE International Geoscience and Remote Sensing Symposium (IGARSS), Vancouver, BC, Canada, 24–29 July 2011; pp. 2033–2036.
10. Voronovich, A.G.; Zavorotny, V.U. Full-polarization modeling of monostatic and bistatic radar scattering from a rough sea surface. *IEEE Trans. Antennas Propag.* **2014**, *62*, 1362–1371. [CrossRef]
11. Hwang, P.A.; Stoffelen, A.; Zadelhoff, G.J.; Perrie, W.; Zhang, B.; Li, H.; Shen, H. Cross-polarization geophysical model function for C-band radar backscattering from the ocean surface and wind speed retrieval. *J. Geophys. Res. Oceans* **2015**, *120*, 893–909. [CrossRef]
12. Mouche, A.A.; Chapron, B.; Zhang, B.; Husson, R. Combined co-and cross-polarized SAR measurements under extreme wind conditions. *IEEE Trans. Geosci. Remote Sens.* **2017**, *55*, 6746–6755. [CrossRef]
13. Hwang, P.A.; Zhang, B.; Perrie, W. Depolarized radar return for breaking wave measurement and hurricane wind retrieval. *Geophys. Res. Lett.* **2010**, *37*, L01604. [CrossRef]
14. Zhang, B.; Perrie, W.; He, Y. Wind speed retrieval from RADARSAT-2 quad-polarization images using a new polarization ratio model. *J. Geophys. Res. Oceans* **2011**, *116*, C08008. [CrossRef]
15. Vachon, P.W.; Wolfe, J. C-band cross-polarization wind speed retrieval. *IEEE Trans. Geosci. Electron. Lett.* **2011**, *8*, 456–459. [CrossRef]
16. Zhang, B.; Perrie, W.; Zhang, J.A.; Uhlhorn, E.W.; He, Y. High-resolution hurricane vector winds from C-band dual-polarization SAR observations. *J. Atmos. Ocean. Technol.* **2014**, *31*, 272–286. [CrossRef]
17. Horstmann, J.; Falchetti, S.; Wackerman, C.; Maresca, S.; Caruso, M.J.; Graber, H.C. Tropical cyclone winds retrieved from C-band cross-polarized synthetic aperture radar. *IEEE Trans. Geosci. Remote Sens.* **2015**, *53*, 2887–2898. [CrossRef]
18. Zhang, B.; Perrie, W. Cross-Polarized Synthetic Aperture Radar: A New Potential Measurement Technique for Hurricanes. *Bull. Am. Meteorol. Soc.* **2012**, *93*, 531–541. [CrossRef]
19. Zadelhoff, G.J.V.; Stoffelen, A.; Vachon, P.W.; Wolfe, J.; Horstmann, J.; Rivas, M.B. Scatterometer hurricane wind speed retrievals using cross polarization. *Atmos. Meas. Tech. Discuss.* **2013**, *6*, 7945–7984. [CrossRef]
20. Zhang, G.; Li, X.; Perrie, W.; Hwang, P.A.; Zhang, B.; Yang, X. A hurricane wind speed retrieval model for C-band RADARSAT-2 cross-polarization ScanSAR images. *IEEE Trans. Geosci. Remote Sens* **2017**, *55*, 4766–4774. [CrossRef]
21. Huang, L.; Liu, B.; Li, X.; Zhang, Z.; Yu, W. Technical evaluation of Sentinel-1 IW mode cross-pol radar backscattering from the ocean surface in moderate wind condition. *Remote Sens.* **2017**, *9*, 854. [CrossRef]
22. Zhou, X.; Yang, X.; Li, Z.; Yu, Y.; Bi, H.; Ma, S.; Li, X. Estimation of tropical cyclone parameters and wind fields from SAR images. *Sci. China Earth Sci.* **2013**, *56*, 1977–1987. [CrossRef]
23. Shao, W.; Li, X.; Hwang, P.; Zhang, B.; Yang, X. Bridging the gap between cyclone wind and wave by C-band SAR measurements. *J. Geophys. Res. Oceans* **2017**, *122*, 6714–6724. [CrossRef]
24. Park, J.-W.; Korosov, A.A.; Babiker, M.; Sandven, S.; Won, J.-S. Efficient Thermal Noise Removal for Sentinel-1 TOPSAR Cross-Polarization Channel. *IEEE Trans. Geosci. Remote Sens.* **2018**, *56*, 1555–1565. [CrossRef]
25. Meissner, T.; Ricciardulli, L.; Wentz, F.J. Capability of the SMAP Mission to Measure Ocean Surface Winds in Storms. *Bull. Am. Meteorol. Soc.* **2017**, *98*. [CrossRef]
26. Fore, A.; Yueh, S.; Tang, W.; Stiles, B.; Hayashi, A. Validation of SMAP radiometer extreme wind speed data product with rapid scatterometer and stepped frequency microwave radiometer. In Proceedings of the 2017 IEEE International Geoscience and Remote Sensing Symposium, Fort Worth, TX, USA, 23–28 July 2017; pp. 398–401.
27. Powell, M.D. Boundary layer structure and dynamics in outer hurricane rainbands. Part I: Mesoscale rainfall and kinematic structure. *Mon. Weather Rev.* **1990**, *118*, 891–917. [CrossRef]

remote sensing

MDPI

Article

Ocean Wind Retrieval Models for RADARSAT Constellation Mission Compact Polarimetry SAR

Tianqi Sun [1,2], Guosheng Zhang [2,3,*], William Perrie [2], Biao Zhang [3], Changlong Guan [1], Shahid Khurshid [4], Kerri Warner [4] and Jian Sun [1]

[1]   College of Oceanic and Atmospheric Sciences, Ocean University of China, Qingdao 266100, China;
      tianqisun.ocean@yahoo.com (T.S.); clguan@ouc.edu.cn (C.G.); sunjian77@ouc.edu.cn (J.S.)
[2]   Fisheries and Oceans Canada, Bedford Institute of Oceanography, Dartmouth, NS B2Y 4A2, Canada;
      william.perrie@dfo-mpo.gc.ca
[3]   School of Marine Sciences, Nanjing University of Information Science and Technology, Nanjing 210044,
      China; zhangbiao@nuist.edu.cn
[4]   Earth Applied Science & Development Section Observation & Geomatics, Meteorological Service of Canada,
      Environment and Climate Change Canada, Ottawa, ON K1A 0H3, Canada;
      shahid.khurshid@canada.ca (S.K.); kerri.warner@canada.ca (K.W.)
*   Correspondence: zgsheng001@gmail.com; Tel.: +1-902-426-7797

Received: 30 October 2018; Accepted: 29 November 2018; Published: 2 December 2018

Abstract: We propose two new ocean wind retrieval models for right circular-vertical (RV) and right circular-horizontal (RH) polarizations respectively from the compact-polarimetry (CP) mode of the RADARSAT Constellation Mission (RCM), which is scheduled to be launched in 2019. For compact RV-polarization (right circular transmit and vertical receive), we build the wind retrieval model (denoted CoVe-Pol model) by employing the geophysical model function (GMF) framework and a sensitivity analysis. For compact RH polarization (right circular transmit and horizontal receive), we build the wind retrieval model (denoted the CoHo-Pol model) by using a quadratic function to describe the relationship between wind speed and RH-polarized normalized radar cross-sections (NRCSs) along with radar incidence angles. The parameters of the two retrieval models are derived from a database including wind vectors measured by in situ National Data Buoy Center (NDBC) buoys and simulated RV- and RH-polarized NRCSs and incidence angles. The RV- and RH-polarized NRCSs are generated by a RCM simulator using C-band RADARSAT-2 quad-polarized synthetic aperture radar (SAR) images. Our results show that the two new RCM CP models, CoVe-Pol and CoHo-POL, can provide efficient methodologies for wind retrieval.

Keywords: compact polarization (CP); RADARSAT Constellation Mission (RCM); geophysical model function (GMF); wind retrieval; CoVe-Pol and CoHo-Pol models; right circular horizontal polarization model; right circular vertical polarization model

---

1. Introduction

The Canadian RADARSAT Constellation Mission (RCM) is scheduled for launch in early 2019 and will provide Compact-Polarimetry (CP) synthetic aperture radar (SAR) data. RCM is the evolution of the RADARSAT Program and the successor of RADARSAT-2, which is a satellite constellation carrying three identical C-band SAR satellites. The RCM CP mode consists of a right hand circular transmit and linear/circular receive radar signal, namely right circular-vertical (RV) polarization, right circular-horizontal (RH) polarization, right circular-right circular (RR) polarization and right circular-left circular (RL) polarization. The CP configuration is designed for Earth observation; compared with conventional linear dual-polarization SAR, compact polarimetry SAR can obtain abundant high-resolution information with wider swath [1].

Researchers have shown that CP SAR is an efficient imaging mode for ocean surface observation and, therefore, elucidation of algorithms and models for retrieval of ocean surface features and marine variables like wind is critically important [2]. Compact polarization SAR is a dual-polarization radar system which transmits circular (or 45°) components and receives two orthogonal polarization components (V or H) with relative phase [3,4]. Accurate information of ocean surface can be obtained from C-band radar data. The main process of physical mechanism is the interactions between the microwaves and related surface water waves. The major interaction is denoted as Bragg scattering when the wavelengths of ocean surface waves are of the same order of radar wavelength. For practical application, wind retrievals from SAR images can be accomplished by C-band geophysical model functions (GMFs), because radar measurements are sensitive to the ocean-surface roughness which is determined by the surface wind field [5]. The GMF is a set of functions between wind vectors and radar backscatter signals denoted as the normalized radar cross section (NRCSs) including dependence on the radar incidence angles.

The C-band GMF model series (CMOD4, CMOD5 and CMOD5.N [5,6]) are used for vertical-vertical (VV) polarization data. For horizontal-horizontal (HH) polarization, there are generally two ways to achieve ocean wind retrieval: (1) conversion of the HH backscatter to VV by a polarization ratio (PR) which is a ratio of VV-polarized NRCS to HH-polarized NRCS [7], or (2) construction of a new relationship between wind vectors and radar backscatters [8]. In terms of C-band linear cross-polarization (VH and HV) ocean backscatter data, C-2PO and C-3PO models are available for wind retrieval [9,10]. For wind retrieval from CP SAR mode, recent studies have attempted to retrieve wind speed by converting right circular-vertical (RV) polarization data into linear dual-polarization data, taking advantage of CMOD5 and CMOD5.N models [11,12]. However, the configurations for RV and VV polarizations are different and the CMOD model series were originally developed for C-Band VV-polarized scatterometers rather than SARs. In view of this situation, we propose two new models in this paper to retrieve wind from C-band RV-pol and RH-pol measurements.

CMOD is a well-behaved parameterization to retrieve wind speed, allowing the NRCS to be dependent on the principal parameters, such as radar incidence angle, relative wind direction and wind speed [5]. However, as there are 28 CMOD coefficients, the process of optimizing the CMOD model to CP data sets to generate a new GMF (for CP SAR) is not feasible using normal computing clusters, because of the requirement to adjust such large number of CMOD coefficients to represent the CP parameters. In order to avoid excessive computation, a sensitivity analysis of the coefficients can play a significant role in optimizing the generation of a new GMF for RV-polarization. The sensitivity analysis is a method to adjust the models by changing the coefficients within a specific range of variations in order to optimize the parameterization and simplify the computation [13,14]. Moreover, another common method to generate a practical empirical algorithm is to fit a function relating the CP variables and the wind parameters. In this study, we utilize these two methods to construct the wind speed retrieval models for RV- and RH-polarization data.

RADARSAT-2 is beyond its 7-year design life, as it was launched in 2007. Thus the launch of RCM is necessary, and application of CP SAR is a new application offering the advantages of full polarimetry SAR mode, with the possibility of better wide-swath coverage. Based on the simulated CP parameters, this study is a preparation for possible ocean wind retrievals from RCM which will be available in the next year. The remainder of this paper is organized as follows: Section 2 describes the database consisting of CP SAR NRCSs simulated by the RCM simulator using quad-polarized RADARSAT-2 SAR images, and collocated wind vectors observed by NCBC buoys. A new CMOD function for RV-polarization data is proposed based on a sensitivity analysis [13], followed by a performance evaluation, and we present a new model for RH polarization wind speed retrieval. Results and the validations of the two new models are shown in Section 3. Discussion and conclusions are given in Sections 4 and 5, respectively.

2. Materials and Methods

In this section, two wind retrieval models are developed based on the collocated data sets. For the compact RV polarization, we employ the CMOD framework to derive a new GMF by optimizing each coefficient. The resultant formulation is denoted the CoVe-Pol model. In this derivation, sensitivity analysis is used to avoid unnecessarily huge calculations [13]. Through the sensitivity analysis, the computation efficiency of the process to generate the new GMF generation is increased by more the 10^{20} times and the accuracy of computed NRCS reaches 10^{-2}. For the compact RH polarization, we propose the CoHo-Pol wind retrieval model which we derive by using a quadratic regression function.

2.1. Datasets

The CP parameters were simulated from RADARSAT-2 quad-polarized data using the RCM simulator, which is provided by the Canadian Space Agency [12]. To develop two new wind retrieval models for RV- and RH-polarization data, we build a database consisting of simulated RV- and RH-polarized NRCSs from RADARSAT-2 fine quad-pol SAR images and collocated wind vectors measured by in situ buoys. The RADARSAT-2 quad-pol SAR images over the selected buoys are acquired. Then, these RADARSAT-2 images are converted to RCM CP mode SAR images by using the CP simulator. Finally, wind speed measured by buoy and the collocated simulated NRCS are paired. The distributions of NRCSs and wind speeds are shown in Figure 1.

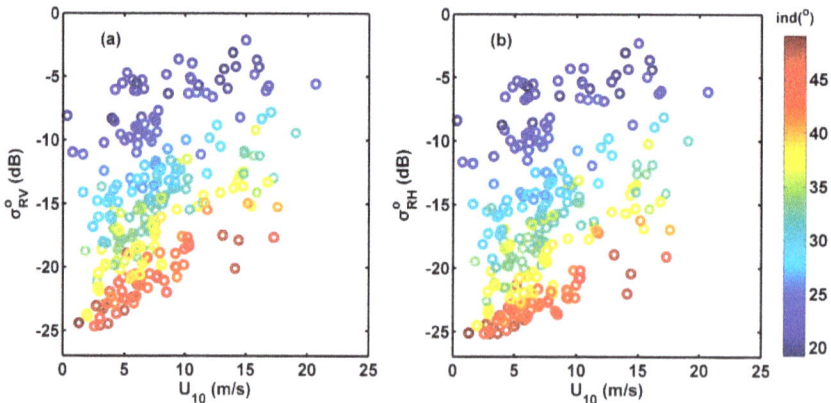

Figure 1. Simulated compact-polarimetry (CP)-polarized normalized radar cross-sections (NRCSs) vs. in situ buoy-measured U_{10}: (**a**) right circular-vertical (RV) polarization; (**b**) right circular-horizontal (RH) polarization.

In this database, we have 267 RADARSAT-2 fine quad-pol SAR images, which are processed by a RCM CP simulator to re-construct CP mode images [1]. The results of simulated CP configurations are used as "ground truth" in this study. The RADARSAT Constellation Mission has several polarization configurations: linear mono-polarized, dual linear-polarized (HH/HV, VV/VH, or VV/HH); dual circular transmit-linear receive; and fully polarimetric [12]. The compact polarimetry SAR mode provides four polarimetric datasets, which are RV, RH, right circular transmit and right circular receive (RR) and right circular transmit and left circular receive (RL). In the medium resolution mode, the pixel spacings in azimuth and range directions are about 100 m and the associated noise floor is about −25 dB [15]. Data from eight National Data Buoy Center (NDBC) buoys collocated with the SAR data are collected, at locations off the east and west coasts of Canada [16]. The buoy locations are shown in Figure 2. At each buoy, the mean wind speed (at 10 m reference height, hereafter U_{10}) and direction are measured by two sensors, averaged over 8-min periods and reported hourly. The wind speeds observed by the buoys are converted to winds at 10 m reference height above the ocean surface using the Tropical Ocean and Global Atmosphere Coupled Ocean-Atmosphere Response Experiment (TOGA COARE) bulk flux algorithm [17], and the winds can be considered as neutral winds [6,18]. The temporal separation between the SAR data and the buoy data is restricted to less than 30 min [11]. The distribution of months is shown in Figure 3. As the SAR is active microwave, the effects of weather and seasons are both almost negligible.

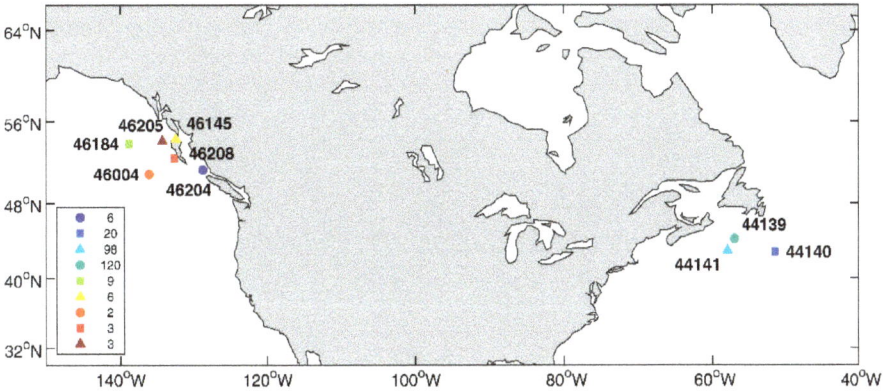

Figure 2. Distribution of buoy locations used in this study. We indicate the number of synthetic aperture radar (SAR) images overlaying the respective buoys in the legend.

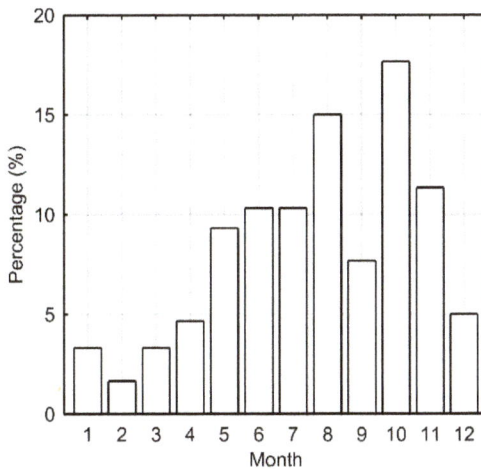

Figure 3. Distribution of months of the collocated data used in this study.

We divide the collected data randomly into two data groups (2/3 and 1/3 of the total number). One contains 178 data which are used to generate the new models, and the other contains 89 data, reserved for model testing.

2.2. CoVe-Pol Model for Right Circular-Vertical (RV) Polarization

Figure 4 shows the flowchart for the derivation process for the CoVe-Pol model for RV polarization data. As with all GMF functions, an empirical functional relation is used to establish the dependency of the normalized backscatter on wind speed, wind direction, and the incidence angle [5]. The general form of the CMOD function is summarized as:

$$\sigma^0(\theta, U_{10}, \varphi) =$$
$$B0(c_0, U_{10}, \theta)[1 + B1(c_1, U_{10}, \theta)\cos(\varphi) + B2(c_2, U_{10}, \theta)\cos(2\varphi)]^{1.6} \tag{1}$$

where σ^0 is the NRCS in linear units, φ is the relative wind direction, which is the angle between local wind direction and radar look direction (both relative to north), U_{10} is the statistically neutral wind referenced to 10 m height, θ is the incidence angle and $B0$, $B1$, $B2$ are coefficients depending on U_{10}, θ, the radar frequency and polarization. The dominant term, $B0$, sets the speed scale for a given measurement. The upwind–crosswind asymmetry term $B2$ allows for a determination of the wind direction, and $B1$ is used to resolve the remaining 180° ambiguity in the wind direction. Coefficients c_i complete the definition of the terms $B0$, $B1$, $B2$. Detailed expressions are shown in Appendix A.

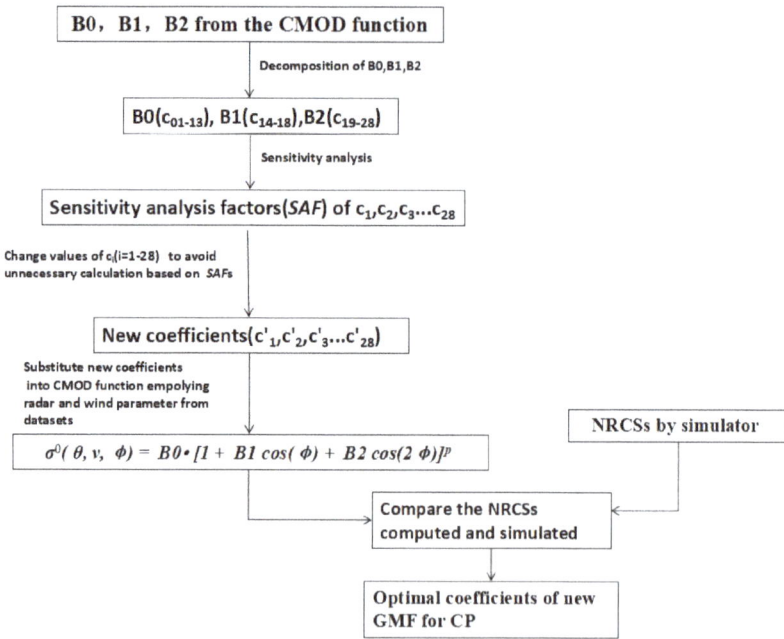

Figure 4. Flowchart for building the CoVe-Pol model through adjustment of the C-band geophysical model function (GMF) model series (CMOD) coefficients obtained by application of a sensitivity analysis.

2.2.1. Sensitivity Analysis

As there are 28 CMOD coefficients ($c_1 \sim c_{28}$) in the empirical model formulation, the sensitivity analysis is used to reduce the computations and to allow a determination of the coefficients. We define the sensitivity analysis factor (*SAF*) of the CMOD coefficients to make the adjustment process more efficient:

$$SAF = \left|(\delta\sigma^0/\sigma^0)/(\delta c_i/c_i)\right| \tag{2}$$

where σ^0 represents the RV-polarized backscatter value (NRCS), the independent variable, and the c_i coefficients ($c_1 \sim c_{28}$) are dependent variables. In this approach, *SAF*s indicate the degree of influence of each coefficient on the CMOD parameterization. From equation (2), we know that for any particular coefficient c_i, the corresponding NRCS (σ^0) can have multiple values, and in each case the ratio is indicative of the degree of influence, *SAF*. If *SAF* cannot provide the required computational accuracy, the corresponding coefficients c_i can be ignored. For any particular coefficient, the basis for this decision is the magnitude of the degree of influence, *SAF*.

Average values of the *SAFs* under various wind speeds range from 1 m/s to 25 m/s as shown in Figure 5, assuming typical conditions for radar incidence angles and wind directions. Thus, it is shown that *SAF* values range widely from 0 to 10^1, which means the influence of different coefficients, c_i, vary greatly. As shown in Figure 5, very close to 0, most *SAF* values are less than 1 and only a few have values exceeding 1, for three typical radar incidence angles (25°, 35° and 45°). Thus, we only focus on coefficients c_i, where *SAF* values indicate greater influence (>0.1) so that the parameterization is simplified and a huge calculation can be avoided.

In this study, we can classify the coefficients by the orders of magnitude of the corresponding *SAFs*, namely 10^0, 10^{-1} and less than 10^{-1}. Thus, when we attempt to get the accuracy of computed NRCS values to 10^0 order of magnitude, we don't need to consider the coefficients with *SAF* values less than 10^0, because their influence is negligible. Likewise, when we focus on the 10^{-1} order of magnitude, the coefficients with *SAF* values lower than 10^{-1} can be ignored. Therefore, based on the *SAFs* of each coefficient, c_i, we firstly tune the coefficients whose *SAF* values have orders of magnitude higher than 10^0, without considering other coefficients. Secondly, we tune the coefficients whose *SAF* values have orders of magnitude higher than 10^{-1}, without considering other coefficients. Then, we tune the coefficients whose SAF values have progressively higher orders of magnitude than 10^{-1}. The reason for this approach is that coefficients whose *SAF* values have orders of magnitude lower than 10^{-1} have essentially no effect on computed NRCSs derived from coefficients whose SAF values have higher orders of magnitude.

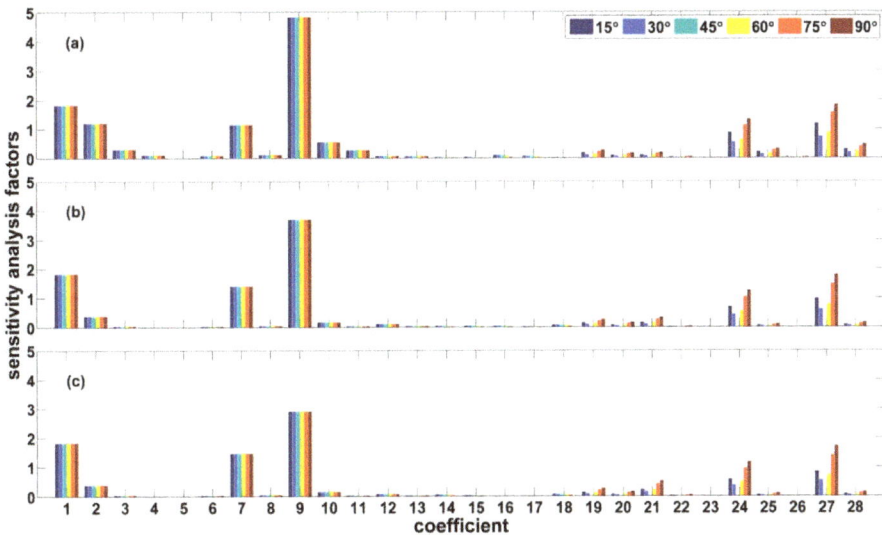

Figure 5. The mean sensitivity factors vs. coefficients: (**a**) radar incidence angle is 25 degree; (**b**) radar incidence angle is 35 degree; (**c**) radar incidence angle is 45 degree. The colors represent different wind directions (15°, 30°, 45°, 60°, 75°, 90°).

We note that although the radar incidence angles can vary as shown in Figure 5, their influence on the coefficients c_i, and on the resultant NRCSs is quite minor. Thus, we combine the average *SAFs* under different radar incidence angles, as shown in Figure 6. The extent of influence for incidence angles on the coefficients c_i for the *SAF* orders of magnitude 10^0 and 10^{-1} are shown in Figure 6a,b, respectively. It is apparent that the coefficients can be divided into three groups according to their degrees of influence. The *SAF* values corresponding to coefficients c_1, c_7, c_9 are more than 1. The *SAF* values corresponding to coefficients c_2, c_3, c_{10}, c_{11}, c_{19}, c_{21}, c_{27}, and c_{28} are in the range from 0.1 to 1, and the remaining c_i values are under 0.1. It is notable that the coefficients with higher *SAF* magnitudes play a part in the adjustment process of the coefficients c_i, for lower orders of magnitude; but this influence does not work, if we put things the other way around. For example, assuming a *SAF* value of 1.3219, the corresponding c_i coefficients cannot influence the computed NRCS values above 10^0 order of magnitude, but these c_i coefficients do influence NRCS values below 10^1 magnitude such as 10^0 and 10^{-1} magnitudes, as reported in this study. Moreover, because *SAF* values for typical wind directions have similar orders of magnitude, we give the average *SAF* values for the c_i coefficients in Table 1.

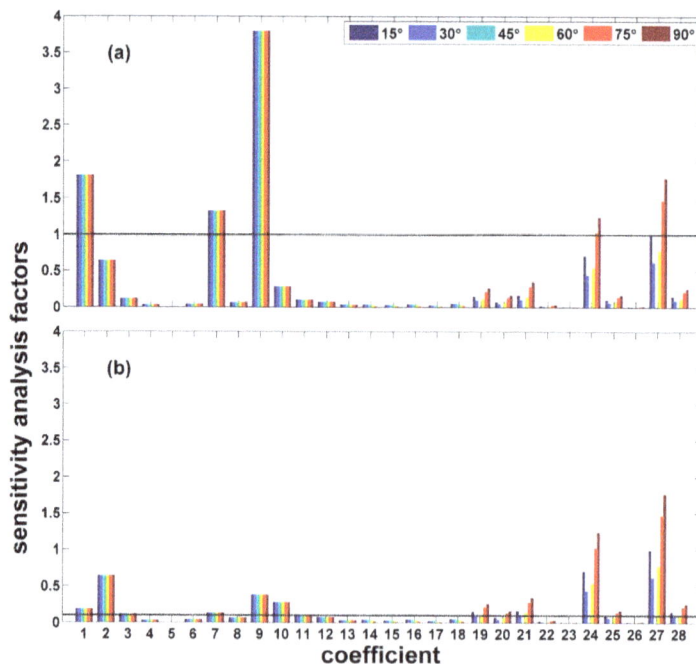

Figure 6. The sensitivity analysis factor (*SAF*) values for different orders of magnitude for computed NRCSs: (**a**) *SAF* over 1; (**b**) *SAF* over 0.1. The colors represent different wind directions (15°, 30°, 45°, 60°, 75°, 90°).

Table 1. Average *SAF* values of the c_i coefficients.

SAF of c_1	1.8081	*SAF* of c_8	0.0630	*SAF* of c_{15}	0.0200	*SAF* of c_{22}	0.0171
SAF of c_2	0.6366	*SAF* of c_9	3.7931	*SAF* of c_{16}	0.0276	*SAF* of c_{23}	0.0024
SAF of c_3	0.1145	*SAF* of c_{10}	0.2767	*SAF* of c_{17}	0.0158	*SAF* of c_{24}	0.6617
SAF of c_4	0.0308	*SAF* of c_{11}	0.1012	*SAF* of c_{18}	0.0332	*SAF* of c_{25}	0.0895
SAF of c_5	0	*SAF* of c_{12}	0.0724	*SAF* of c_{19}	0.1381	*SAF* of c_{26}	0.0056
SAF of c_6	0.0399	*SAF* of c_{13}	0.0335	*SAF* of c_{20}	0.0774	*SAF* of c_{27}	0.9411
SAF of c_7	1.3219	*SAF* of c_{14}	0.0262	*SAF* of c_{21}	0.1697	*SAF* of c_{28}	0.1344

2.2.2. Determination of the Coefficients for CoVe-Pol Model

We divide the coefficients into three groups according to the magnitude of corresponding *SAF*s, which represent the degree of influence of every coefficient c_i on the computed NRCSs. Thus, we adjust the coefficients, proceeding from higher magnitudes of their degree of influence, to lower orders of magnitude. Through this method, the computational accuracy of the CoVe model reaches 10^{-2}, and calculation of the required adjustments in the coefficients c_i can be achieved using a common computer cluster because the computation has been significantly reduced.

As a first step, we change the values of coefficients c_1, c_7, c_9, which determine the accuracy of computed NRCSs at *SAF* order of magnitude 10^0. Thus, the NRCSs are computed by using the conventional GMF formulation with input of the wind speed, incidence angles and wind direction observed by buoys. Comparing the root mean square errors (RMSEs) between the resulting CMOD parameterization and the simulated RCM data (computed and simulated values for NRCSs), there is an optimal set of computed NRCSs with corresponding c_1, c_7, c_9 values that minimize the RMSE, as defined in (3) below.

Secondly, in order to achieve the CMOD adjustment at 10^{-1} order of magnitude coefficients, we change the values of coefficients c_2, c_3, c_{10}, c_{11}, c_{19}, c_{21}, c_{27}, c_{28} and c_1, c_7, c_9 based on the new CMOD obtained in the first step above. The optimal values of the coefficients are obtained in the same way as previously. For the 10^{-2} order of magnitude, there are 25 sensitive coefficients which must be adjusted, making the calculation too huge to be practical. Thus, we tune the c_i coefficients empirically. Thus, a GMF with new coefficients for compact RV-polarization SAR is proposed following this adjustment procedure, as displayed in Table A1 in the Appendix A.

The RV-polarized NRCSs computed by the new GMF are in good agreement with simulated RCM data, shown in Figure 7a. Additional details are given in Appendix A.

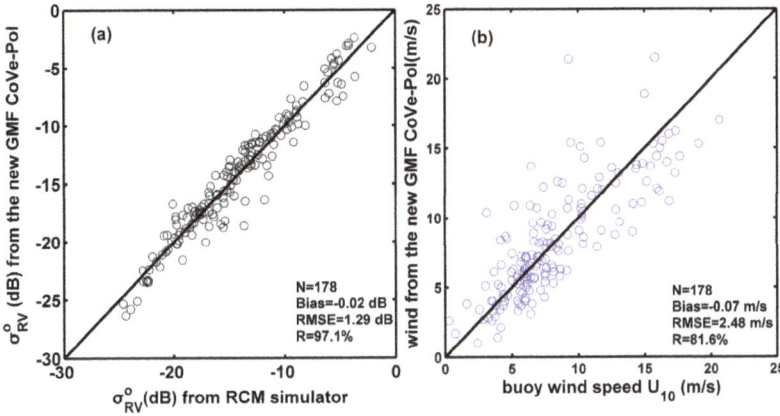

Figure 7. Comparisons of: (**a**) RV-polarized NRCSs between RADARSAT Constellation Mission (RCM) simulated data and results computed by the new compact RV polarization GMF; and (**b**) SAR-retrieved wind speeds from CoVe-Pol model and data measured by buoys.

The definitions of bias, RMSE, and correlation coefficient (R) are,

$$RMSE = \sqrt{\frac{1}{n}\sum_{i=1}^{n}(G_i - D_i)} \tag{3}$$

$$bias = \frac{(G - D)}{n} \tag{4}$$

$$R = \frac{Cov(G, D)}{\sqrt{Cov(G, D)Cov(G, D)}} \tag{5}$$

where G represents the computed results from the GMF, D is the wind speed from the data sets, n is the number of measurements.

2.3. CoHo-Pol Model for Right Circular-Horizontal (RH) Polarization

To build the compact RH polarized wind speed retrieval model, denoted CoHo-Pol, we employ the parameterization method from Komarov et al. [8]. Thus, we use a quadratic relationship in a regression model between buoy wind speed, as a dependent variable, and RH-polarized NRCS, along with incidence angles as independent variables:

$$V = a_0 + a_1\sigma_{RH}^0 + a_2\theta + a_3\left(\sigma_{RH}^0\right)^2 + a_4\theta^2 + a_5\sigma_{RH}^0\theta \tag{6}$$

where V is the wind speed (m/s), θ is the radar incidence angle (degree), and σ_{RH}^0 is the RH-polarized NRCS (dB). Table 2 presents the parameters for the model.

Table 2. Regression coefficients for the horizontal-horizontal (HH) model.

a_0	a_1	a_2	a_3	a_4	a_5
−17.8296	0.9490	1.8640	0.0447	−0.0034	0.0525

2.4. Validation

As the coefficients CoVe-Pol models and CoHo-Pol models are obtained by training the first group of data sets, additional verification tests are performed using the part of the collected data reserved for model testing. We test the CoVe-Pol model for the NRCS values and wind speed. Thus, we substitute the variables (NRCSs simulated, wind speed observed by buoys, radar incidence angles and wind directions) into the CoVe-Pol model, and we compare the simulated NRCS values with the wind speeds observed by the buoys. To provide additional testing of the RH model, CoHo-Pol, the retrieved wind speeds are obtained by substituting σ_{RH}^0 and θ as given by the reserved data sets into the RH model.

3. Results

For wind retrieval, we use simulated compact polarization SAR data and parameters from buoy observations to validate the new RV-polarized GMF, CoVe-Pol. Thus, the wind speed will be determined after we substitute values for the RV-polarized NRCS, incidence angles and wind directions. We compare the wind speeds retrieved by the CoVe-Pol model with the wind speeds observed by buoys in Figure 7b. The bias is 0.07 m/s, the RMSE is 2.48 m/s and the correlation coefficient is 81.3%. Although the accuracy of CoVe-Pol model appears encouraging, additional tests and validation are still needed in the future, when RCM data is available.

The performance of the regression model for compact RH polarization data, CoHo-Pol, is shown in Figure 8, which indicates that the model is an effective methodology for wind retrieval from RH polarized data. Comparing model results to wind speeds observed by buoys, the RMSE is 2.37 m/s.

The wind retrieval models for RV and RH data are presented for compact polarimetry measurements. We test CoVe-Pol model for the NRCS values and wind speed, in Figure 9a,b. The RMSE for σ_{RV}^0 is only 1.28 dB and for wind speed, 2.36 m/s, and the values of the correlation coefficients are 97.9% and 82.4%. These results indicate that CoVe-Pol model is a potentially good method for wind retrieval; the computed NRCSs and retrieved winds are in good agreement with the simulated NRCSs and the independently measured buoy winds.

Comparing the retrieved wind speeds and the buoy observations, the RMSE for winds retrieved by the CoHo-Pol model is 2.39 m/s, and the correlation coefficient is 81.5%, which is shown in Figure 10.

Validations demonstrate that the CoVe-Pol and CoHo-Pol models are reliable and useful retrieval models for RV and RH polarized SAR data, respectively.

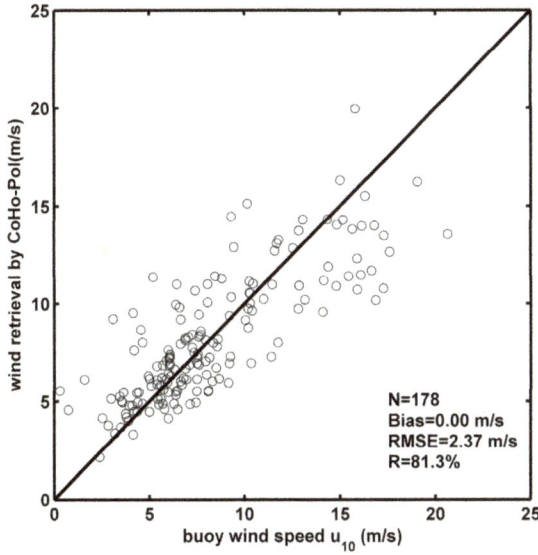

Figure 8. SAR-retrieved wind speeds from RH polarization mode model denoted CoHo-Pol vs. buoy-measured U_{10} for the reserved training subset.

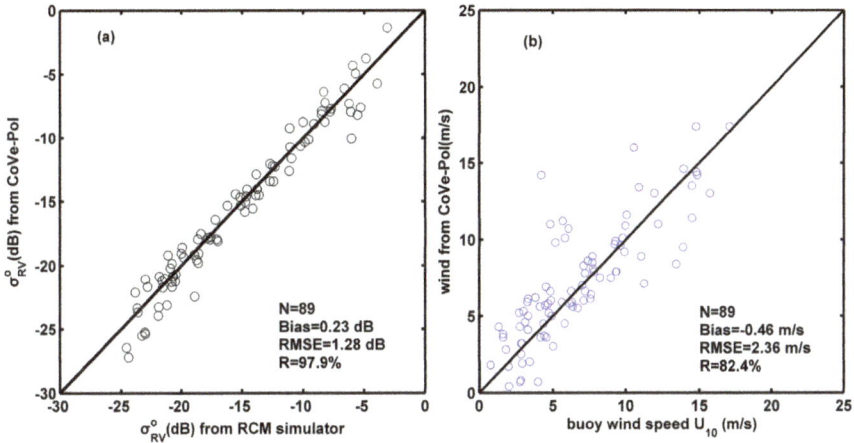

Figure 9. Validation of CoVe-Pol model by inputting wind speed and RV-polarized NRCSs: (**a**) σ^0_{RV} from CoVe-Pol model vs. σ^0_{RV} simulated by the RCM simulator; (**b**) wind speed retrieved by CoVe-Pol model vs. wind speed measured by buoys.

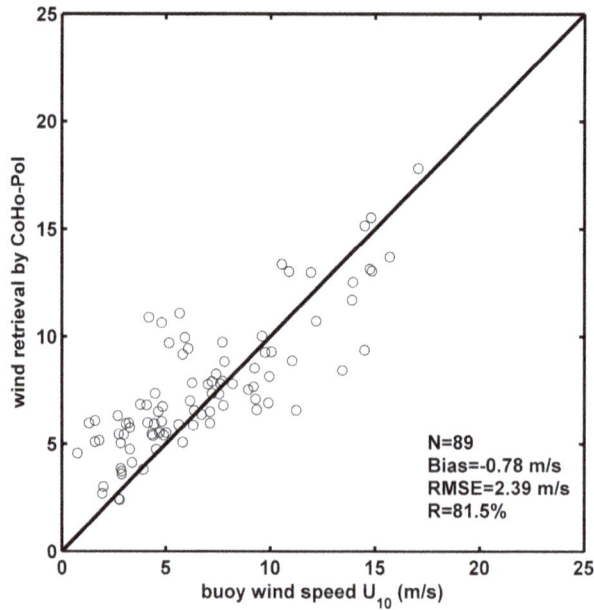

Figure 10. Wind speed retrieved by CoHo-Pol, the RH polarization mode model, vs. reserved wind speeds observed by buoys; the test data subset.

4. Discussion

Based on the almost linear relationships between NRCSs of VV and RV polarizations, the C-band RV-polarized wind retrieval model has been simply proposed using the C-band VV-polarized wind retrieval model [11,12]. However, there are two factors should be further discussed: (1) the VV-polarized wind retrieval models, routinely CMODs, are developed for scatterometer but not SAR, and (2) the relationships between the two polarized NRCSs are almost linear but not accurately. Therefore, this study aims to improve the CP mode wind retrieval accurate by tuning each parameter in the CMOD frame. To reduce the large computations, we employ the sensitivity analysis. Lu et al. (2018) developed a new wind retrieval model for C-band VV-polarization [19]. In the CMOD series, including CMOD4, CMOD5, CMOD5.N, CMOD6, CMOD7 and et al., the equations are the same but with different parameters [20,21]. We compare the results from CoVe-Pol models here and the method proposed by Geldsetzer et al. (2015) based on the first data group [11], which is shown in Figure 11. It is obvious that the new model CoVe-POL present better wind retrieval results.

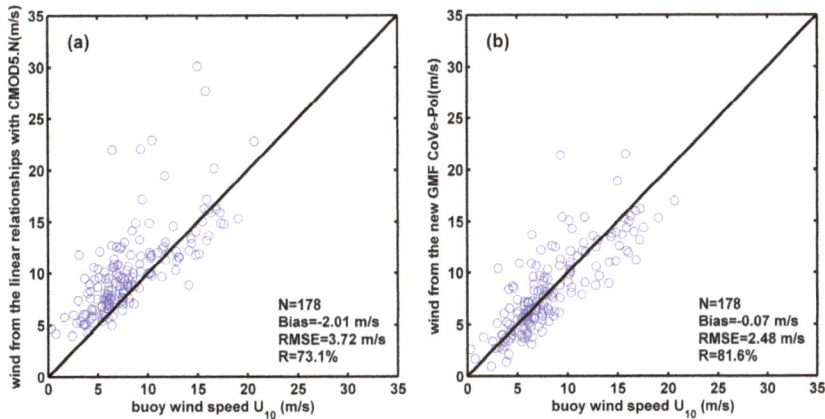

Figure 11. The comparsion of wind speed estimation results between using the linear realtionships with CMOD5.N and using CoVe-Pol model: (**a**) wind speed estimation through the linear relationships with CMOD5.N vs. wind speed measured by buoys; (**b**) wind speed retrieved by CoVe-Pol model vs. wind speed measured by buoys.

Zhang et al. (2018) proposed a semi-empirical ocean surface model for RCM CP mode and the simulated results suggest that the RV-polarization has a better potential capability for the ocean wind retrieval than the RH-polarization [22]. This result is consistent with the wind retrieval results here. As indicated by Zhang et al. (2018) [22], the noise floor should be a challenge for sea surface wind retrieval. As shown in Figure 9a, the NRCSs are under-estimated by the new proposed wind retrieval model when the values are around −25 dB. On one hand, this may be due to the errors of the model. However, on the other hand, this should be caused by the noise floor which is designed as −25 dB in the RCM. Therefore, the low wind retrieval from RCM RV and RH polarizations will be a challenge. Moreover, the sea surface ice or oil spill can be detected by SAR due to covering the ocean surface and changing the backscattering features [22]. The determination of oil spill from low wind condition is a current issue from the existing SAR observations. As the noise floor for RADARSAT-2 is much lower than the RCM, this would also be a problem for the sea ice or oil spill study using RCM in the future.

5. Conclusions

The estimation of ocean surface winds by SAR is an important research field of satellite remote sensing. Because RCM will provide CP products which differ from the conventional polarimetry SAR, the establishment of new specific models for potential wind retrieval from CP SAR parameters is an urgent need.

In this paper, we propose two wind retrieval models for C-band RCM SAR CP model: (1) CoVe-Pol model for RV polarization data, and (2) CoHo-Pol model for RH polarization data. The two models are derived from collected data consisting of 267 RADARSAT-2 SAR quad-polarized images and collocated buoy data. These two models can be applied to the real CP data when RCM will have been launched and succeeds in providing data. The CP-polarized data sets are generated from quad-polarized SAR data by a RCM simulator. We have divided these data randomly into two data groups: one for building new models and the other reserved for model testing.

To develop CoVe-Pol model, we carried out a sensitivity analysis in the process of creating the new GMF coefficients. We separate the derivation process for the coefficients into several steps. These steps are designed according to a sequence determined by the orders of magnitude of the CMOD coefficients, in order to reduce the required computations, so that the numerical process can be possible in terms of available computer resources. Utilizing sensitivity analysis factors (*SAFs*) for the coefficients provides an efficient methodology for building a new GMF for the RV polarization data, by adjustment of these coefficients. In this approach, when the new GMF is derived, the correlation coefficient between RV-polarized computed NRCSs and simulated NRCSs reaches 97.1%, and the RMSE is only 1.29 dB. The wind retrieval by the associated CoVe-Pol model parameterization is shown to have a good performance based on the RV polarized SAR data. The RMSE is 2.48 m/s, and the bias is 0.07 m/s.

To produce the RH model for wind retrieval, we use a quadratic function in a regression model to relate buoy wind speed to the RH NRCS data along with the radar incidence angles. Comparing results with the winds measured by buoys, the RH model, denoted CoHo-Pol, is shown to behave well in wind retrievals, with RMSE of 2.37 m/s. The model results indicate that the RH model is a useful way to retrieve wind speed as a fast inversion methodology.

We test the two new models, CoVe-Pol and CoHo-Pol, with the reserved test data set, and show that there is strong agreement between both the models and the data. Thus, these two new models can potentially be applied to retrieve wind from CP C-band SAR measurements. In February 2019, the three satellites of RCM are scheduled for launching together. Compared to what we have now with the separated single SAR satellite (e.g., RADARSAT-2 or Sentinel-1), the three continuous observations make the temporal studies of oceanography and/or atmosphere possible. Therefore, the ocean wind retrieval models developed here would be important for temporal oceanography or atmosphere dynamic research based on RCM SAR data.

Author Contributions: Writing-Original Draft Preparation and processing datasets, T.S.; Designing methodology, proposing the model and designing this study, G.Z.; Supervision, writing-Review and editing, W.P.; Validation, B.Z.; Formal Analysis, C.G.; Resources, S.K.; Data Curation, K.W; Software, J.S.

Funding: This research was funded by National Natural Science Youth Foundation of China under Grant 41706193, in part by the Canadian Data Utilization and Application Program "Winds from SAR" RCM Readiness Project between ECCC and DFO, in part by the Canadian Space Agency SWOT and Office of Energy Research and Development (OERD) Programs, and Open Fund of Key Laboratory of Geographic Information Science (Ministry of Education), East China Normal University under Grant No. KLGIS2017A06, National Natural Science Foundation of China under Grant Nos. 41376010), and the Excellent Youth Science Foundation of Jiangsu Province under Grant BK20160090.

Acknowledgments: The authors thank the China Scholarship Council for supporting this scientific research cooperation under the sponsorship [2015] 3022, the Canadian Space Agency for providing RADARSAT-2 data, NOAA NDBC for supplying buoy data (http://www.ndbc.noaa.gov/).

Conflicts of Interest: The authors declared that they have no conflicts of interest to this work.

Abbreviations

CP	compact-polarimetry
GMF	geophysical model function
HH	horizontal-horizontal
HV	horizontal-vertical
NDBC	National Data Buoy Center
NRCSs	normalized radar cross-sections
PR	polarization ratio
RCM	RADARSAT Constellation Mission
RH	right circular transmit and horizontal receive
RL	right circular transmit and left circular receive
RMSEs	root mean square errors
RR	right circular transmit and right circular receive
RV	right circular transmit and vertical receive
SAF	sensitivity analysis factor
SAR	synthetic aperture radar
TOGA COARE	Tropical Ocean and Global Atmosphere Response Experiment
VH	vertical-horizontal
VV	vertical-vertical

Appendix A Cove-Pol Model Formulation and Coefficients

The form of the CoVe-Pol model parameterization:

$$\sigma^0(\theta, U_{10}, \varphi) =$$
$$= B0(c_0, U_{10}, \theta)[1 + B1(c_1, U_{10}, \theta)\cos(\varphi) + B2(c_2, U_{10}, \theta)\cos(2\varphi)]^{1.6} \tag{A1}$$

where $B0$, $B1$ and $B2$ are functions of wind speed U_{10} and incidence angle θ, or alternatively, $\chi = (\theta - 40)/25$. The $B0$ term is defined as:

$$B0 = 10^{a_0 + a_1 U_{10}} f(a_2 U_{10}, s_0) \tag{A2}$$

where,

$$f(s, s_0) = \begin{cases} (s_0)^\alpha g(s_0), & s < s_0 \\ g(s), & s > s_0 \end{cases} \tag{A3}$$

where,

$$g(s) = 1/(1 + \exp(-s)), \text{ and } \alpha = s_0(1 - g(s_0)) \tag{A4}$$

The functions a_0, a_1, a_2, γ and s_0 depend on incidence angle only:

$$a_0 = c_1 + c_2 x + c_3 x^2 + c_4 x^3$$
$$a_1 = c_5 + c_6 x \tag{A5}$$
$$a_2 = c_7 + c_8 x$$

$$\gamma = c_9 + c_{10} x + c_{11} x^2$$
$$s_0 = c_{12} + c_{13} x \tag{A6}$$

The $B1$ term is modeled as follows:

$$B1 = \frac{c_{14}(1 + x) - c_{15}v(0.5 + x - \tanh[4(x + c_{16} + c_{17}v)])}{1 + \exp(0.34(v - c_{18}))} \tag{A7}$$

The $B2$ term was chosen as,

$$B2 = (-d_1 + d_2 v_2)\exp(-v_2) \tag{A8}$$

Here v_2 is given by,

$$v_2 = \begin{cases} a + b(y - 1)^n & , \quad y < y_0 \\ y & , \quad y \geq y_0 \\ y = \frac{v + v_0}{v_0} \end{cases} \tag{A9}$$

where,

$$y_0 = c_{19}, \quad n = c_{20} \tag{A10}$$

$$a = y_0 - (y_0 - 1)/n, \quad b = 1/\left[n(y_0 - 1)^{n-1}\right] \tag{A11}$$

The quantities v_0, d_1 and d_2 are functions of incidence angle only,

$$v_0 = c_{21} + c_{22} x + c_{23} x^2$$
$$d_1 = c_{24} + c_{25} x + c_{26} x^2 \tag{A12}$$
$$d_2 = c_{27} + c_{28} x$$

The coefficients are given in Table A1.

Table A1. CoVe-Pol coefficients.

c_1	−0.9200	c_8	0.0159	c_{15}	0.0064	c_{22}	−3.2592
c_2	−1.1935	c_9	5.4536	c_{16}	0.3141	c_{23}	1.2905
c_3	0.0321	c_{10}	0.2633	c_{17}	0.0117	c_{24}	6.0876
c_4	0.3421	c_{11}	−2.2313	c_{18}	45.4000	c_{25}	2.3296
c_5	0	c_{12}	0.0472	c_{19}	2.0293	c_{26}	0.3168
c_6	0.0040	c_{13}	−0.0689	c_{20}	2.9350	c_{27}	4.0550
c_7	0.0882	c_{14}	0.0043	c_{21}	16.7318	c_{28}	1.5237

References

1. Charbonneau, F.T.; Brisco, B.; Raney, R.K.; McNairn, H.; Liu, C.; Vachon, P.W.; Shang, J.; DeAbreu, R.; Champagne, C.; Merzouki, A.; et al. Compact Polarimetry Overview and Applications Assessment. *Can. J. Remote Sens.* **2010**, *36*, S298–S315. [CrossRef]
2. Cloude, S.R.; Goodenough, D.G.; Chen, H. Compact Decomposition Theory. *IEEE Geosci. Remote Sens. Lett.* **2012**, *9*, 28–32. [CrossRef]
3. Souyris, J.C.; Imbo, P.; Fjortoft, R.; Mingot, S.; Lee, J.S. Compact polarimetry based on symmetry properties of geophysical media: The /spl pi//4 mode. *IEEE Trans. Geosci. Remote Sens.* **2005**, *43*, 634–646. [CrossRef]
4. Keith, R.K. Hybrid-Polarity SAR Architecture. *IEEE Trans. Geosci. Remote Sens.* **2007**, *45*, 3397–3404. [CrossRef]
5. Hersbach, H.; Stoffelen, A.; De Haan, S. An improved C-band scatterometer ocean geophysical model function: CMOD5. *J. Geophys. Res.* **2007**, *112*, C03006. [CrossRef]
6. Hersbach, H. Comparison of C-Band Scatterometer CMOD5.N Equivalent Neutral Winds with ECMWF. *J. Atmos. Ocean. Technol.* **2010**, *27*, 721–736. [CrossRef]
7. Vachon, P.W.; Dobson, F.W. Wind Retrieval from RADARSAT SAR Images: Selection of a Suitable C-Band HH Polarization Wind Retrieval Model. *Can. J. Remote Sens.* **2000**, *26*, 306–313. [CrossRef]
8. Komarov, A.S.; Zabeline, V.; Barber, D.G. Ocean Surface Wind Speed Retrieval from C-band SAR Images Without Wind Direction Input. *IEEE Trans. Geosci. Remote Sens.* **2013**, *52*, 980–990. [CrossRef]
9. Zhang, B.; Perrie, W. Cross-Polarized Synthetic Aperture Radar: A New Potential Measurement Technique for Hurricanes. *Bull. Am. Meteor. Soc.* **2012**, *93*, 531–541. [CrossRef]
10. Zhang, G.S.; Li, X.F.; Perrie, W.; Hwang, P.A.; Zhang, B.; Yang, X.F. A Hurricane Wind Speed Retrieval Model for C-band RADARSAT-2 Cross-polarization ScanSAR Images. *IEEE Trans. Geosci. Remote Sens.* **2017**, *55*, 4766–4774. [CrossRef]
11. Denbina, M.; Collins, M.J. Wind Speed Estimation using C-band compact polarimetric SAR for wide swath imaging modes. *ISPRS J. Photogramm. Remote Sens.* **2016**, *113*, 75–85. [CrossRef]
12. Geldsetzer, T.; Charbonneau, F.; Arkett, M.; Zagon, T. Ocean Wind Study Using Simulated RCM Compact-Polarimetry SAR. *Can. J. Remote Sens.* **2015**, *41*, 418–430. [CrossRef]
13. Hamby, D.M. A Review of Techniques for Parameter Sensitivity Analysis of Environmental Models. *Environ. Monit. Assess.* **1994**, *32*, 135–154. [CrossRef] [PubMed]
14. Holvoet, K.; van Griensven, A.; Seuntjens, P.; Vanrolleghem, P.A. Sensitivity analysis for hydrology and pesticide supply towards the river in SWAT. *Phys. Chem. Earth* **2005**, *30*, 518–526. [CrossRef]
15. Canadian Space Agency. Available online: http://www.asc-csa.gc.ca/eng/satellites/radarsat/ (accessed on 20 October 2018).
16. National Data Buoy Center. Available online: http://www.ndbc.noaa.gov (accessed on 20 October 2018).
17. Fairall, C.W.; Bradley, E.F.; Hare, J.E.; Grachev, A.A.; Edson, J.B. Bulk Parameterization of Air-Sea Fluxes: Updates and Verification for the COARE Algorithm. *J. Clim.* **2003**, *16*, 571–591. [CrossRef]
18. Smith, S.D. Coefficients for sea surface wind stress, heat flux, and wind profiles as a function of wind speed and temperature. *J. Geophys. Res.* **1988**, *93*, 15467–15472. [CrossRef]
19. Lu, Y.R.; Zhang, B.; Perrie, W.; Mouche, A.A.; Li, X.F.; Wang, H. A C-Band Geophysical Model Function for Determining Coastal Wind Speed Using Synthetic Aperture Radar. *IEEE J. Sel. Top. Appl. Earth Obs. Remote Sens.* **2018**, *11*, 2417–2428. [CrossRef]
20. Elyouncha, A.; Neyt, X.; Stoffelen, A.; Verspeek, J. Assessment of the corrected CMOD6 GMF using scatterometer data. *Remote Sens. Ocean Sea Ice Coast. Waters Large Water Reg.* **2015**, *9638*. [CrossRef]
21. Stoffelen, A.; Verspeek, J.; Vogelzang, J.; Verhoef, A. The CMOD7 Geophysical Model Function for ASCAT and ERS Wind Retrievals. *IEEE J. Sel. Top. Appl. Earth Obs. Remote Sens.* **2017**, *10*, 2123–2134. [CrossRef]
22. Zhang, G.S.; Perrie, W.; Zhang, B.; Khurshid, S.; Warner, K. Semi-empirical ocean surface model for compact-polarimetry mode SAR of RADARSAT Constellation Mission. *Remote Sens. Environ.* **2018**, *217*, 52–60. [CrossRef]

remote sensing

MDPI

Article

Comparison of C-Band Quad-Polarization Synthetic Aperture Radar Wind Retrieval Models

He Fang [1,2], Tao Xie [3,4,*], William Perrie [2], Guosheng Zhang [2], Jingsong Yang [5] and Yijun He [1]

[1]   School of Marine Sciences, Nanjing University of Information Science and Technology,
      Nanjing 210044, China; fanghe_doc@163.com (H.F.); yjhe@nuist.edu.cn (Y.H.)
[2]   Fisheries & Oceans Canada, Bedford Institute of Oceanography, Dartmouth, NS B2Y 4A2, Canada;
      william.perrie@dfo-mpo.gc.ca (W.P.); zgsheng001@gmail.com (G.Z.)
[3]   School of Remote Sensing and Geomatics Engineering, Nanjing University of Information Science and
      Technology, Nanjing 210044, China
[4]   Laboratory for Regional Oceanography and Numerical Modeling, Qingdao National Laboratory for Marine
      Science and Technology, Qingdao 266237, China
[5]   State Key Laboratory of Satellite Ocean Environment Dynamics, Second Institute of Oceanography,
      State Oceanic Administration, Hangzhou 310012, China; jsyang@sio.org.cn
*   Correspondence: xietao@nuist.edu.cn; Tel.: +86-25-5869-5696

Received: 13 July 2018; Accepted: 5 September 2018; Published: 11 September 2018

Abstract: This work discusses the accuracy of C-2PO (C-band cross-polarized ocean backscatter) and CMOD4 (C-band model) geophysical model functions (GMF) for sea surface wind speed retrieval from satellite-born Synthetic Aperture Radar (SAR) images over in the Northwest Pacific off the coast of China. In situ observations are used for comparison of the retrieved wind speed using two established wind retrieval models: C-2PO model and CMOD4 GMF. Using 439 samples from 92 RADARSAT-2 fine quad-polarization SAR images and corresponding reference winds, we created two subset wind speed databases: the *training* and *testing* subsets. From the *training* data subset, we retrieve ocean surface wind speeds (OSWSs) from different models at each polarization and compare with reference wind speeds. The RMSEs of SAR-retrieved wind speeds are: 2.5 m/s: 2.11 m/s (VH-polarized), 2.13 m/s (HV-polarized), 1.86 m/s (VV-polarized) and 2.26 m/s (HH-polarized) and the correlation coefficients are 0.86 (VH-polarized), 0.85(HV-polarized), 0.87(VV-polarized) and 0.83 (HH-polarized), which are statistically significant at the 99.9% significance level. Moreover, we found that OSWSs retrieved using C-2PO model at VH-polarized are most suitable for moderate-to-high winds while CMOD4 GMF at VV-polarized tend to be best for low-to-moderate winds. A hybrid wind retrieval model is put forward composed of the two models, C-2PO and CMOD4 and sets of SAR test data are used in order to establish an appropriate wind speed threshold, to differentiate the wind speed range appropriate for one model from that of the other. The results show that the OSWSs retrieved using our hybrid method has RMSE of 1.66 m/s and the correlation coefficient are 0.9, thereby significantly outperforming both the C-2PO and CMOD4 models.

Keywords: ocean surface wind speed retrieval; synthetic aperture radar (SAR); quad-polarized SAR

---

1. Introduction

Ocean surface wind speed (OSWS) plays a significant role in the global climate, directly influencing energy transport between ocean basins, ocean water mass formations and circulation. As a result, observations and monitoring of OSWS can improve our understanding of the physical mechanisms of oceanic-atmospheric interactions, hurricane and severe storm predictions and decision making and numerical weather predication (NWP) and marine forecasts [1].

In recent decades, with the development of satellite remote sensing, the reliability of OSWS retrieved from various satellite sensors has matured and improved. Among various satellites, the microwave scatterometers (SCAT) play a vital role in getting coverage over the entire global ocean. However, a major drawback for SCAT-derived wind speeds is the coarse resolution of the data (12.5–50 km), which limits our ability to get a better understanding of the coastal oceans and to study related processes in the lower atmospheric and oceanic boundary layers, such as surface currents, waves, winds and their interactions [2]. Spaceborne synthetic aperture radar (SAR) can mitigate this difficulty, because of its ability to retrieve OSWSs, day or night, in almost all-weather conditions, at high spatial resolution and large areal coverage [3]. At this time, the retrieval of OSWS at high (<1 km) resolution from quad-polarized spaceborne SAR images is a mature geophysical application. Many efforts have been devoted to developing optimal reliable methodologies to elucidate the geophysical relationship between the normalized radar cross section (NRCS) and OSWS and to apply this relationship to accurately compute wind speeds [4–8].

OSWS retrieved from co-polarized (HH-and VV-polarized; horizontal transmit, horizontal receive and vertical transmit, vertical receive, respectively) SAR data are normally computed employing various empirical geophysical model functions (GMFs). For VV polarized SAR data, these GMFs were initial developed from C-band scatterometer measurements. They are routinely called CMOD (C-band model) GMFs and they relate the wind speeds and directions to the local incidence angle and NRCS. Using radar incidence angle and wind direction, along with the NRCS at each pixel in the VV-polarized channel from C-band SAR, the associated OSWS can be retrieved from various CMOD GMFs, such as CMOD4 [9], CMOD-IFR2 [10], CMOD5 [11] and CMOD5.N [12]. Recently, the latest CMOD GMF, called CMOD7, was developed in several steps from CMOD5.N for application to intercalibrate ERS (ESCAT) and ASCAT scatterometers [13]. Although CMOD GMFs for VV-polarized have been widely used, based on a large number of SAR data, however, no similarly well-developed GMF exists for HH-polarized SAR imagery. To remedy this difficulty, hybrid model functions, called the polarization ratios (PRs), were proposed to map the expected NRCS at VV-polarized mode to the HH-polarized value for the same wind direction and speed. When these CMOD GMFs, as mentioned above, are applied to HH-polarized SAR images, various PR models have to be used to convert HH-NRCS to VV-NRCS before application for wind retrieval [14–18].

In conclusion, OSWSs retrieved at co-polarized channel are a mature technical achievement that has been widely validated in different SAR systems. However, the NRCS value for co-polarized SAR imagery exhibits data saturation when wind speeds exceed about 16 m/s for local incidence angle under 35° [19]. Moreover, available experimental and theoretical evidence suggest that dampening or single saturation of co-polarized channel radar backscatter occurs across a wide range of wind speeds and radar frequencies [20,21]. In recent years, C-band cross-polarized (HV-and VV-polarized, horizontal transmit, vertical receive and vertical transmit, horizontal receive, respectively) ocean backscatter has been shown to be almost independent of incidence angle and wind direction and to be quite linear with respect to the OSWS. This unique sensitivity for cross-polarized data is mainly attributed to the contribution of breaking waves [22]. The relationship between the cross-polarized NRCS and OSWS can directly provide wind speeds from SAR images, without the requirement of wind direction or incidence angles. Some cross-polarized OSWS retrieval models have been developed based on this relationship [23,24]. And more critically, the measured NRCS values for cross-polarized SAR seem to be almost not saturated, even at very high speeds (up to 50 m/s), which indicates that they can potentially be used to retrieve hurricane-generated winds [25–28].

Based on the above developments, it is apparent that OSWSs can be retrieved from cross- and co-polarized SAR data using these established methods. The differing sensitivity between contemporaneous cross and co-polarized SAR signals can be advantageously exploited to infer local information about the wind fields. However, each model has its own wind speed application range; for example, cross-polarized NRCS does not suffer from saturation effects at high wind speeds, which are evident in the co-polarized NRCS. Moreover, retrieved OSWSs from cross-polarized SAR data have better accuracy than winds retrieved from co-polarized NRCS at high wind speed regimes, especially at wind speeds above 20 m/s [27,28]. La et al. [29] compare different retrieval models for

OSWS based on empirical (EP) and theoretical electromagnetic (EM) approaches. They show that OSWS estimates from CMOD5.N GMF and two-scale EM models (small-slope and resonant curvature approximation) are very close, for low and moderate incidence angles, whereas retrieved OSWS from EM models give overestimates for high wind speeds. Our objective in this paper is to validate and elucidate the advantages and disadvantages of retrieval models for the cross- and co-polarized data and their respective wind speed ranges for reliable applications. Based on our results, we attempt to propose a new hybrid OSWS retrieval model, which can more accurately retrieve wind speed data from C-band RS-2 quad-polarized SAR data. Two C-band OSWS retrieval GMFs, namely C-2PO and CMOD4 and the SAR data are introduced at Section 2. Section 3 introduces the OSWS retrieval results for quad-polarized RS-2 images. Discussion is given in Section 4 and conclusions, in Section 5.

2. Materials and Methods

2.1. RADARSAT-2 Quad-Polarized SAR Images

In this study, in order to retrieve OSWS from cross- and co-polarized SAR images, 92 RS-2 fine quad-polarized model single-look complex (SLC) SAR images were used. The area covered by the selected SAR images includes Chinese waters: the Yellow Sea, the East China Sea and the South China Sea, covering the area from 14°N to 38°N and 110°E to 130°E between February 2011 and December 2015. The RS-2 satellite transit times for our research area are about 10:00 (Ascending) and 22:00 (Descending) Coordinate Universal Time (UTC). The range of local radar angles is between 20° and 49° and the nominal incidence angles vary by about 1.5° across a swath of 25 km. RS-2 fine quad-polarized model SLC data have the capability to provide C-band VV, HH, HV and HV polarized images with a low noise floor. In addition, inter-channel cross talk is corrected in the processor to better than −35 dB, which is appropriate for cross-polarized backscatter measurements, without contamination from the co-polarized (HH- and VV-polarized) data.

For each individual SAR image with a specific beam mode, the spatial resolution (pixel spacing) in range and azimuth is about 5 m. Direct calculation of OSWS from the original scene can result in noisy patterns due to the presence of speckle noise in the raw SAR image. Therefore, the raw RS-2 SAR data is preprocessed in order to obtain the orthorectified NRCS images at each polarization. *Firstly*, the OSWS can only be retrieved from SAR images that are independent of ocean surface features that are not due to the local wind. In order to exclude SAR scenes that contain features not associated with the local wind, a filter is applied in this study. The filter is used to distinguish between inhomogeneous and homogeneous SAR images and, additionally, to retrieve ocean waves and wind speeds [30,31]. In the *next step*, the calibrated 5-m spatial resolution image is degraded to 100-m resolution image. To achieve this, we perform 20 × 20 pixel boxcar averaging of the NRCS in each polarization, so that the reconstructed pixel spacing is 100 m. *Finally*, the radiometric correction method is used to transform NRCS values from intensity units to decibel units. The conversion formula is expressed as follows:

$$\sigma_0 \ (\text{dB}) = 10 \cdot \log_{10} \sigma_0 (\text{no units}) \tag{1}$$

where σ_0 is the NRCS value in decibels (dB).

2.2. ECMWF ERA-Interim Reanalysis Winds

Carvalho et al. [32] evaluated the performance of different reanalysis wind datasets and found that ERA-Interim reanalysis provides the most realistic initial and boundary data for oceanographic applications, therefore allowing the possibly for development of reliable retrieval models of OSWS from spaceborne SAR data. Therefore, we select ERA-Interim wind field data as the reference wind data for this study.

The ERA-Interim global atmospheric reanalysis *daily* wind speed of components U (east-west direction) and V (south-north direction) at 10-m height with a high spatial resolution (up to 0.125°) at 6-h intervals are provided by the European Center for Medium Range Weather Forecasts (ECMWF) [33].

These data are available for the period from 1979 to the present. Figure 1 shows the wind speed in the study areas in this paper. In order to validate the wind retrievals from SAR images with collocated in situ measurements, ERA-Interim wind speeds were taken as reference values. However, the swaths of RS-2 SAR scenes are about 25 km, which means that there are about 5–6 ERA-Interim grid cells in each SAR image. In our study, the acquisition time for RS-2 SAR images is 1~2 h earlier than the ERA-Interim data. For a viable comparison, the ERA-Interim wind grid cells at 12:00:00 and 24:00:00 UTC are interpolated to generate the wind vectors at the RS-2 acquisition time [34]. In addition, conventional co-polarized CMOD GMFs have two unknown parameters, that is, wind direction and speed which must be provided by external sources, prior to wind speed retrieval. Therefore, auxiliary data for wind directions are necessary as inputs to these GMFs because wind directions are difficult to directly measure from SAR images.

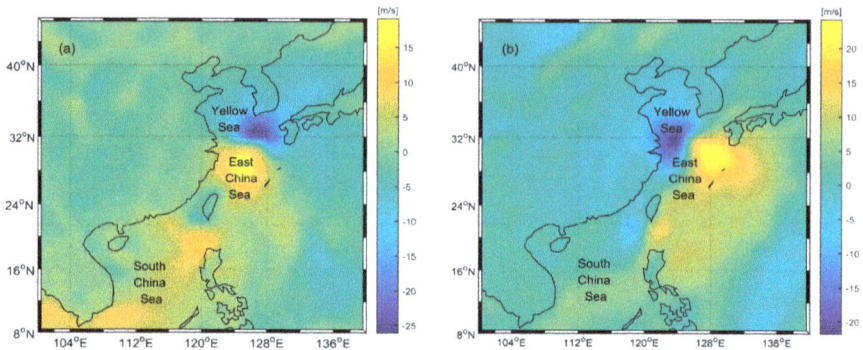

Figure 1. ERA-Interim wind speeds: (**a**) east-west direction and (**b**) south-north direction in Chinese waters. The ERA-Interim wind products are acquired on 20 August 2012 at 12:00:00 UTC from ECWMF website.

2.3. Creation of Wind Speed Databases

A general flowchart (Figure 2) for the establishment of the wind speed database for the developing wind speed retrieval method is described in this section. *Firstly*, all available quad-polarized RS-2 SAR images are preprocessed. *Subsequently*, we identify all existing ERA-Interim grid cells that are located inside the corresponding SAR scenes. In our study, the ERA-Interim wind field data are available at 0.125°-resolution grids whereas our RS-2 imagery has a higher spatial resolution: 5.4 m in the range direction and 8.0 m in the azimuth direction. The swath widths of RS-2 SAR scenes are about 25 km, which means that the number of ERA-Interim grid cells inside one SAR scene varies from 5 to 6. Thus, 439 samples are extracted from 92 SAR images. As the resolution of the ERA-Interim wind data is 0.125°, there are about one hundred pixels between the adjacent SAR measurements we selected. These selected measurements can be treated as essentially independent. It is notable that the locations (latitude and longitude) of ERA-Interim grid cells and SAR pixels are misalignment in most cases. To obtain the radar incidence angles, NRCS at each polarization and other parameters, we use the bilinear interpolation method rather than other downscaling approaches such as the nearest grid-points [35].

Finally, we created two subset wind speed databases: the *training* subset and the *testing* subset. Each subset contains SAR parameters (including NRCS at each polarization, radar incidence angles and external wind directions) and ERA-Interim wind speeds at the same location. The *training* subset contains 285 samples that are used for training different OSWS retrieval models at each polarization. Based on the analyzed retrieval results, we build a hybrid wind speed retrieval method and employ the *testing* subset (154 samples) to validate these algorithms.

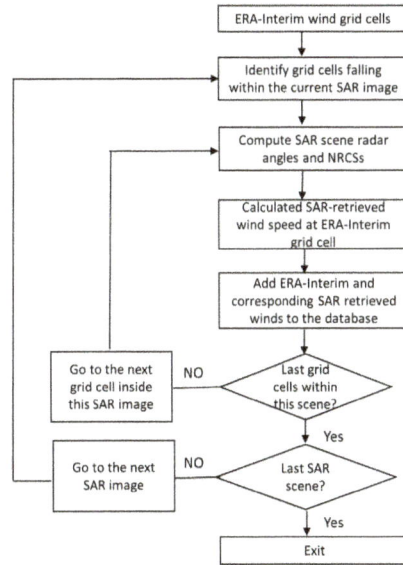

Figure 2. Flowchart for building a database for the developing wind speed retrieval method.

2.4. Quad-Polarized SAR Wind Speed Retrieval Algorithm

For the instruments operating at C-band and VV-polarized channel, several empirical GMFs, for example, CMOD4, CMOD-IFR2, CMOD5 and CMOD5.N have been developed and validated through a series of satellite scatterometer missions. Figure 3a shows the variations of SAR-simulated VV-polarized NRCSs from CMOD GMFs with wind speed at a local radar incidence angle of 30°, with relative wind direction of 180°. Clearly, all CMOD functions produce very similar results for wind speeds below 20 m/s. In fact, C-band VV-polarized OSWS retrieval will always become saturated under high wind conditions. Han et al. [36] retrieved OSWS based on CMOD4, CMOD-IFR2 and CMOD5 using RS-2 SAR images of the East China Sea and the results suggested that CMOD4 is the most promising GMF of these formulations.

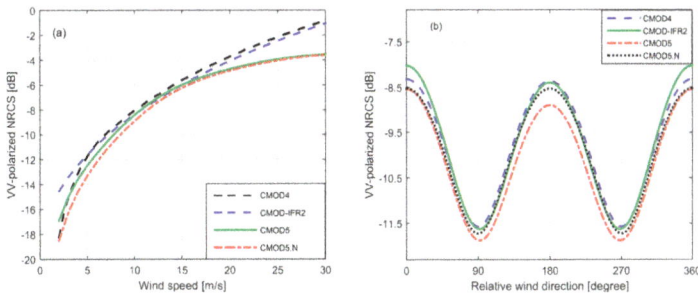

Figure 3. (**a**) Dependence of VV-polarized normalized radar cross section (NRCS) from CMOD geophysical model functions (GMFs) versus ocean surface wind speeds (OSWSs) at a local incidence angle of 30° and wind direction relative radar look angle of 180°. (**b**) Dependence of VV-polarized NRCS from CMOD GMFs versus wind direction relative radar look angle at a local incidence angle of 30° and wind speed of 10 m/s.

In this study, we selected CMOD4 GMF as the VV-polarized wind speed retrieve algorithm and the general form is expressed as follows:

$$\sigma_{VV}^0(\theta, U_{10}, \phi) = A_0(\theta, U_{10})[1 + A_1(\theta, U_{10})\cos\phi + A_2(\theta, U_{10})\cos 2\phi]^{1.6} \tag{2}$$

Here, σ_{VV}^0 is the VV-polarized NRCS, ϕ is the external wind direction ψ relative radar look angle. The other parameters A_0, A_1 and A_2 are coefficients which are dependent on the local radar incidence angle θ. and OSWS at 10-m reference height U_{10}. Moreover, the external wind direction, defined as ψ, should be obtained from ERA-Interim data, whereas the local incidence angle θ and the NRCS can be directly computed from the corresponding RS-2 SAR image.

To date, although many CMOD GMFs have been presented for VV-polarized data, no similar well-developed, verified OSWS retrieval models exist to extract wind speed from HH-polarized SAR images. To overcome this deficiency, the empirical PR models, which are related to local radar incidence angles, were developed for application in retrieving OSWS using the HH-polarized SAR channel. Following the usual notation, we define PR as

$$PR = \frac{\sigma_0^{VV}}{\sigma_0^{HH}} \tag{3}$$

where σ_0^{VV} and σ_0^{HH} are the NRCSs of VV and HH polarizations, respectively. Various PR models have been proposed as functions of incidence angles and several PR models are compared in the Figure 3. Recently, using a nonlinear least squares algorithm to fit the collocated 877 RS-2 fine quad-polarization PR and incidence angles, a new C-band PR incidence angle dependent model was proposed by Zhang et al. [20], given as

$$PR = B_1 \exp(B_2\theta) + B_3 \tag{4}$$

where $B_1 = 0.2828$, $B_2 = 0.0451$, $B_3 = 0.2891$ and R represent the polarization ration value, respectively. In our study, we have selected Zhang's PR model to retrieve OSWS from HH-polarized data, because this PR model is based on RS-2 fine quad-polarization data which is the same as the data used in this study. Moreover, for convenience, the OSWS retrieved from CMOD4 GMF using VV-polarized data is denoted CMOD4+VV and the alternative, using converted HH-polarized data by the PR model, is denoted CMOD4+HH+PR.

Generally speaking, co-polarized scattering is the result of ocean surface scattering, whereas cross-polarized scattering results from sea surface tilts or by volume scattering; thus, cross-polarized scattering is less correlated than co-polarized scattering data. Studies show that the NRCS in cross-polarized mode is essentially independent of radar incidence angle and wind direction but has a linear relationship with respect to OSWS and thus generates a new potential capability to monitor marine wind speed [24]. Using RS-2 fine quad-polarized mode SAR images and collocated buoy data, via a nonlinear least squares method, two C-band cross-polarized ocean backscatter models (C-2PO) relating to the equivalent neutral OSWS at 10 m height were presented by Vachon et al. [23] and Zhang et al. [24], although only the latter specifically denote their model by the acronym, 'C-2PO.' These two C-2PO models are as follows:

$$\sigma_0(cross - pol) = 0.580u_{10} - 35.652 \quad \text{(Zhang\_model)} \tag{5}$$

$$\sigma_0(cross - pol) = 0.595u_{10} - 35.60 \quad \text{(Vachon\_model)} \tag{6}$$

where $\sigma_0, cross - pol$ is the HV- or VH-polarized NRCS and u_{10} is OSWS at the 10-m height. Figure 4b shows the NRCSs simulated by both the Zhang_model and the Vachon_model, with OSWS. It can be noticed that the NRCSs increase with increasing OSWS and there is little difference between these two models. Here, we select Zhang's C-2PO model as the cross-polarized OSWS retrieval algorithm in this study. In addition, for the sake of consistency and convenience, the C-2PO model at VH-polarization is denoted VH-C2PO and the other, at HV-polarization is denoted HV-C2PO.

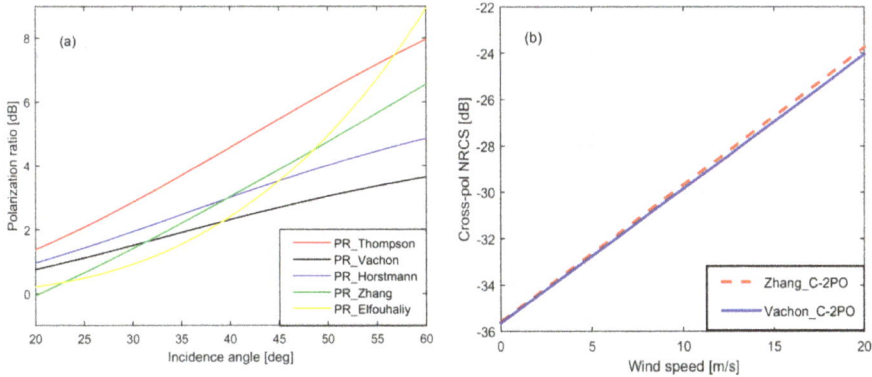

Figure 4. (**a**) Polarization ratio (PR) as a function of the radar incidence angle from the literature; (**b**) The relationship between wind speed and cross-polarized NRCS.

3. Results

3.1. OSWS Retrieval Case

In the following discussion, we apply the above-mentioned quad-polarized SAR OSWS retrieval algorithm to one case and we describe the calculation process in detail. This case is a RS-2 fine quad-polarized SLC SAR image acquired on 27 August 2012, at 10:25:24 UTC. *First*, we reconstruct the spatial resolution at 100 m and extract the corresponding NRCS and incidence angle from each pixel at each polarization. *Next*, we compute how many ERA-Interim grid cells fall within the SAR scene and then we interpolate the NRCSs and incidence angles at the coordinates for each grid cell. With this calculation, the example shown in Figure 5 has 5 grid cells, indicated by numbers S1 to S5, which fall within this SAR scene. For the cross-polarized channel, the procedure to retrieve OSWSs from the C-2PO models is relatively simple because they are only related to VH- and HV-polarized NRCS values. Figure 5 shows the SAR-retrieved wind speeds from VH- and HV-polarized image, and the corresponding NRCS distribution, on a 100-m resolution scale, without need for radar incidence angle or any external wind-direction inputs. Wind speed retrieval results can be seen in Table 1.

For co-polarized data, we first extract the external wind directions from ERA-Interim reanalysis data on 27 August 2012, at 10:00:00 UTC. Specifically, the NRCS values of the HH-polarized data need to be converted by the PR model before being input to the CMOD4 GMF. In this case, the radar incidence angles are in the range from 41.04° to 42.42° and thus, the PR value can be directly computed from Equation (4) at coordinates S1 to S5. Figure 6 shows the SAR-retrieved wind speeds from VV- and HH-polarized image, and the corresponding NRCS distribution. Based on the above parameters, wind speeds in the VV- and HH-polarized data can be calculated from CMOD4 and the results can be seen in Table 1.

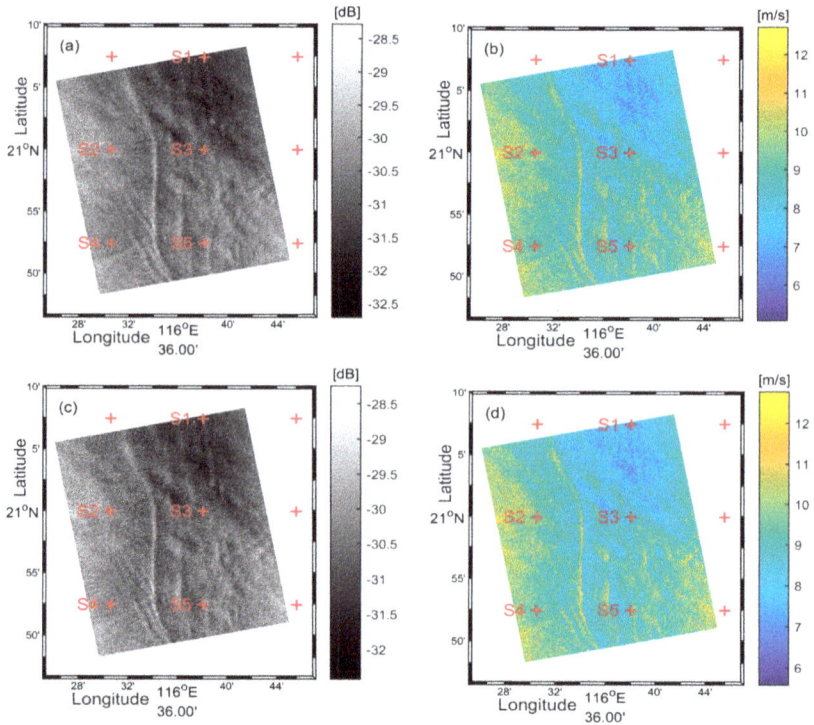

Figure 5. C-band (**a**) VH- and (**c**) HV-polarized SAR images in the South China Sea waters from RADARSAT-2 fine quad-polarization mode SLC SAR data acquired on 27 August 2012, at 10:25:24 UTC (grayscale color bar denoted NRCS). OSWS retrieved from (**b**) VH- and (**d**) HV-C2PO model. Symbol '+' denotes winds grid cells from ERA-Interim data. RADARSAT-2 Data and Product MacDonald, Detweiler and Associates Ltd., All Rights Reserved.

Table 1. OSWSs retrieved from the RS-2 fine quad-polarized mode SAR images compared with corresponding wind acquired from ERA-Interim daily (in units of m/s).

Sample	Coordinate	ERA-Interim	VH	HV	VV	HH
S1	116.625°E 21.125°N	7.3682	6.28	6.45	6.96	5.47
S2	116.50°E 21.000°N	7.9413	9.60	9.87	10.72	10.13
S3	116.625°E 21.000°N	7.3628	8.16	8.31	8.82	7.98
S4	116.50°E 20.875°N	7.2097	9.39	9.66	8.75	8.11
S5	116.625°E 20.875°N	7.3583	9.78	9.89	8.26	8.73

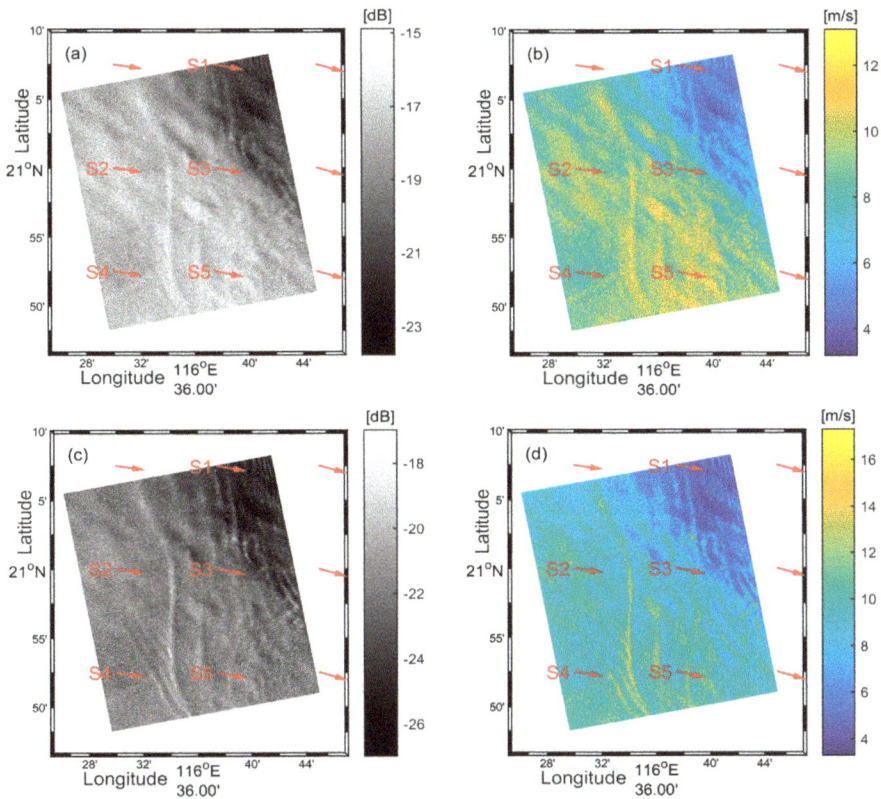

Figure 6. C-band (**a**) VV- and (**c**) HH-polarized SAR images in the South China Sea waters from RADARSAT-2 fine quad-polarization mode SLC SAR data acquired on 27 August 2012, at 10:25:24 UTC (grayscale and color bar denote NRCS). OSWS retrieved from CMOD using (**b**) VV and (**d**) HH-polarized imagery. Red arrows denote wind directions from ERA-Interim data. RADARSAT-2 Data and Product MacDonald, Detweiler and Associates Ltd., All Rights Reserved.

3.2. OSWS Retrieval Using Training Database

The comparison results for OSWS are computed for each ERA-Interim grid cell at each polarization using the *training* database. The overall *training* data are included in Figure 7, which compares the results between SAR-retrieved winds and in situ ERA-Interim winds. As can be seen, winds from all of the OSWS retrieval models exhibit a good agreement with ERA-Interim reanalysis winds at both validations sites. The RMSEs of SAR-retrieved wind speeds are all below 2.5 m/s: 2.11 m/s (VH-polarized), 2.13 m/s (HV-polarized), 1.86 m/s (VV-polarized) and 2.26 m/s (HH-polarized) and the correlation coefficients are 0.86 (VH-polarized), 0.85 (HV-polarized), 0.87 (VV-polarized) and 0.83 (HH-polarized) which are statistically significant at the 99.9% significance level, respectively. For VH- and HV-C2PO retrieval results (Figure 7a,b), the scatter plot and RMSE results are quite similar because the NRCS values in these two polarizations are quite similar. In fact, because of the monostatic property of the RADARSAT-2 SAR and the reciprocity theorem, the VH-polarized component is equal to the HV-polarized component of the Polarimetric Scattering Matrix (PSM), specifically SVH = SHV. Moreover, the NRCSs in the C-2PO model are calculated from the dual-channel intensity information of the cross-polarized SAR images and thus, $NRCS_{VH} = NRCS_{HV}$. This also means that the wind speed retrieval results from the C-2PO model for VH- and HV-polarized modes are consistent. Therefore, in the next section, we only focus on VH-polarized OSWS retrieval in cross-polarized SAR data.

In terms of OSWS retrieval from co-polarized SAR data, the VV-polarized retrieved wind speed significantly outperforms the HH-retrieved winds, as the former produces a smaller RMSE value of 1.86 m/s and correlation coefficient of 0.87 m/s, which is statistically significant at the 99.9% significance level. In addition, the RMSE value of the HH-retrieved wind speed is much larger than that reported in previous studies [20]. This fact indicates that the SAR OSWSs retrieved at these locations in the Northwest Pacific are slightly inaccurate. The most important factor in this process is the PR model. Theoretically, the most accurate PR model is that which can convert $NRCS_{HH}$ values to *exact* $NRCS_{VV}$ values. The empirical PR model that we selected in this study is empirically fit to 877 RS-2 observed PR values and incidence angles off the East and West Coasts of USA and the Gulf of Mexico, while the study area in this work is the Northwest Pacific near China. Another factor to consider is that the reference wind speed data in previous studies [20] are from National Data Buoy Center (NDBC) buoys whereas our reference wind speeds are from ERA-Interim reanalysis wind data. To sum up, for co-polarized channel, whether VV-polarized or HH-polarized, both can use CMOD4 GMF to calculate OSWS from the SAR images. Therefore, in order to obtain better accuracy for OSWS retrieval results, for co-polarized SAR data, we recommend retrieving OSWS using VV-polarized data. Additional discussion is given in the next section.

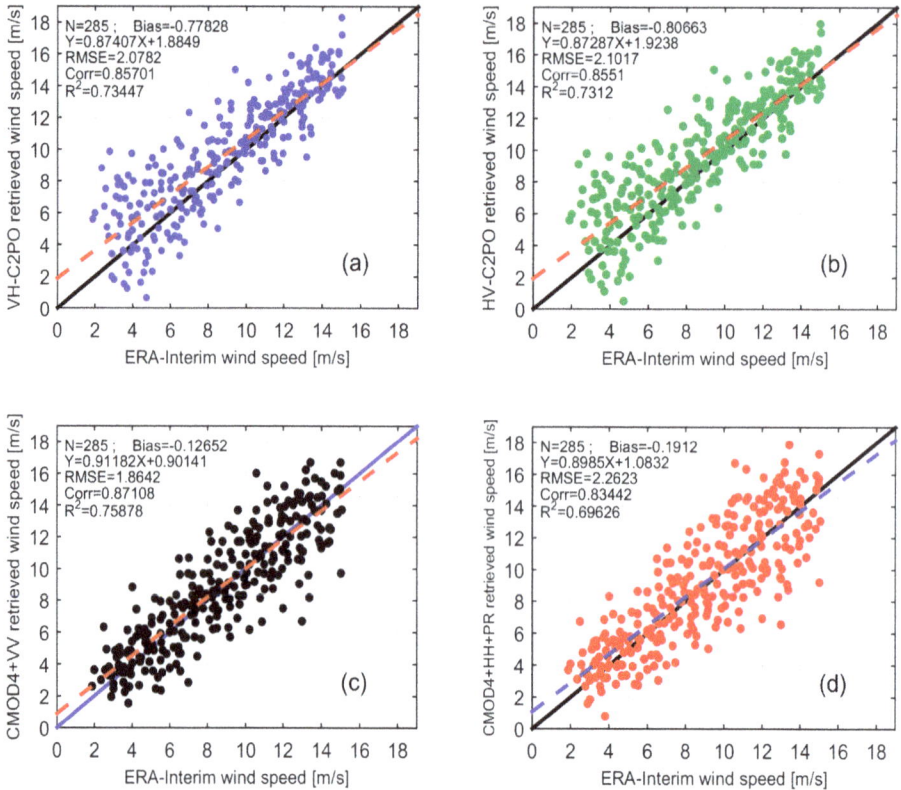

Figure 7. (**a**) VH-, (**b**) HV-, (**c**) VV- and (**d**) HH-polarized OSWS values retrieved from SAR images using corresponding GMF models compared to ERA-Interim winds data as the *training* database.

3.3. Different between C-2PO and CMOD4 GMF OSWS Retrievals

In previous studies, Vachon et al. [23] used 'their C-2PO model' to compute OSWSs using the VH- and HV-polarized channels to achieve a good agreement with wind data from NDBC buoys. However, these good results benefit from relatively high wind speed observations, to a certain degree. Similarly, Zhang et al. [24] retrieved OSWS from their C-2PO function and then compared their results with NDBC buoy measurements. The retrieved wind speeds have essentially no bias (0.04 m/s) with an RMSE of 1.39 m/s. Nevertheless, the OSWS values less than about 6 m/s are excluded in this retrieval experiment. Thus, the C-2PO model using cross-polarized data seems to be not suitable for low wind speeds but performs relatively well at moderate-to-high wind speeds. Thus far in this study, we have assumed that 8 m/s is the wind speeds threshold according to the *training* dataset. Based on this assumption, we retrieved OSWSs from C-2PO model for OSWSs greater than 8 m/s; and from CMOD4 GMF for OSWSs less than or equal to 8 m/s. Results show that the RMSEs of the SAR-retrieved OSWSs are 2.53 m/s in the former case using VH-C2PO and 1.61 m/s in the latter case using CMOD4-VV, when winds are less than or equal to 8 m/s. Similarly, the RMSEs of the SAR-retrieved OSWSs are 1.86 m/s (VH-C2PO) in the former case and 2.31 m/s (CMOD4+VV) in the latter case, when winds are less than or equal 8 m/s.

4. Discussion

4.1. A Hybrid OSWS Retrieval Algorithm Using Quad-Polarized RS-2 SAR Data

4.1.1. Methodology for Precise OSWS Threshold Based on the *Training* Dataset

The preliminary, estimated OSWS threshold (8 m/s) in the above discussion was selected as an *empirical* or *test* value. However, to better understand the appropriate scope of applications for C-2PO and CMOD4 GMF, an accurate OSWS threshold value is essential. In the next section, we put forward a method to find the best threshold, from the perspective of a quantitative analysis of the *training* dataset. The procedure is as follows:

(1) Create three one-dimensional arrays of wind speeds (OSWS): ERA-Interim, retrieved from C-2PO and retrieved from CMOD4. These three arrays have the same number of elements and one-to-one correspondence to the ERA-Interim OSWS.
(2) Calculate the maximum, minimum and length of the ERA-Interim array and denote as max_ERA, min_ERA and n, respectively;
(3) Set up OSWS threshold array from min_ERA to max_ERA in steps of 0.05 and with m as the length of these arrays.
(4) Design a double loop program. The outer loop variable is j from 1 to m and the inner loop variable is i from 1 to n;
(5) In the outer loop, the OSWS threshold value ranges from min_ERA to max_ERA in steps of 0.05 m/s. In the inner loop, we compute a new one-dimensional array when the threshold is a constant, called the hybrid OSWS array, depending on the follow rule: we select CMOD4 retrieved OSWS when ERA-Interim OSWS *less than or equal* to the reference OSWS; otherwise, we select the C-2PO retrieved OSWS, when ERA-Interim OSWS *greater than* the reference array; then, compute RMSEs between ERA-Interim OSWS and the hybrid OSWS array.
(6) Find the position of the minimum RMSE value. The reference array element corresponding to this position is the best threshold value. Figure 8 shows a sketch of this method.

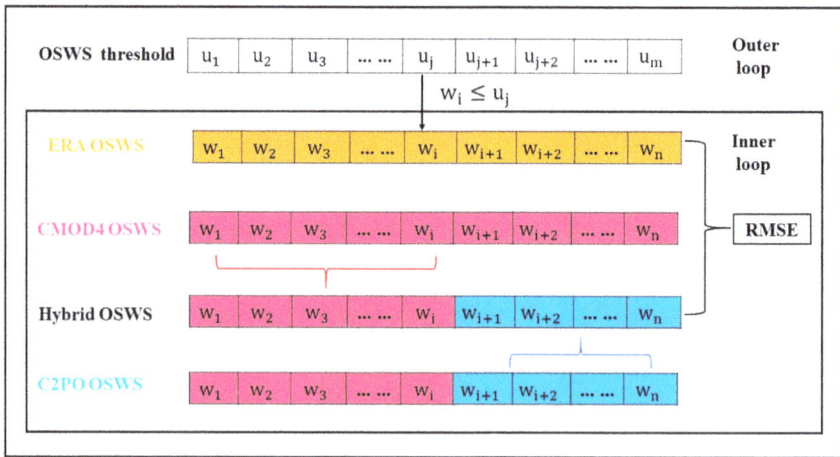

Figure 8. Design of a double loop program to find the best OSWS threshold value based on the *training* data.

Next, we calculate the most appropriate OSWS threshold value using the above method, based on our *training* dataset. Figure 9 shows the variation in RMSE between the ERA-Interim OSWS array and the hybrid OSWS array. In addition, when the loop variable is equal to 1, the hybrid OSWS array is the C2PO-retrieved OSWS array with RMSE of 2.07 m/s. Similarly, when the loop variable is equal to 285, the hybrid array is the CMOD4-retrieved OSWS array with RMSE of 1.86 m/s. These results are in complete conformity to Figure 7a,c. Finally, from Figure 9, we can estimate that the RMSE reaches a minimum of 1.59 m/s when the loop variable equals 156. The corresponding wind speed element is 9.4 m/s. Therefore, the most appropriate OSWS threshold value is 9.4 m/s based on our *training* dataset.

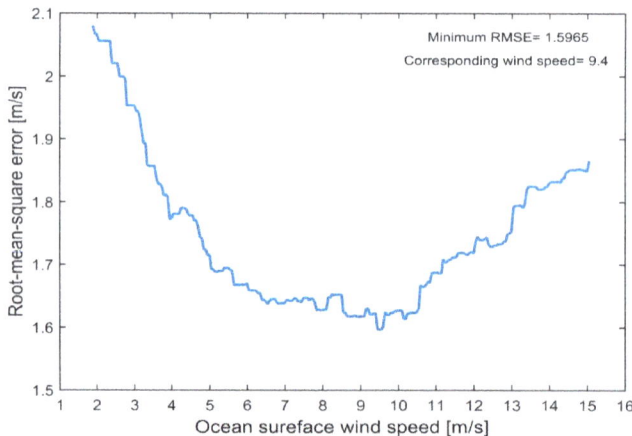

Figure 9. Variation of RMSE with different wind speeds using our hybrid OSWS retrieval model.

4.1.2. Establishment and Validation of the Hybrid OSWS Retrieval Model

According to the analysis presented previously, we computed the most appropriate OSWS threshold (9.4 m/s) based on our *training* dataset. It is important to note that the C-2PO model provides a relationship between the VH-polarized NRCS and OSWS which can be simplified as an empirical linear equation. Thus, we can use the $NRCS_{vh}$ values as a discriminant, meaning that when the OSWS is less than or equal to 9.4 m/s (corresponding $NRCS_{vh}$ is -30.2 dB), we select CMOD4+VV as our OSWSs retrieval model. By contrast, when the wind speeds are higher than 9.4 m/s, we use VH-C2PO as our OSWSs retrieval algorithm. Figure 10 shows the flowchart for the new hybrid wind speed retrieval method.

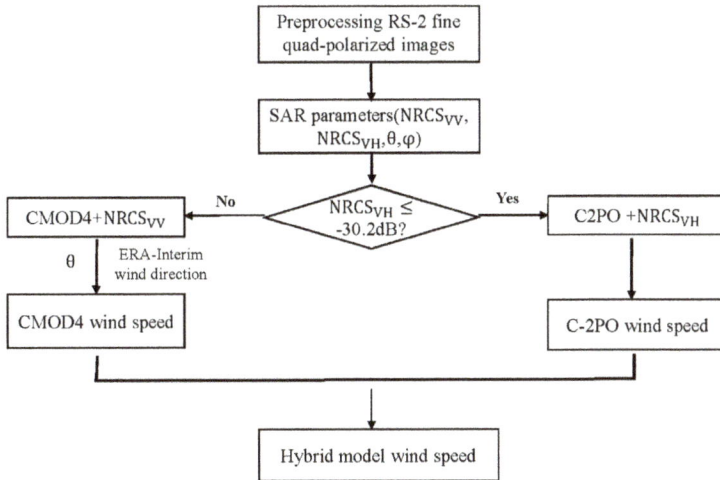

Figure 10. Flowchart for the hybrid wind speed retrieval algorithm.

In the next test, we *first* use VH-C2PO and CMOD4+VV models to retrieve OSWSs based on our empirical test database. The results are shown in Figure 11. *Subsequently*, the wind speeds retrieved from our hybrid model using the *same* data are shown in Figure 12. The RMSEs of the SAR-retrieved wind speeds are 1.92 m/s (VH-C2PO), 1.80 m/s (CMOD4+VV) and 1.66 m/s (Hybrid model), respectively. Clearly, our hybrid OSWS model has the smallest RMSE and thus it can be considered to be most suitable for wind speed retrievals at winds within the range from 1 to 16 m/s.

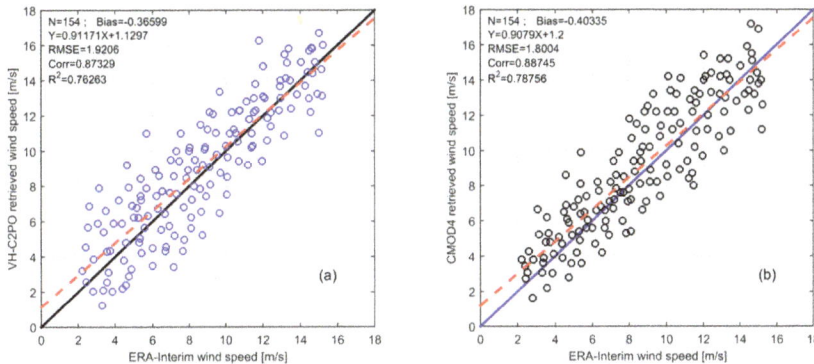

Figure 11. OSWSs retrieved from (**a**) VH-C2PO and (**b**) CMOD4+VV models from quad-polarized RADARSAT-2 images.

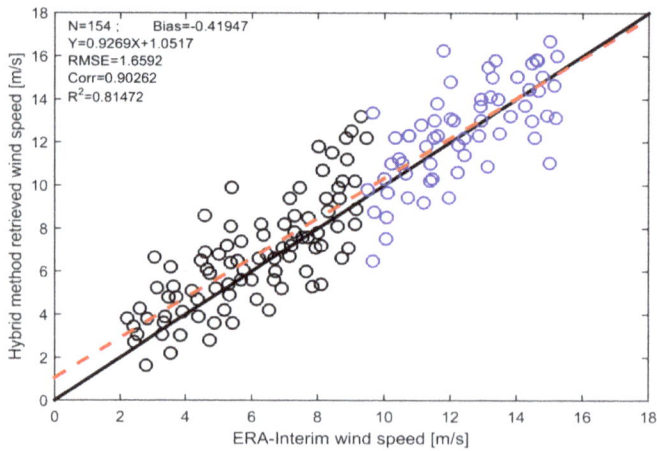

Figure 12. OSWSs retrieved from hybrid model in this study from quad-polarized RADARSAT-2 images.

4.2. Error Analysis of OSWS Retrieved Using C-2PO Model at Low-to-Moderate Winds

For low-to-moderate winds, OSWSs retrieved from the C-2PO model for VH- and HV-polarized data have a relatively large RMSE which indicates that this model has a relatively poor retrieval performance. However, SAR systems are quite complicated and thus the OSWS retrievals from SAR images can depend on a number of factors. In this section, on the basis of the underlying mechanisms for the SAR imagery, we propose three possible reasons to explain why the C-2PO model based on cross-polarized data might have a poor performance for the retrieval of low-to-moderate winds.

4.2.1. Effect of Modeling the Data from C-2PO

Based on 546 RS-2 fine quad-polarized mode SAR images and in situ weather buoys maintained by Environment Climate Change Canada (ECCC) and Fisheries and Oceans Canada (DFO) off the east and west coasts of Canada, Vachon and Wolf [23] first proposed a new C-band cross-polarized empirical model. This model, as yet unnamed, suggested that the relationship between NRCS and OSWS is independent of wind direction and incidence angle and that there is no saturation effect at high wind speeds and that it can directly retrieve OSWS. Note that the data source for the establishment of this model is from higher wind speed observations which simplifies wind speed retrieval from SAR imagery for sufficiently high wind speeds.

Within the following year, independent of Vachon's work, Zhang and Perrie [24] developed a C-band cross-polarization ocean model, which they denoted as C-2PO, using the RS-2 fine quad-polarized mode SAR measurements for high (>20 m/s) wind retrievals. Zhang and Perrie selected 534 RS-2 SAR images collocated with NDBC buoy measurements under different sea states and retrieved wind speeds from C-2PO model. The retrieved wind speeds have essentially no bias (0.04 m/s) with an RMS error of 1.39 m/s. However, these good results exclude the wind speeds less than about 6 m/s. Thus, from the point of the modeling, the C-2PO model may not be suitable for low wind speed retrievals from SAR images.

4.2.2. Effect of the Noise Level

For the low wind speed retrieved from SAR data, VV-polarized CMOD4 GMF performs better than the VH-polarized C-2PO model. The reasons are related to noise level are as follows: (1) the noise level (floor) is the same value at VV- and VH-polarized mode in the same pixels, (2) VV-polarized NRCS values are much stronger than the VH-polarized NRCS values under the same wind conditions and (3) VH-polarized NRCS values are close to the noise level whereas VV-polarized NRCS values are much larger that the noise level and thus the VH-polarized NRCS values are sometimes annihilated by the noise level.

In terms of cross-polarized SAR wind speed retrieval, the C-2PO model is applied to the NRCSs without removing the noise level because of the complicated relationship of the Signal-to Noise Ratio (SNR) [24,26]. In fact, the NRCS values induced by local winds are close to the noise level values, especially under low wind speed conditions (Figure 13). One reason is that the actual noise level for RADARSAT-2 fine quad-polarized SAR data in an individual pixel is unknown and only the estimated noise level is provided. The other reason is that the cross-polarized NRCSs induced by the low wind might be above the actual noise level. However, sometimes, these NRCSs may be beneath the noise level. If we remove the estimated noise level for all pixels, we cannot apply C-2PO model to retrieve the wind speed.

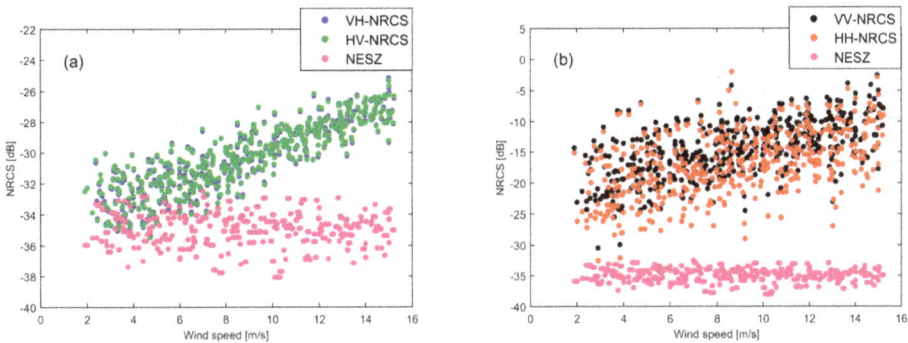

Figure 13. Sigma naught versus in situ ERA-Interim wind speed for (**a**) cross- and (**b**) co-polarized data.

4.2.3. Effect of the Wind-Roughness Relationship

In previous studies focusing on linear polarizations [37], several EM mechanisms are relevant to the radar backscatter denoted as the NRCS from the ocean surface: (1) Bragg resonance scattering, (2) quasi-specular reflection and (3) diffraction of radio waves on sharp wedges. Generally speaking, Bragg resonance scattering mechanisms, as related to ocean surface roughness and quasi-specular reflection and diffraction of radio waves, are considered in relation to wave breaking.

Bragg resonance plays a main role in the VV-polarized NRCS but is negligible for the VH-polarization NRCS [38]. In addition, non-Bragg scattering dominates the VH-polarized NRCS but is negligible for the VV-polarized NRCS. A summary is shown in Table 2 for the roles of Bragg and non-Bragg resonance scattering mechanisms with respect to the VH and VV polarizations. For SAR imaging under the low wind speed conditions, VV-polarized NRCS values mainly depend on the ocean surface roughness, which can be described by the Bragg resonance. The relationship among winds, roughness and NRCS values is 'stable' and thus the VV-polarized NRCSs (CMOD4) apply to low wind speed retrieval. The VH-polarized imaging depends on the wave breaking mechanism. However, waves induced by winds are not easily broken at low wind speeds. The relationship among winds, roughness and NRCS values is 'instable' and thus the VH-polarized NRCSs model (C-2PO) is not suitable for low wind speed retrieval.

Table 2. The roles of the two mechanisms with respect to the VH and VV polarizations.

Mechanisms	VH-Polarized	VV-Polarized
Bragg Resonance	Negligible	Main
Non-Bragg	Main	Negligible

4.2.4. Effect of the Reconstructed Spatial Resolution

Direct calculation of OSWSs from the original SAR scene can result in noisy patterns due to the presence of speckle noise in the raw SAR images. Therefore, the raw RS-2 SAR data needs to be preprocessed to reconstruct an appropriate spatial resolution. For this reason, we make a 20 × 20 pixel boxcar averaging of the NRCS, in each polarization, so that the reconstructed pixel spacing is 100 m. Previous research has suggested that the reconstructed spatial resolution has an effect on the accuracy of OSWS retrieval with the C-2PO model for the cross-polarized channels from RS-2 fine quad-polarized images [39]. To investigate this phenomenon thoroughly, we selected one SAR image under very low wind speed conditions. This case is a RS-2 quad-polarized SLC SAR image acquired on 29 April 2012, at 05:33:24 UTC and collocated with a NDBC buoy (#46035; 57°1′33″N, 177°44′16″W) in the Bering Sea. In addition, most of the NDBC buoy anemometers are installed at a height of 5 m. Therefore, the OSWS from the 5-m anemometers is converted to OSWS at 10-m height using the power-law wind profile method under near-neutral stability conditions [40]. Finally, the NDBC buoy-measured 10-m OSWS is 2.93 m/s on 29 April 2012, at 05:30:00 UTC.

In the next step, we first reconstruct the spatial resolution at VH- and VV-polarized mode at 100, 600, 1100, 1600, 2100 and 3100 m, respectively. Various SAR data including NRCSs, incidence angles and external wind directions are interpolated to the buoy location. VH-polarized wind speeds retrieved from the C-2PO model can be directly computed from Equation (5). The results of the SAR-retrieved wind speeds are 7.72 m/s (100 m), 6.18 m/s (600 m), 5.14 m/s (1100 m), 4.67 m/s (1600 m), 6.46 m/s (2100 m) and 6.69 (3100 m) and the differences are 4.79 m/s (100 m), 3.25 m/s (600 m), 2.21 m/s (1100 m), 1.64 m/s (1600 m), 2.33 m/s (2100 m) and 3.19 m/s (3100 m), respectively. A more intuitive and straightforward assessment of the influence of influence of different spatial scales is given in Figure 14., which shows the reconstructed SAR images and corresponding retrieved OSWS at spatial resolutions of 100, 600, 1100, 1600, 2100 and 3100 m for the VH-polarized channel.

Obviously, the retrieved OSWS values from C-2PO model at 1600 m spatial resolution have the smallest difference and thus the optimal resolution from VH-polarized data is 1600 m, in this case. This phenomenon can be explained by the fact that the speckles in the SAR images have been suppressed by averaging the pixel spacing to coarser resolution and the wind field is smoothed by the increase in the wind cell spacing.

For VV-polarized data, the ERA-Interim reanalysis wind field are used as external wind directions acquired on 29 April 2012, at 00:00:00 UTC. SAR retrieved wind speeds from CMOD4 GMF can be taken from Equation (2). The results of the SAR-retrieved wind speeds are 6.6 m/s (100 m), 4.69 m/s (600 m), 4.91 m/s (1100 m), 5.25 m/s (1600 m), 5.6 m/s (2100 m) and 5.95 m/s (3100 m) and the differences are 3.67 m/s (100 m), 1.76 m/s (600 m), 1.98 m/s (1100 m), 2.32 m/s (1600 m), 2.67 m/s (2100 m) and 3.02 m/s (3100 m), respectively. Similarly, Figure 15 shows the reconstructed SAR images and corresponding retrieved OSWS values at spatial resolutions of 100, 600, 1100, 1600, 2100 and 3100 m for the VV-polarized channel. The CMOD4 GMF retrieves OSWS at 600 m spatial resolution has the smallest difference and thus, the optimal resolution in VV-polarized mode is 600 m, in this case.

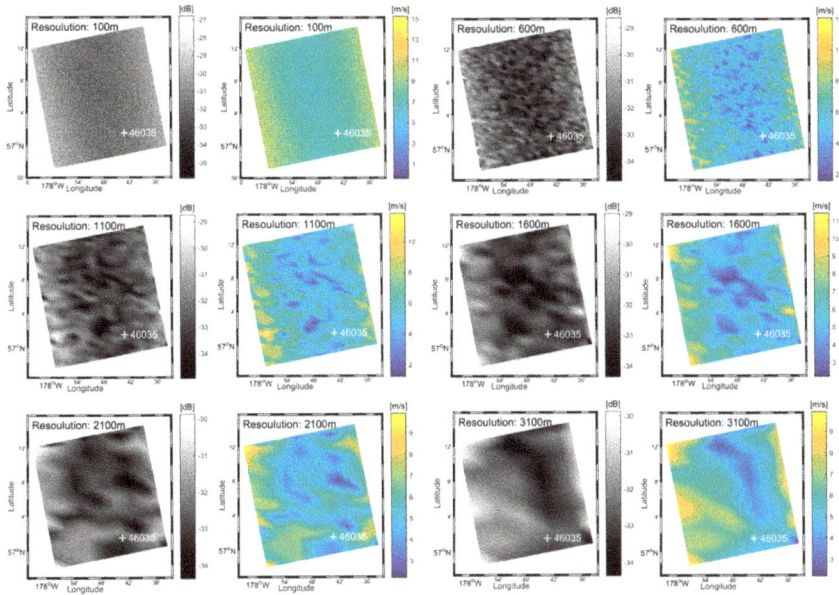

Figure 14. VH-polarized SAR images at reconstructed spatial resolutions of 100, 600, 1100, 1600, 2100 and 3100 m and corresponding OSWSs retrieved from C2PO model. The in-situ buoy (#46035, 57°1′33″N 177°44′16″W) wind speed is 2.92 m/s. RADARSAT-2 Data and Product MacDonald, Detteiler and Associates Ltd., All Rights Reserved.

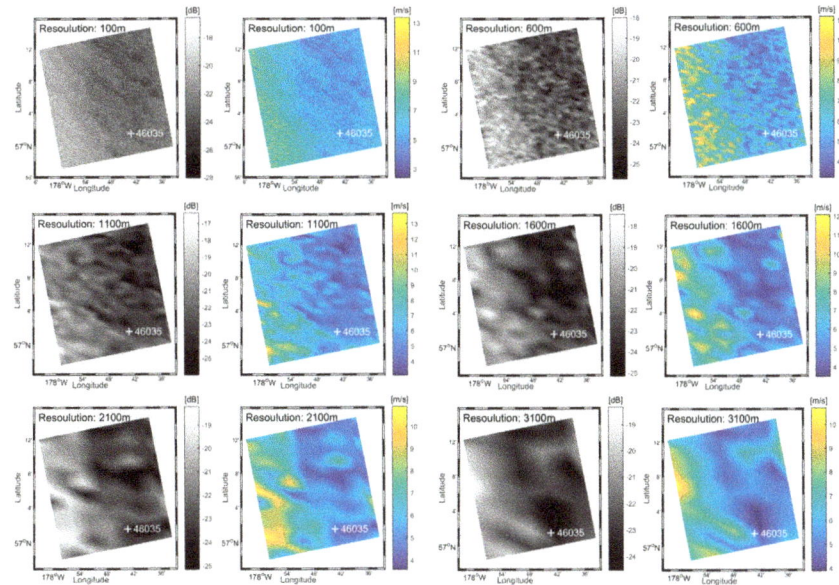

Figure 15. VV-polarized SAR images at reconstructed spatial resolutions of 100, 600, 1100, 1600, 2100 and 3100 m and corresponding OSWSs retrieved from the C2PO model. The in-situ buoy (#46035, 57°1′33″N 177°44′16″W) wind speed is 2.92 m/s. RADARSAT-2 Data and Product MacDonald, Detteiler and Associates Ltd., All Rights Reserved.

Compared with OSWS retrieval from VV-polarized SAR mode data, OSWS can be retrieved with better accuracy (at 1600 m) for the VH-polarized channel for low winds. However, the results appear to exhibit 'instability' for cross-polarized OSWS retrieval with the C-2PO model for the RS-2 fine quad-polarization data. The simple explanation is as follows: OSWSs retrieved from the SAR image are based on the Bragg scattering theory and thus the accuracy of retrieved wind speeds are closely related to the radar backscatter signal. In terms of the backscatter signal, the intensity at VH-polarized channel is far less than the intensity at the VV-polarized channel. With increased pixel averaging, the speckle is reduced. Although the SNR for the VH-polarized channel becomes stronger when the noise is reduced, the changes are small because the backscatter signal intensity is itself quite weak. As a result, OSWSs retrieved from C-2PO model for VH-polarized mode have 'instability', with the change of spatial resolution (due to averaging) in low winds. For the VV-polarized channel, the signal intensity is strong and thus the speckle noise has little effect, with the decrease in spatial resolution. Therefore, OSWSs retrieved from the CMOD4 GMF for VV-polarized mode data are more accurate with the change of spatial resolution due to appropriate averaging, in relatively low winds.

5. Conclusions

In this paper, ocean surface wind speed measurements made by RADARSAT-2 in cross-polarized and co-polarized modes were analyzed, using 439 samples from 92 fine quad-polarization SAR images and corresponding ERA-Interim winds in the Northwest Pacific in waters off the coast of China. We *first* created two subset wind speed databases: the *training* and *testing* subsets. From the *training* data subset, we retrieve OSWSs from different GMF models for different polarizations, as appropriate and we compared the results with corresponding ERA-Interim winds. The RMSEs of SAR-retrieved wind speeds are all below 2.5 m/s: specifically, 2.11 m/s (VH-polarized), 2.13 m/s (HV-polarized), 1.86 m/s (VV-polarized) and 2.26 m/s (HH-polarized) and the correlation coefficients are 0.86 (VH-polarized), 0.85 (HV-polarized), 0.87 (VV-polarized) and 0.83 (HH-polarized), which are statistically significant at the 99.9% significance level, respectively.

Through analysis of the SAR data considered in this study, we have presented the advantages and disadvantages for SAR wind retrieval models for cross- and co-polarized data and the respective wind speed ranges for reliable application. We found that OSWS retrieved using the C-2PO model for VH-polarized data are most suitable for moderate-to-high winds while the CMOD4 GMF at VV-polarized tends to be best for low-to-moderate winds. In addition, under higher wind conditions, such as generated by hurricanes, many studies have suggested that the NRCS in cross-polarization mode essentially does not saturate [19,23,25,26,28]. Thus, the cross-polarized channel is more appropriate for retrieval of high winds, as may be generated by hurricanes. To better understand the appropriate scope of applications for C-2PO and CMOD4 GMF, an accurate OSWS threshold value algorithm is proposed based on our *training* dataset, from the perspective of a quantitative analysis. According to the analysis results, a hybrid methodology is put forward and applied to the test data subset. The results show that the accuracy of the retrieved OSWSs from our hybrid method can significantly outperform C-2PO or the CMOD4 models, producing an RMSE of 1.66 m/s and correlation coefficient of 0.9. Finally, we proposed four possible reasons to explain why the C-2PO model based on the cross-polarized retrieved wind speeds has a rather poor performance at low wind speeds. They are modeling, noise level, wind-roughness relationship and reconstructed spatial resolution.

From the perspective of data analysis and physical mechanism, we put forward the hybrid OSWS retrieval model, which provide readers a new idea to retrieve OSWS from C-band quad-polarized SAR images. However, for OSWS retrieved from quad-polarized SAR images, there are still some deficiencies in using reference wind speeds from the ERA-Interim winds data in this paper. For example, the spatial resolution retrieved from SAR images may be not comparable to the resolution of wind grid cells from the reference data. In future work, in order to set up a more accurate OSWS retrieval model, we will take real in-situ (i.e., buoys) as our reference wind data. Besides, from a model perspective, whether C-2PO or CMOD GMFs are used, they are empirical formulations based on relationships between the NRCSs and wind speed. In the future, we plan to make a more thorough comparison of all the model functions and include comparisons using multiple remotely sensed datasets and additional buoy measurements to improve results. There is hope that a new C-band wind speed retrieval model will provide improved retrievals, without having to consider differing wind intensities, such as low, moderate, high and so forth.

Author Contributions: The first author of this paper, H.F., wrote the manuscript. T.X. conceived and designed the experiments. H.F. and G.Z. performed the experiments. J.Y. and Y.H. supervised the work. W.P. reviewed, edited and provided valuable suggestions and inputs for the final manuscript. H.F. and W.P. wrote the paper.

Funding: This research received no external funding other than what is acknowledging below.

Acknowledgments: This work was supported by the National Key R&D Program of China (2016YFC1401007), the Global Change Research Program of China (2015CB953901), the National Natural Science Foundation of China project (41776181), the Jiangsu Meteorological Bureau 2014 modernization project "Jiangsu Ocean meteorological Integrated Service System", the Postgraduate Research & Practice Innovation Program of Jiangsu Province (KYCX18_1012), the China Scholarship Council for 1 year's study at the Bedford Institute of Oceanography, the National Natural Science Youth Foundation of China (41706193), the Canadian Space Agency DUAP Program "Winds from SAR" RCM Readiness Project and the Canadian Office of Energy Research and Development.

Conflicts of Interest: The authors declare no conflict of interest.

References

1. Monaldo, F.M.; Li, X.; Pichel, W.G.; Jackson, C.R. Ocean wind speed climatology from spaceborne SAR imagery. *Bull. Am. Meteorol. Soc.* **2014**, *95*, 565–569. [CrossRef]
2. Duan, B.; Zhang, W.; Yang, X. Assimilation of Typhoon Wind Field Retrieved from Scatterometer and SAR Based on the Huber Norm Quality Control. *Remote Sens.* **2017**, *9*, 987. [CrossRef]
3. Xie, T.; Perrie, W.; He, Y.; Li, H.; Fang, H.; Zhao, S.; Yu, W. Ocean surface wave measurements from fully polarimetric SAR imagery. *Sci. China Earth Sci.* **2015**, *58*, 1849–1861. [CrossRef]
4. Shen, H.; Perrie, W.; He, Y.; Liu, G. Wind speed retrieval from VH dual-polarization RADARSAT-2 SAR images. *IEEE Trans. Geosci. Remote Sens.* **2014**, *52*, 5820–5826. [CrossRef]
5. Fang, H.; Xie, T.; Perrie, W.; Zhao, L.; Yang, J.; He, Y. Ocean Wind and Current Retrievals Based on Satellite SAR Measurements in Conjunction with Buoy and HF Radar Data. *Remote Sens.* **2017**, *9*, 1321. [CrossRef]
6. Denbina, M.; Collins, M.J. Wind speed estimation using C-band compact polarimetric SAR for wide swath imaging modes. *ISPRS J. Photogramm. Remote Sens.* **2016**, *113*, 75–85. [CrossRef]
7. Liu, G.; Yang, X.; Li, X.; Zhang, B.; Pichel, W.; Li, Z.; Zhou, X. A systematic comparison of the effect of polarization ratio models on sea surface wind retrieval from C-band synthetic aperture radar. *IEEE J. Sel. Top. Appl. Earth Obs. Remote Sens.* **2013**, *6*, 1100–1108. [CrossRef]
8. Jagdish; Kumar, S.; Chakraborty, A.; Kumar, R. Validation of wind speed retrieval from RISAT-1 SAR images of the North Indian Ocean. *Remote Sens. Lett.* **2018**, *9*, 421–428. [CrossRef]
9. Stoffelen, A.; Anderson, D. Scatterometer data interpretation: Derivation of the transfer function CMOD4. *J. Geophys. Res.* **1997**, *102*, 5767–5780. [CrossRef]
10. Quilfen, Y.; Chapron, B.; Elfouhaily, T.; Katsaros, K.; Tournadre, J. Observation of tropical cyclones by high-resolution scatterometry. *J. Geophys. Res.* **1998**, *103*, 7767–7786. [CrossRef]
11. Hersbach, H.; Stoffelen, A.; Haan, S. An improved C-band scatterometer ocean geophysical model function: CMOD5. *J. Geophys. Res. Oceans* **2007**, *112*, C03006. [CrossRef]
12. Mouche, A.; Chapron, B.; Zhang, B.; Husson, R. Combined Co- and Cross-Polarized SAR Measurements under Extreme Wind Conditions. *IEEE Trans. Geosci. Remote Sens.* **2017**, *55*, 6746–6755. [CrossRef]
13. Stoffelen, A.; Verspeek, J.A.; Vogelzang, J.; Verhoef, A. The CMOD7 geophysical model function for ASCAT and ERS wind retrievals. *IEEE J. Sel. Top. Appl. Earth Obs. Remote Sens.* **2017**, *10*, 2123–2134. [CrossRef]
14. Thompson, D.R.; Elfouhaily, T.M.; Chapron, B. Polarization ratio for microwave backscattering from the ocean surface at low to moderate incidence angles. In Proceedings of the 1998 IEEE International Geoscience and Remote Sensing Symposium Proceedings (IGARSS'98), Seattle, WA, USA, 6–10 July 1998; Volume 3, pp. 1671–1673.

15. Vachon, P.W.; Dobson, F.W. Wind retrieval from RADARSAT SAR images: Selection of a suitable C-band HH polarization wind retrieval model. *Can. J. Remote Sens.* **2000**, *26*, 306–313. [CrossRef]

16. Horstmann, J.; Koch, W.; Lehner, S.; Tonboe, R. Wind retrieval over the ocean using synthetic aperture radar with C-band HH polarization. *IEEE Trans. Geosci. Remote Sens.* **2000**, *38*, 2122–2131. [CrossRef]

17. Mouche, A.A.; Hauser, D.; Daloze, J.F.; Guerin, C. Dual-polarization measurements at C-band over the ocean: Results from airborne radar observations and comparison with ENVISAT ASAR data. *IEEE Trans. Geosci. Remote Sens.* **2005**, *43*, 753–769. [CrossRef]

18. Johnsen, H.; Engen, G.; Guitton, G. Sea-surface polarization ratio from Envisat ASAR AP data. *IEEE Trans. Geosci. Remote Sens.* **2008**, *46*, 3637–3646. [CrossRef]

19. Shen, H.; Perrie, W.; He, Y. On SAR wind speed ambiguities and related geophysical model functions. *Can. J. Remote Sens.* **2009**, *35*, 310–319. [CrossRef]

20. Zhang, B.; Perrie, W.; He, Y. Wind speed retrieval from RADARSAT-2 quad-polarization images using a new polarization ratio model. *J. Geophys. Res. Atmos.* **2011**, *116*. [CrossRef]

21. Hwang, P.A.; Zhang, B.; Toporkov, J.V.; Perrie, W. Comparison of composite Bragg theory and quad-polarization radar backscatter from RADARSAT-2, with applications to wave breaking and high wind retrieval. *J. Geophys. Res. Oceans* **2010**, *115*. [CrossRef]

22. Hwang, P.A.; Perrie, W.; Zhang, B. Cross-polarization radar backscattering from the ocean surface and its dependence on wind velocity. *IEEE Trans. Geosci. Remote Sens. Lett.* **2014**, *11*, 2188–2192. [CrossRef]

23. Vachon, P.W.; Wolfe, J. C-band cross-polarization wind speed retrieval. *IEEE Trans. Geosci. Remote Sens. Lett.* **2011**, *8*, 456–458. [CrossRef]

24. Zhang, B.; Perrie, W.; Vachon, P.W.; Li, X.; Pichel, W.G.; Guo, J.; He, Y. Ocean vector winds retrieval from C-band fully polarimetric SAR measurements. *IEEE Trans. Geosci. Remote Sens.* **2012**, *50*, 4252–4261. [CrossRef]

25. Zhang, B.; Perrie, W. Cross-polarized synthetic aperture radar: A new potential technique for hurricanes. *Bull. Am. Meteorol. Soc.* **2012**, *93*, 531–541. [CrossRef]

26. Horstmann, J.; Falchetti, S.; Wackerman, C.; Maresca, S.; Caruso, M.J.; Graber, H.C. Tropical cyclone winds retrieved from C-band cross-polarized synthetic aperture radar. *IEEE Trans. Geosci. Remote Sens.* **2015**, *53*, 2887–2898. [CrossRef]

27. Zhang, G.; Li, X.; Perrie, W.; Hwang, P.A.; Zhang, B.; Yang, X. A Hurricane Wind Speed Retrieval Model for C-Band RADARSAT-2 Cross-Polarization ScanSAR Images. *IEEE Trans. Geosci. Remote Sens.* **2017**, *55*, 4766–4774. [CrossRef]

28. Touzi, R.; Vachon, P.W.; Wolfe, J. Requirement on antenna cross-polarization isolation for the operational use of C-band SAR constellations in maritime surveillance. *IEEE Trans. Geosci. Remote Sens.* **2010**, *7*, 861–865. [CrossRef]

29. La, T.V.; Khenchaf, A.; Comblet, F.; Nahum, C. Assessment of Wind Speed Estimation from C-Band Sentinel-1 Images Using Empirical and Electromagnetic Models. *IEEE Trans. Geosci. Remote Sens.* **2018**, *56*, 4075–4087. [CrossRef]

30. Lehner, S.; Schulz-Stellenfleth, J.; Schättler, B.; Breit, H.; Horstmann, J. Wind and wave measurements using complex ERS-2 SAR wave mode data. *IEEE Trans. Geosci. Remote Sens.* **2000**, *38*, 2246–2257. [CrossRef]

31. Horstmann, J.; Schiller, H.; Schulz-Stellenfleth, J.; Lehner, S. Global wind speed retrieval from SAR. *IEEE Trans. Geosci. Remote Sens.* **2003**, *41*, 2277–2286. [CrossRef]

32. Carvalho, D.; Rocha, A.; Gómez-Gesteira, M. Ocean surface wind simulation forced by different reanalyses: Comparison with observed data along the Iberian Peninsula coast. *Ocean Model.* **2012**, *56*, 31–42. [CrossRef]

33. Dee, D.P.; Uppala, S.M.; Simmons, A.J.; Berrisford, P.; Poli, P.; Kobayashi, S.; Andrae, U.; Alonso-Balmaseda, M.; Balsamo, G.; Bauer, P.; et al. The ERA-Interim reanalysis: Configuration and performance of the data assimilation system. *Q. J. R. Meteorol. Soc.* **2011**, *137*, 553–597. [CrossRef]

34. Xu, Q.; Lin, H.; Li, X.F.; Zheng, Q.; Pichel, W.; Liu, Y. Assessment of an analytical model for sea surface wind speed retrieval from spaceborne SAR. *Int. J. Remote Sens.* **2010**, *31*, 993–1008. [CrossRef]

35. Komarov, A.; Zabeline, V.; Barber, D. Ocean surface wind speed retrieval from C-band SAR images without wind direction input. *IEEE Trans. Geosci. Remote Sens.* **2014**, *52*, 980–990. [CrossRef]

36. Han, B.; Xu, X.; Li, H. Wind Speed Retrieval of Ocean Surface Using Radarsat-2 Co-polarization Data. *Remote Sens. Technol. Appl.* **2017**, *32*, 419–426.

37. Kudryavtsev, V.; Hauser, D.; Caudal, G.; Chapron, B. A semiempirical model of the normalized radar cross-section of the sea surface 1. Background model. *J. Geophys. Res.* **2003**, *108*. [CrossRef]
38. Zhang, G.; Perrie, W.; Li, X.; Zhang, J.A. A hurricane morphology and sea surface wind vector estimation fmodel based on C-band cross-polarization SAR imagery. *IEEE Trans. Geosci. Remote Sens.* **2017**, *55*, 1743–1751. [CrossRef]
39. Zhang, K.; Xu, X.; Han, B.; Mansaray, L.R.; Guo, Q.; Huang, J. The Influence of Different Spatial Resolutions on the Retrieval Accuracy of Sea Surface Wind Speed with C-2PO Models Using Full Polarization C-Band SAR. *IEEE Trans. Geosci. Remote Sens.* **2017**, *55*, 5015–5025. [CrossRef]
40. Ren, Y.; Li, X.; Zhou, G. Sea surface wind retrievals from SIR-C/X-SAR data: A revisit. *Remote Sens.* **2015**, *7*, 3548–3564. [CrossRef]

remote sensing

MDPI

Article

Group Line Energy in Phase-Resolved Ocean Surface Wave Orbital Velocity Reconstructions from X-band Doppler Radar Measurements of the Sea Surface

Andrew J. Kammerer [1,*] and Erin E. Hackett [2]

[1] Marine Meteorology Division, Naval Research Laboratory, Monterey, CA 93943, USA
[2] Department of Coastal and Marine Systems Science, Coastal Carolina University, Conway, SC 29528, USA;
    ehackett@coastal.edu
*   Correspondence: Andrew.Kammerer@nrlmry.navy.mil; Tel.: +1-831-656-4780

Received: 2 November 2018; Accepted: 24 December 2018; Published: 2 January 2019

Abstract: The wavenumber-frequency spectra of many radar measurements of the sea surface contain a linear feature at frequencies lower than the first order dispersion relationship commonly referred to as the "group line". Plant and Farquharson, showed numerically that the group line is at least partially caused by wave interference-induced breaking of steep short gravity waves. This paper uses two wave retrieval techniques, proper orthogonal decomposition (POD) and FFT-based dispersion curve filtering, to examine two X-band radar datasets, and compare wave orbital velocity reconstructions to ground truth wave buoy measurements within the field of view of the radar. POD allows group line energy to be retained in the reconstruction, while dispersion curve filtering removes all energy not associated with the first order dispersion relationship. Results show that when group line energy is higher or comparable to dispersion curve energy, the inclusion of this group line energy in phase-resolved orbital velocity reconstructions increases the accuracy of the reconstruction. This increased accuracy is demonstrated by higher correlations between POD reconstructed time series with buoy ground truth measurements than dispersion curve filtered reconstructions. When energy lying on the dispersion relationship is much higher than the group line energy, the FFT and POD reconstruction methods perform comparably.

Keywords: Doppler radar; radar; sea surface roughness; air-sea interaction; proper orthogonal decomposition; ocean surface waves; dispersion curve filtering; marine X-band radar; phase-resolved wave fields

---

1. Introduction

In the last several decades, sea clutter has transitioned from a nuisance of operating radar systems in marine environments to a useful tool for quantitative measurements of ocean surface waves. This transition began with the seminal paper by Young et al. in 1985 [1], establishing the Fourier-based dispersion curve filtering method. In the years since this paper was published, techniques derived from Young et al. [1] for calculation of wave statistics have become well established [2–5]. However, methods and validation for production of phased-resolved wave fields remains an open area of research [6–9].

Most wavenumber – frequency (k-ω) spectra of radar images of the sea surface exhibit a "group line" feature: a low frequency feature below the first order dispersion relationship that passes through the origin. Numerous explanations exist for the origin of this feature that are wave-related and non-wave related, such as: shadowing, non-linear wave-wave interactions, contamination (e.g., hard targets in the images), turbulence advected by winds, interference-induced wave breaking, and non-linear scattering effects [10–16]. Recently, in their 2012 study, Plant and Farquharson provide numerical evidence that interference induced wave breaking from the interaction of linear wave fields are a primary source of the group line feature [10]. They demonstrate that the superposition of wind waves and swell can generate steep short gravity waves that break near the local maxima of surface slope, resulting in Doppler measurements of the phase speed of these steep short gravity waves. Such effects as noted by Plant and Farquharson as well as non-linear second order wave-wave interactions may be features that should be accounted for in the generation of instantaneous (phase-resolved) sea surface elevation maps produced by radar [9,13]. However, it is difficult to separate these wave contributions to the group line feature to validate whether they are important to the reconstruction of sea surface elevation maps because other non-wave related effects may also occupy the same frequency-wavenumber space (e.g., shadowing).

In this study, we use a non-Fourier based method, proper orthogonal decomposition (POD), to reconstruct ocean surface orbital velocities from X-band Doppler measurements of the sea surface [17]. This method permits the inclusion of some of the spectral energy in the group line feature in the reconstruction of instantaneous orbital velocities. The inclusion of the group line energy is based on how much it contributes to the overall variance of the measured spatial series [17]. In order to evaluate the importance of group line associated energy to accurate phase-resolved reconstructions of ocean surface wave orbital velocities, the POD results are compared to orbital velocity maps produced using the conventional Fourier-based method (FFT), which filters the k-ω spectra on the linear dispersion relationship for surface gravity waves, and therefore removes all group line features.

We show that inclusion of some portion of the energy in the group line does improve correlation with GPS wave buoy ground truth orbital velocity time series measurements, although this energy does not greatly impact wave statistics (e.g., significant wave height) when computed over approximately 20 min time periods [17]. The results show higher correlation with buoy time series when including group line energy, provided that group line energy is comparable or higher than the energy on the dispersion relationship. These results are demonstrated with experimental data of bimodal seas. The results of this study support the numerical findings of Plant and Farquharson [10] with experimental data, and show that at least a portion of the group line energy is wave field related and contributes to accurate instantaneous phase-resolved sea surface orbital velocities. Presumably by inference, the interference pattern referred to by Plant and Farquharson [10] impacts the wave phasing, and should be included for accurate instantaneous sea surface wave retrievals.

2. Experimental Data

Experimental data for this study was collected during the *R/V Melville* experiment, which took place from 14 September to 17 September 2013. Details of this experiment are described in Kammerer and Hackett [17]; only relevant details are provided here. The data used in this study was collected using a ship-mounted rotating radar system with a center frequency of 9.41 GHz. It was vertically (VV) polarized, coherent-on-receive [18], and rotated at 24 RPM. Doppler velocity is calculated using the pulse-pair processing method [19]. For noise reduction, the mean of 12 pulse-pairs are taken, generating a Doppler velocity range distribution every 0.86° of rotation (or every 0.006 s). The resulting Doppler velocity distributions are a function of range (r), time (t), and azimuth (φ) ($D(r,t,\varphi)$), and cover a range of 960 m at a resolution of 3.75 m with a blanking range of 100 m around the vessel. One "frame" of Doppler data is produced with every full revolution of the radar system.

GPS mini wave buoys developed by the *Coastal Observing and Research and Development Center* at the *Scripps Institution of Oceanography* [20] were deployed during the R/V Melville experiment to record ground truth wave data for comparison and validation of radar reconstructions. These buoys were free drifting in and around the area of operation. Only wave buoys in the field of view (FOV) of the radar are used for this study. Figure 1 shows an example of buoy tracks during collection of a radar dataset. The GPS wave buoys have a sample rate of 1 Hz. Prior to time series processing, the buoy data are high pass filtered with a cut off frequency of 0.05 Hz (period of 20 seconds) and de-trended to eliminate non-wave low frequency signals mostly related to the transitioning of GPS satellites. To facilitate comparisons with the radar data, the buoy time series are then low pass filtered and down-sampled to match the temporal resolution of the radar data. Measurement uncertainty and other information about the GPS wave buoys can be found in Drazen [20].

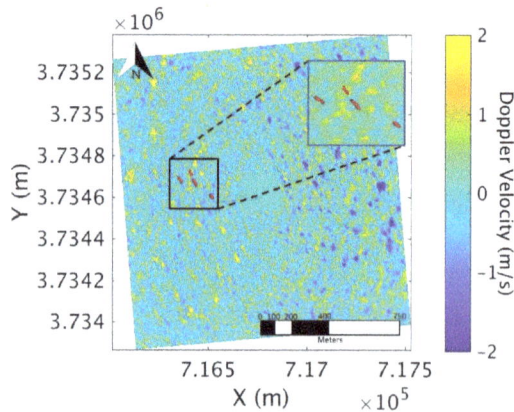

Figure 1. Tracks from the four GPS wave buoys in the radar field of view during collection of radar dataset 1 are shown in red, overlaid on an example frame of Doppler velocity data. The black box shows a zoom-in of that area.

Two radar datasets from this experiment are used for this study. Datasets were selected based on availability of wave buoys in the field of view of the radar (Table 1) (and for their unique group line features). Dataset 2 was taken approximately 25 min after dataset 1 and during this time winds were decreasing in magnitude (decaying seas). Both datasets were collected under bimodal seas with wind-waves and swell present. Dataset 1 was collected in higher wind speed (15 m/s), and dataset 2 was collected while wind speed was declining (7 m/s). Figure 2 (panels (a) and (b)) shows the k-ω spectrum for each dataset. Dataset 1 has a strong group line feature with a high magnitude of group line energy relative to dispersion curve associated energy, while dataset 2 shows a weaker group line feature relative to dispersion curve associated energy. The primary difference between dataset 1 and 2 is the amount of group line energy relative to dispersion curve associated energy because they were obtained in such close proximity in time to each other. Table 1 shows general environmental statistics for both datasets.

Table 1. Environmental statistics for both radar datasets used in this study. Shown in the table is the date and time the dataset was collected, as well as the significant wave height (H_s) as measured by the wave buoys, wind speed (U_w) as measured by a shipboard anemometer, root mean squared surface velocity (V_{rms}) as measured by the wave buoys (averaged over all buoys in the radar FOV), angle between the two wave systems ($\Delta\Theta$) during dataset collection based on radar derived directional wave spectra, and the number of buoys in the radar FOV.

Dataset	Date	Time	H_s (m)	U_w (m/s)	V_{rms} (m/s)	$\Delta\Theta$ (°)	*Buoys*
1	17 September 2013	00:08:45	1.65	15	0.54	41	4
2	17 September 2013	00:33:34	1.48	7	0.49	46	3

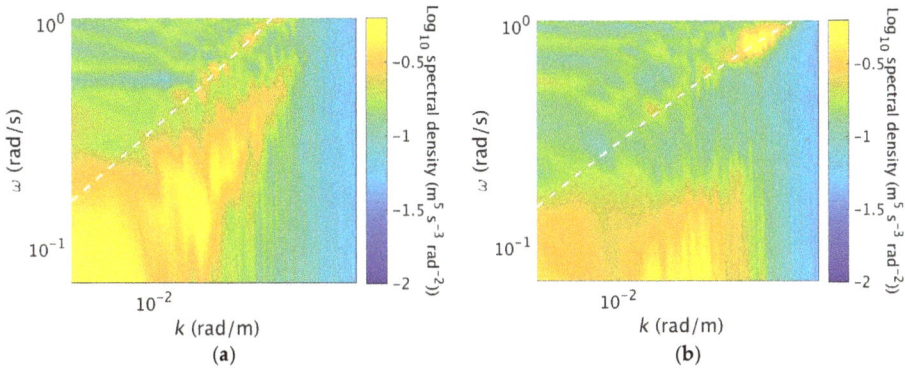

Figure 2. (a) k-ω Doppler velocity spectrum for dataset 1; (b) k-ω Doppler velocity spectrum for dataset 2. The white dashed line in both panels shows the linear dispersion relationship (adjusted for currents and ship forward speed).

3. Phased Resolved Orbital Velocity Maps

Ocean surface wave orbital velocity maps are produced for all possible frames of radar data using both the POD and dispersion curve filtering methods. The time series of these maps are used to generate time series of orbital velocity by extracting the orbital velocities from each frame at the location where a wave buoy was present. Buoy measured wave orbital velocity time series are compared to orbital velocity time series derived from these radar-based wave orbital velocity maps. Correlation (c) and root mean squared error (E_{rms}) are calculated between the buoy velocity time series and the velocity map derived time series for each method for all available buoys in the FOV of the radar, and these statistics are then averaged across all buoys (see Table 1).

3.1. Pre-Processing

Prior to the POD or FFT wave orbital velocity retrieval methods being performed, pre-processing steps were applied to the radar data. Both datasets were de-trended along range, converted to a Cartesian grid ($D(r,t,\phi)$ to $D(x,y,t)$), and georeferenced. Geo-referencing relative to the ship GPS position is performed for each cell of the Cartesian grid so that an accurate cell location of the wave buoys can be identified. Due to sensitivity of the POD method to wave direction [17], the data is rotated such that the dominant wave system is propagating along the direction of the x-axis.

3.2. POD Based Wave Field Extraction

The POD method is applied as described in Kammerer, and Kammerer and Hackett [17,21], which was adapted from Hackett et al. [22]. Only relevant details are repeated here for the reader's convenience. This method takes one frame of Doppler velocity radar data, $D(x,y)$, and decomposes the signal into a series of orthonormal basis functions (or modes) and spatial coefficients. Mode functions are determined by the best fit to the variance of the data as opposed to being assumed *a priori*, as in Fourier analysis. As mode number (n) increases, the amount of signal variance accounted for in that mode decreases. The summation of the product of all the mode functions and spatial coefficients results in reconstruction of the original signal.

Because the variance of the Doppler velocity is dominated by the ocean surface wave orbital velocities, a summation of the product of the leading mode functions and spatial coefficients results in a reconstruction of the ocean wave field orbital velocities. This reconstruction method allows energy associated with the group line as well as the dispersion curve to be maintained depending on how much they contribute to the overall variance of the map. POD is performed on frames 16 to N-16, where N is the total number of radar frames in the dataset. This set is selected to maintain the same time series length as the FFT-method, which is described subsequently.

3.3. Conventional FFT Based Wave Field Extraction

The FFT based wave field extraction method is applied as described in Kammerer [21], which was based on the method outlined by Young et al. [1]. After the pre-processing is complete the Doppler radar data is in the form $D(x,y,t)$ with 141 samples in the x and y directions (i.e., a matrix of size 141 × 141 × N). The first 32 frames of radar data are extracted and the x and y dimensions are zero-padded such that the size of D is 256 × 256 × 32. A 3-dimensional FFT is then applied to D. The subsequent Fourier coefficients are then multiplied by a binary dispersion relationship filter ($d_k(k_x,k_y,\omega)$):

$$\begin{aligned} d_k &= 1 \quad \text{for } \sigma - W\Delta\omega < \omega < \sigma + W\Delta\omega \\ d_k &= 0 \quad \text{otherwise} \end{aligned} \tag{1}$$

where ω is radian frequency, k_x is radian wave number in the x-direction, k_y is radian wave number in the y-direction, $\Delta\omega$ is the radian frequency resolution (0.07 rad/s), W is discrete filter width ($W = 1$ is used for this study), and σ is the linear deep-water dispersion relationship including current (U):

$$\sigma = U \cdot k + \sqrt{g|k|} \tag{2}$$

where, U is the current, k is radian wavenumber, and g is gravitational acceleration.

An inverse Fourier transform (IFFT) returns the filtered Fourier coefficients from the spectral domain back to the spatial domain ($V(x,y,t)$). Only the middle frame of the 3D data stack is extracted and saved as the phase-resolved wave orbital velocity map for the center time of the stack. The 3D stack is then shifted forward by one frame in time for the next set of 32 frames and the process repeats until the last frame in the stack is frame N. The resulting FFT dispersion-filtered time series of phase-resolved wave orbital velocity maps consists of frames 16 to N-16.

4. Time Series Extraction for Buoy Comparisons

For each frame in the time series of wave orbital velocity maps, the spatial coordinates of each buoy in the radar field of view are identified. In order to account for uncertainty in the GPS measurements of the buoy and ship, the velocity of the identified range cell as well as all eight adjacent cells are compared to the buoy wave orbital velocity, and the value closest to the buoy orbital velocity is extracted as the velocity for the radar time series. This process is repeated for each buoy in the field of view of the radar, and for each frame of the radar time series for both the FFT and POD derived ocean surface orbital velocity maps. Figure 1 (Section 2) shows an example of buoy tracks overlaid on one frame of radar data for dataset 1.

Finally, Pearson's correlation coefficient (c) between each buoy and radar-based time series is computed. Additionally, the root mean squared error (E_{rms}) between the buoy velocity time series, and the POD and FFT reconstructed time series are computed. Both statistics are evaluated for each available wave buoy.

5. Results and Discussion

Figure 3 shows the time series comparisons between the wave buoy, POD, and FFT reconstructions for dataset 1, and Figure 4 shows the time series comparisons for dataset 2. Note, dataset 1 has four buoys in the radar field of view, and dataset 2 has three buoys in the radar FOV (see Table 1, Section 2), with one of the three buoys only being in the field of view for part of the dataset. The number of leading POD modes used for the example reconstructions shown in Figures 3 and 4 was selected to be representative of peak correlation (see Figure 5 and Table 2). Dataset 1 is shown using an $n = 32$ mode reconstruction and dataset 2 is shown using an $n = 6$ mode reconstruction. It can be seen in Figures 3 and 4 that the POD reconstructed orbital velocity time series is generally in-phase, and of comparable magnitude to the wave buoy measured velocities for both dataset 1 and 2. The dispersion filtered time series for dataset 1 seems to be of lower magnitude than the wave buoy measured velocities, and out of phase with the buoy measurements at times. However, for dataset 2, the dispersion filtered time series seems to perform in a visually similar way to the POD time series.

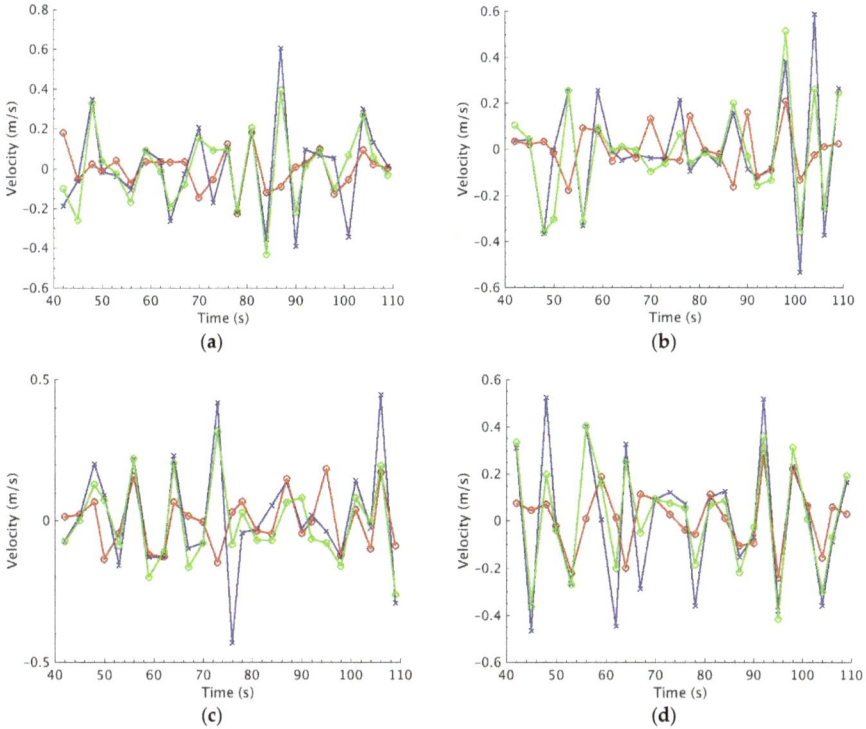

Figure 3. Time series comparisons between the GPS wave buoys (blue lines), POD reconstructions using the leading 32 modes ($n = 32$) (green lines), and dispersion filtered reconstructions (red line) for dataset 1. (**a**) are the time series for wave buoy 279; (**b**) are the time series for wave buoy 283; (**c**) are the time series for buoy 286; and (**d**) are the time series for buoy 289.

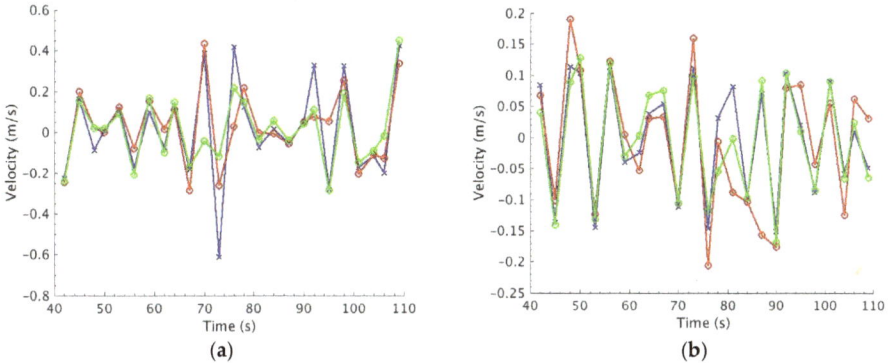

Figure 4. *Cont.*

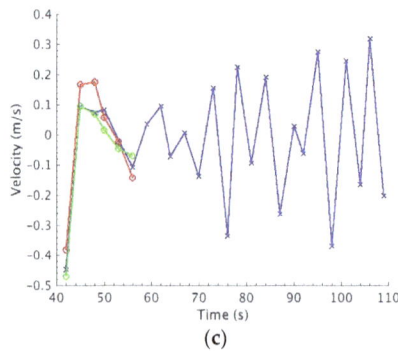

(c)

Figure 4. Time series comparisons between the GPS wave buoys (blue lines), POD reconstructions using the leading 6 modes ($n = 6$) (green lines), and dispersion filtered reconstructions (red line) for dataset 2. (**a**) are the time series for wave buoy 279; (**b**) are the time series for wave buoy 280; and (**c**) are the time series for buoy 286.

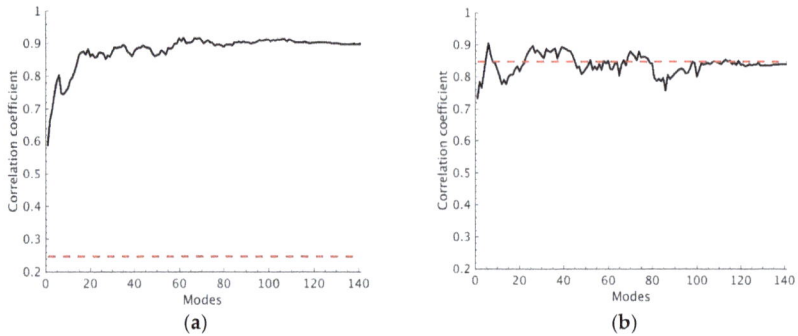

(a)   **(b)**

Figure 5. Average correlation coefficient (*c*) between POD reconstructed orbital velocity time series and wave buoy orbital velocity time series for each (1 to *n*) mode reconstruction (black line) and average correlation coefficient between FFT-based dispersion curve orbital velocity time series and wave buoy time series (red dashed line). Correlation coefficients are averaged over all available buoys. (**a**) shows dataset 1 and (**b**) shows dataset 2.

Table 2. Correlation coefficient (*c*) and E_{rms} for both wave retrieval methods: proper orthogonal decomposition (POD) and dispersion filtering (FFT) for each dataset. For POD, the peak *c* and minimum E_{rms} are provided with the mode selection (modes 1-*n*) shown parenthetically.

Dataset	POD: *c*	POD: E_{rms} (*m/s*)	FFT: *c*	FFT: E_{rms} (*m/s*)
1 (group line)	0.93 (modes 1–32)	0.085 (modes 1–33)	0.25	0.23
2 (no group line)	0.90 (modes 1–6)	0.072 (modes 1–43)	0.85	0.092

Figure 5 shows the averaged correlation coefficient between the GPS wave buoy orbital velocity time series and POD reconstructed velocity time series for POD reconstructions using various numbers of modes (1 to *n*) as well as for the FFT–based reconstructed time series for dataset 1 (a) and dataset 2 (b). Note that correlation coefficients are calculated separately for each available buoy time series and averaged over all buoys. For both dataset 1 and dataset 2, POD reconstructed time series attain above 0.8 correlation coefficient with buoy time series within the leading 10 modes, and peak in correlation at ~0.9. For dataset 1, when the group line energy is stronger relative to dataset 2, POD reconstructions achieve a significantly higher correlation with ground truth wave buoy measurements then the FFT method regardless of the number of modes used in the POD reconstruction. In contrast, for dataset 2, in which dispersion curve energy is higher relative to group line energy, the FFT and POD methods result in similar correlation coefficients regardless of the number of modes used for the POD reconstruction. Furthermore, when spectral energy is limited to primarily dispersion curve associated energy (as in dataset 2), the POD method reaches a correlation "plateau" in fewer modes than when the energy spectra is more complex (i.e., containing group line energy in addition to dispersion curve energy). In summary, POD reconstructed orbital velocity time series correlate highly with buoy-measured wave orbital velocity time series for both datasets, regardless of the group line to dispersion curve energy ratio; when the ratio of group line energy relative to dispersion curve energy is high, the POD method attains significantly higher correlations with buoy measurements than the conventional dispersion curve filtering method. This difference in the correlation coefficients is attributed to the inclusion of group line energy in the POD reconstructions. Nevertheless, the selection of the optimal number of leading modes is non-trivial, but any mode selection achieved higher correlations than dispersion-filtered time series.

The averaged E_{rms} between the reconstructed time series and the ground truth wave buoy time series for both reconstruction methods and datasets is show in Figure 6. Recall, E_{rms} is computed as the root-mean-square of the difference in orbital velocity between the time series. The POD method clearly attains lower E_{rms} than the dispersion curve filtering method for dataset 1 regardless of the number of modes used. In contrast, for dataset 2, E_{rms} is comparable between the methods. The E_{rms} results are consistent with the correlation results shown in Figure 5. When significant group line energy is present, the POD method attains lowers E_{rms} and higher c than the dispersion filtering method (for all mode reconstructions), and when group line energy is low relative to dispersion curve energy, both wave retrieval methods perform comparably. A summary of results for both the correlation and E_{rms} metrics are presented in Table 2.

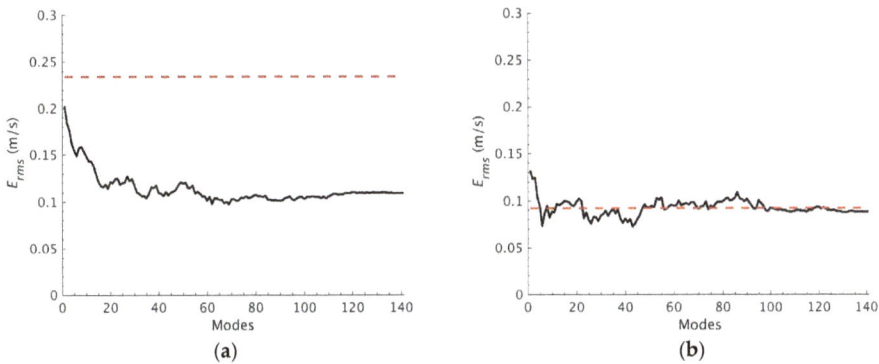

Figure 6. Averaged root mean squared error (E_{rms}) between POD reconstructed wave orbital velocity time series and ground truth wave buoy measured orbital velocity time series for each (1 to n) mode reconstruction (black line), and between FFT-based dispersion curve filtered time series and wave buoy ground truth (red dashed line) for dataset 1 (**a**) and dataset 2 (**b**).

Because the primary difference in the methods is the inclusion of group line energy, and the difference between the datasets is primarily associated with the relative strength of group line energy, we conjecture that the group line energy mostly influences the phasing of the ocean surface wave field and its inclusion in the wave retrieval improves the comparisons with time series buoy data. Figure 7 shows the k-ω Doppler velocity spectrum of the POD reconstructions for datasets 1 and 2 ($n = 32$ and $n = 6$ respectively) as well as the k-ω spectra of the dispersion curve filtered reconstructions. The k-ω spectra of the POD reconstructions for both datasets contain energy both associated with the linear dispersion curve and with the group line feature. Note that dataset 1 (**a**) has high group line energy relative to dataset 2 (**b**), and the dispersion-filtered reconstructions do not contain any group line energy (c and d). Because the POD reconstructed time series correlate more highly with ground truth buoy time series than the dispersion curve filtered time series for dataset 1 and show smaller E_{rms}, and because the k-ω POD reconstruction spectrum contains significant energy associated with the group line, we surmise that the group line contains energy from the wave field, whose inclusion in wave retrieval contributes to more accurate phase-resolved wave orbital velocity reconstructions.

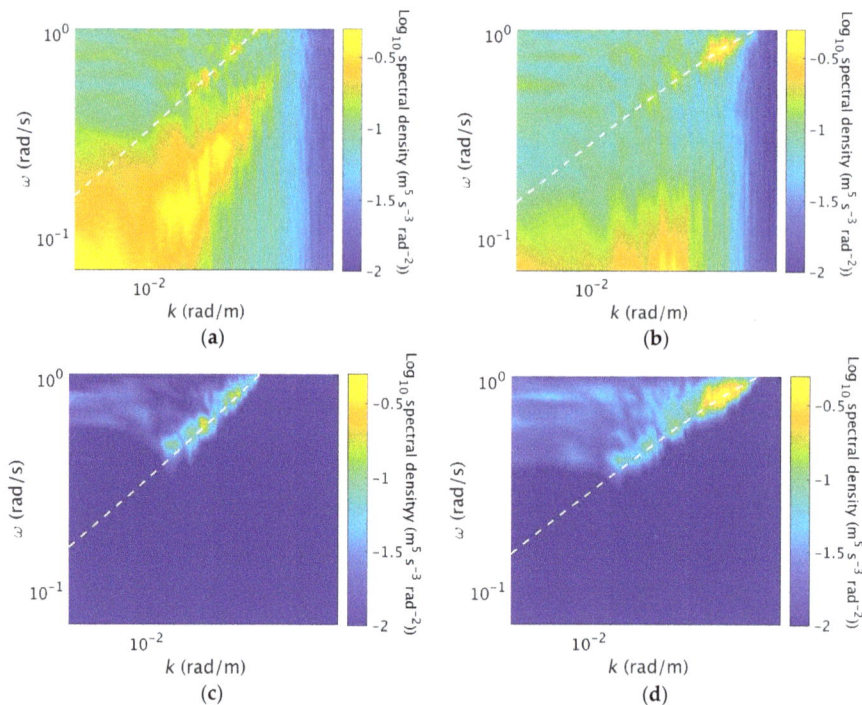

Figure 7. (**a**) an example k-ω POD reconstruction spectrum for dataset 1 using 32 modes ($n = 32$); (**b**) k-ω POD reconstruction spectrum for dataset 2 using 6 modes ($n = 6$); (**c**) k-ω spectrum for the dispersion curve filtered result for dataset 1 and (**d**) k-ω spectrum for the dispersion curve filtered result for dataset 2. Energy that appears above and to the left of the dispersion curve in (**c**) and (**d**) is an artifact of the 3D dispersion cone filtering integrated into 2D for presentation of the figure.

6. Summary and Conclusions

This study has shown that when group line energy is high relative to dispersion curve energy, POD orbital velocity time series attain higher correlation with ground truth wave buoy velocity measurements (and lower E_{rms}) than conventional FFT-based dispersion curve filtering derived wave retrieval. In contrast, when the dispersion curve energy is higher than group line energy, FFT-based dispersion curve filtering attained similar correlation with ground truth measurements than POD; similar E_{rms} is also observed for this dataset. Furthermore, peak correlation between POD-based time series and buoy time series is attained in fewer modes in this case (dataset 2). It is demonstrated that accurate phase-resolved reconstructions of ocean surface wave orbital velocities can be produced from X-band Doppler measurements of the ocean surface using POD and buoy data. More research on automating mode selection is needed in order to use the POD method without optimizing it to buoy data.

The combination of group line energy contained in the POD reconstructed k-ω spectra and the improvement in correlation with ground truth buoy measurements over the conventional dispersion

Remote Sens. **2019**, *11*, 71

curve filtering method (for dataset 1, with high group line to dispersion curve energy ratio) shows experimental evidence that some portion of the energy associated with the group line feature contains contributions from ocean surface waves. The inclusion of this portion of group line energy increases phase accuracy when the ratio of group line to dispersion curve energy is high. When the ratio of group line energy to dispersion curve energy was lower (dataset 2) performance of the FFT-method to produce phase-resolved maps was more comparable to the POD-based reconstruction method. Thus, the POD method can accurately reconstruct phase-resolved ocean surface velocity maps regardless of the ratio of group line to dispersion curve energy, whereas the conventional FFT-filtering method performs sub-optimally when group line energy is high.

The experimental evidence in this paper supports the numerical results of Plant and Farquharson [10], showing that at least a portion of the group line energy is associated with wave field features, and inclusion of these features contributes to more accurate phase-resolved reconstructions of the ocean surface wave field. The wave field statistics (e.g., those derived from 1D wave spectra) do not appear to be significantly impacted by neglect of the group line energy as shown in Kammerer [21], but the phase-resolved wave field does appear to be impacted as shown in this manuscript. This result implies that the group line mostly influences the phase-resolved wave field, which supports the findings of Plant and Farquharson [10] that attribute the group line feature to wave interference effects that would presumably also affect wave phasing due to the influence of superposition. Further research should examine the significance of the group line energy in phase-resolved wave retrieval for a wider range of conditions.

Author Contributions: Conceptualization, E.E.H.; Methodology, E.E.H. and A.J.K.; Formal Analysis, A.J.K.; Investigation, E.E.H. and A.J.K.; Writing-Original Draft Preparation, A.J.K.; Writing-Review & Editing, E.E.H. and A.J.K.; Supervision, E.E.H.; Project Administration, E.E.H.; Funding Acquisition, E.E.H.

Funding: This research was funded by the *Office of Naval Research*, grant number [N00014-15-1-2044].

Acknowledgments: The authors would like to thank Joel Johnson and Shanka Wijesundara from *The Ohio State University* for providing the radar experimental data, and Eric Terrill and Tony DePaulo from the *Coastal Observing Research and Development Center* at the *Scripps Institute of Oceanography* for providing the wave data from GPS mini-buoys.

Conflicts of Interest: The authors declare no conflict of interest.

References

1. Young, I.R.; Rosenthal, W.; Ziemer, F. A three-dimensional analysis of marine radar images for the determination of ocean wave directionality and surface currents. *J. Geophys. Res. Oceans* **1985**, *90*, 1049–1059. [CrossRef]
2. Nieto Borge, J.C.; Reichert, K.; Dittmer, J. Use of nautical radar as a wave monitoring instrument. *Coast. Eng.* **1999**, *37*, 331–342. [CrossRef]
3. Nieto Borge, J.C.; Rodriguez, G.R.; Hessner, K.; González, P.I. Inversion of marine radar images for surface wave analysis. *J. Atmos. Ocean. Technol.* **2004**, *21*, 1291–1300. [CrossRef]
4. Carrasco, R.; Streßer, M.; Horstmann, J. A simple method for retrieving significant wave height from Dopplerized X-band radar. *Ocean. Sci.* **2017**, *13*, 95–103. [CrossRef]
5. Al-Habashneh, A.; Moloney, C.; Gill, E.W.; Huang, W. An adaptive method of wave spectrum estimation using X-band nautical radar. *Remote Sens.* **2015**, *7*, 16537–16554. [CrossRef]
6. Lyzenga, D. Polar Fourier transform processing of marine radar signals. *J. Atmos. Ocean. Technol.* **2017**, *34*, 347–354. [CrossRef]
7. Lyzenga, D.; Nwogu, O.; Beck, R.; O'Brien, A.; Johnson, J.; de Paolo, A.; Terrill, E. Real-time estimation of ocean wave fields from marine radar data. In Proceedings of the IEEE International Geoscience and Remote Sensing Symposium, Milan, Italy, 31 July 2015; pp. 3622–3625.
8. Lyzenga, D.; Nwogu, O.; Trizna, D.; Hathaway, K. Ocean wave field measurements using X-band Doppler radars at low grazing angles. In Proceedings of the IEEE International Geoscience and Remote Sensing Symposium, Honolulu, HI, USA, 25–30 July 2010; pp. 4725–4728.

9.   Nwogu, O.; Lyzenga, D. Surface-wavefield estimation from coherent marine radars. *IEEE Geosci. Remote Sens. Lett.* **2010**, *7*, 631–635. [CrossRef]

10.   Plant, W.J.; Farquharson, G. Origins of features in wave number-frequency spectra of space-time images of the ocean. *J. Geophys. Res. Oceans* **2012**, *117*, C6. [CrossRef]

11.   Dugan, J.P.; Fetzer, G.J.; Bowden, J.; Farruggia, G.J.; Williams, J.Z.; Piotrowski, C.C.; Vierra, K.; Campion, D.; Sitter, D.N. Airborn optical system for remote sensing of ocean waves. *J. Atmos. Ocean. Technol.* **2001**, *18*, 1267–1276. [CrossRef]

12.   Dugan, J.P.; Piotrowski, C.C. Surface current measurements using airborne visible image time series. *Remote Sens. Environ.* **2003**, *83*, 309–319. [CrossRef]

13.   Dugan, J.P.; Piotrowski, C.C. Measuring currents in a coastal inlet by advection of turbulent eddies in airborn optical imagery. *J. Geophys. Res.* **2012**, *117*, 1–15. [CrossRef]

14.   Frasier, S.J.; McIntosh, R.E. Observed wavenumber-frequency properties of microwave backscatter from the ocean surface at near-grazing angles. *J. Geophys. Res.* **1996**, *101*, 18391–18407. [CrossRef]

15.   Stevens, C.L.; Poulter, E.M.; Smith, M.J.; McGregor, J.A. Nonlinear features in wave-resolving microwave radar observations of ocean waves. *IEEE J. Ocean. Eng.* **1999**, *24*, 470–480. [CrossRef]

16.   Rino, C.L.; Eckert, E.; Siegel, A.; Webster, T.; Ochadlick, A.; Rankin, M.; Davis, J. X-band low-grazing-angle ocean backscatter obtained during LOGAN 1993. *IEEE J. Ocean. Eng.* **1997**, *22*, 18–26. [CrossRef]

17.   Kammerer, A.J.; Hackett, E. Use of proper orthogonal decomposition for extraction of ocean surface wave fields from X-band radar measurements of the sea surface. *Remote Sens.* **2017**, *9*, 881. [CrossRef]

18.   Smith, G.E.; O'Brien, A.; Pozderac, J.; Baker, C.J.; Johnson, J.T.; Lyzenga, D.R.; Nwogu, O.; Trizna, D.B.; Rudolf, D.; Schueller, G. High power coherent-on-receive radar for marine surveillance. In Proceedings of the 2013 International Conference on Radar, Adelaide, Australia, 16 August 2013; pp. 434–439.

19.   Miller, K.; Rochwarger, M. A covariance approach to spectral moment estimation. *IEEE Trans. Inf. Theory.* **1972**, *18*, 588–596. [CrossRef]

20.   Drazen, D.; Merrill, C.; Gregory, S.; Fullerton, A. Interpretation of in-situ ocean environmental measurements. In Proceedings of the 31st Symposium on Naval Hydrodynamics, Monterey, CA, USA, 11–16 September 2016; p. 8.

21.   Kammerer, A.J. The Application of Proper Orthogonal Decomposition to Numerically Modeled and Measured Ocean Surface Wave Fields Remotely Sensed by Radar. Master's Thesis, Coastal Carolina University, Conway, SC, USA, June 2017.

22.   Hackett, E.E.; Merrill, C.F.; Geiser, J. The application of proper orthogonal decomposition to complex wave fields. In Proceedings of the 30th Symposium on Naval Hydrodynamics, Hobart, Australia, 2–7 November 2014; p. 9.

remote sensing

MDPI

Article

Analysis on the Effects of SAR Imaging Parameters and Environmental Conditions on the Standard Deviation of the Co-Polarized Phase Difference Measured over Sea Surface

Andrea Buono [1,*,†,‡], Carina Regina de Macedo [1,‡], Ferdinando Nunziata [1,‡], Domenico Velotto [2,‡] and Maurizio Migliaccio [1,‡]

1   Dipartimento di Ingegneria, Universitá degli Studi di Napoli Parthenope, 80143 Napoli NA, Italy; carinaregina.demacedo@uniparthenope.it (C.R.d.M.); ferdinando.nunziata@uniparthenope.it (F.N.); maurizio.migliaccio@uniparthenope.it (M.M.)
2   German Aerospace Center (DLR), Remote Sensing Technology Institute, 82234 Weßling, Germany; domenico.velotto@dlr.de
*   Correspondence: andrea.buono@uniparthenope.it; Tel.: +39-081-547-6706
†   Current address: Centro Direzionale, isola C4, 80143 Napoli, Italy.
‡   These authors contributed equally to this work.

Received: 15 November 2018; Accepted: 19 December 2018; Published: 21 December 2018

Abstract: This study aimed at analyzing the effect of Synthetic Aperture Radar (SAR) imaging parameters and environmental conditions on the standard deviation of the co-polarized phase difference (σ_{φ_C}) evaluated over sea surface. The latter was shown to be an important polarimetric parameter widely used for sea surface target monitoring purposes. A theoretical model, based on the tilted-Bragg scattering, is proposed to predict the behavior of σ_{φ_C} against incidence angle for different roughness conditions. Then, a comprehensive experimental analysis, based on the processing of L-, C- and X-band polarimetric SAR scenes collected over different test areas under low-to-moderate wind conditions and covering a broad range of incidence angle, was carried out to discuss the effects of sensor's and environmental parameters on sea surface σ_{φ_C}. Results show that SAR imaging parameters severely affect σ_{φ_C}, while the impact of meteo-marine conditions, under low-to-moderate wind regime, is almost negligible. Those outcomes have significant relevance to support the design of effective and robust algorithms for marine and maritime applications based on σ_{φ_C}, including the detection of metallic targets (ships and offshore infrastructures as oil/gas platforms, aquacultures, wind farms, etc.) and polluted areas.

Keywords: oceans; Synthetic Aperture Radar; polarimetry; co-polarized phase difference

1. Introduction

The preservation and sustainable management of ocean resources and ecosystems are mandatory goals according to several international strategies, policy programmes and technical reports as the 2030 Agenda, the Blue Growth European Union program and the United Nations report for sustainable development [1–5]. Within this context, remote sensing tools, due to their synoptic view and frequent revisit time, are of paramount importance for a broad range of applications including global weather predictions, storm and hurricane warnings, wave and current forecastings, coastal storm surges, ship routing, commercial fishing and climate change [6].

Microwave imaging of sea surface from space allows the retrieval of several geophysical parameters that, once tailored models are available, can be transformed into added-value products that include, among others, significant wave height from radar altimeters, wave spectrum from Synthetic

Aperture Radars (SARs), sea surface wind field from scatterometers and sea surface temperature and salinity from radiometers [6]. Furthermore, additional information on ocean targets of interest such as sea ice extent, pollutants, metallic infrastructures, icebergs, ships, etc. can be derived.

When dealing with microwave active remote sensing of the oceans, the unprecedented benefits offered by SAR sensors operating in multi-polarization modes is unambiguously recognized for a wide range of marine and maritime applications, including coastline extraction [7–9], metallic target detection [10–12], sea pollution monitoring [13–16] and sea ice observation [17–19]. All the approaches share a similar physical rationale that relies on the exploitation of the different polarimetric properties that characterize the target of interest and the sea clutter. Nonetheless, the performance is significantly affected by the amount of scattering information available that, in turn, depends on [8,10,14,20,21]:

- SAR acquisition parameters, e.g., polarization, Angle Of Incidence (AOI), incident wavelength and Noise-equivalent sigma zero (NESZ);
- target features, e.g., damping properties of the pollutant, ship orientation and ice layer thickness; and
- meteo-marine conditions, e.g., sea state, swell and wave patterns.

Among the different polarimetric SAR architectures, conventional dual-polarimetric SAR imaging modes, i.e., the ones that consist of transmitting a linearly polarized wave (horizontal (H) or vertical (V)) while receiving coherently in an orthogonal linear basis (H–V), are attracting more attention in the perspective of operational services since they offer, for a wide range of ocean applications, a sufficient polarimetric information content over a large swath [7,16,22]. Among the dual-polarimetric SARs, it is worth mentioning the one operated by the German TerraSAR-X (TSX) mission that provides coherent HH-VV SAR measurements routinely. Among the polarimetric features that can be extracted from a coherent dual-polarimetric HH-VV SAR, the standard deviation of the co-polarized phase difference (σ_{φ_C}) has been demonstrated to be a powerful tool for marine and maritime applications. In fact, it is successfully used to perform the observation of metallic targets [23–26], the monitoring of oil pollution [14,16,27,28] and iceberg detection/sea ice classification [29,30].

All those applications share the same physical rationale, i.e., they exploit σ_{φ_C} as a reliable and robust estimator of the correlation between the co-polarized channels. This means that, at least from a theoretical viewpoint, since sea surface scattering is ruled by the Bragg mechanism, the co-polarized channels result in a unitary correlation that makes the co-polarized phase difference statistical distribution resembling a Dirac delta function [31–34]. In real cases, when a low-depolarizing scenario is considered, e.g., a sea surface Bragg scattering, a large correlation between co-polarized channels applies that results in a narrow co-polarized phase difference distribution. Hence, marine targets, e.g., ships, oil slicks, icebergs, etc., that result in departure from Bragg scattering are characterized by a broader co-polarized phase difference distribution. This means that σ_{φ_C} can be successfully exploited to emphasize non-Bragg scattering targets with respect to the Bragg-like sea surface background.

Following this rationale, in the literature, σ_{φ_C} has been mainly investigated in terms of sea/target separability (e.g., mean contrast and average target-to-clutter ratio are usually adopted as figures of merit [17,20,23,26,35]). In [28], σ_{φ_C} values spanning the range $\approx$ 3–18° have been reported, at C-band, over sea surface under low-to-moderate wind conditions over different geographical locations. Dierking and Wesche [29] found that, at C-band, larger σ_{φ_C} values are in place when the measured co-polarized channels approach the sensor's noise floor. In [30], σ_{φ_C} values of about 6° have been measured at C-band with AOI = 40° over sea surface and thin sea ice. In [10], larger σ_{φ_C} values, i.e., about 30°, have been measured over sea surface at X-band. The authors also observed a sensitivity of σ_{φ_C} with both the estimation window's size and the sea state parameters. In [36], the behavior of σ_{φ_C} is investigated with reference to a specific test case, i.e., the Taylor Energy oil seep in the Gulf of Mexico, where a large TSX SAR dataset was considered under limited SAR imaging configuration (X-band, AOI = 26°, 34° and 43°).

Hence, literature studies clearly point out the dependence of σ_{φ_C} on both sensor's and environmental parameters. Nevertheless, the behavior of σ_{φ_C} is analyzed for specific applications only, i.e., for a given target (ships and oil slicks); sensor's configuration, i.e., for a given frequency and/or limited AOI range; and environmental conditions [16,23,29,30,37]. In no case, to the best of our knowledge, the role of sensor's and environmental parameters on $\sigma_{\varphi C}$ is investigated in a systematic way.

In this study, a significant extension of the work carried out in [36] was made. The analysis undertaken in [36] was improved as follows:

- The analysis on the effects of different frequencies, i.e., L- and C-band, and sea surface tilting angle on sea surface $\sigma_{\varphi C}$ was included.
- The analysis on the influence of incidence angle on sea surface $\sigma_{\varphi C}$ was undertaken on a much broader AOI range, i.e., about 20–60°.
- The behavior of sea surface $\sigma_{\varphi C}$ with respect to AOI was investigated by comparing model predictions' with actual SAR measurements over the whole range of Bragg AOIs ($\approx$20–60°).

Hence, a comprehensive analysis on the behavior of $\sigma_{\varphi C}$ over sea surface was provided for the first time. A large polarimetric SAR dataset collected in a wide range of SAR acquisition parameters (frequency, AOI, and NESZ) and meteo-marine conditions (sea surface roughness and wind speed (WS)) was considered to give a more complete understanding on how those parameters affect sea surface $\sigma_{\varphi C}$. Furthermore, to better interpret the experimental results, a theoretical scattering framework, based on the polarimetric X-Bragg model, is proposed that allows giving a physical understanding of the behavior of sea surface $\sigma_{\varphi C}$ under noise-free conditions. It must be explicitly pointed out that this comprehensive study is of paramount importance to support all the operational techniques based on $\sigma_{\varphi C}$, e.g., to find the most suitable SAR configuration in relation with the ocean target to be detected.

The remainder of the paper is organized as follows: in Section 2, the theoretical background that lies at the basis of the sensitivity analysis is presented; in Section 3, SAR dataset and ancillary information are provided; in Section 4, the experiments based on the theoretical model provided in Section 2 are presented and discussed; in Section 5, experiments related to the sensitivity analysis undertaken on the actual SAR dataset are presented and discussed; and conclusions are drawn in Section 6.

2. Polarimetric Framework

The co-polarized phase difference was theoretically predicted over sea surface using a polarimetric sea surface scattering model. The latter, based on the X-Bragg scattering model developed in [36], assumed that sea surface is mainly governed by the tilted-Bragg scattering mechanism. This is a reasonable assumption under low-to-moderate wind conditions (i.e., $\approx$ 2–14 m/s) and under intermediate AOIs (i.e., $\approx$ 20–60°).

From the scattering viewpoint, sea surface is considered as a distributed low-depolarizing scene. Hence, second-order descriptors, i.e., the covariance matrix $\mathbf{C}$, are needed to describe sea surface polarimetric scattering [38]. In the monostatic backscattering case, invoking reciprocity and the reflection symmetry property, the 3 $\times$ 3 Hermitian and semi-definite positive covariance matrix $\mathbf{C}$ can be defined as follows [38]:

$$\mathbf{C} = \begin{pmatrix} C_{11} & C_{12} & C_{13} \\ C_{12}{}^* & C_{22} & C_{23} \\ C_{13}{}^* & C_{23}{}^* & C_{33} \end{pmatrix} = \begin{pmatrix} \langle |S_{hh}|^2 \rangle & 0 & \langle S_{hh}S_{vv}{}^* \rangle \\ 0 & 2\langle |S_{hv}|^2 \rangle & 0 \\ \langle S_{vv}S_{hh}{}^* \rangle & 0 & \langle |S_{vv}|^2 \rangle \end{pmatrix}, \tag{1}$$

where S_{pq} is the complex scattering amplitude with $\{p,q\} = \{h,v\}$, while $\langle \cdot \rangle$, $|\cdot|$ and * stand for ensemble average, modulus and complex conjugate operators, respectively.

According to the X-Bragg scattering theory, the sea surface covariance matrix can be predicted as follows [39–41]:

$$
\mathbf{C_X} = \begin{pmatrix} C_{X_{11}} & 0 & C_{X_{13}} \\ 0 & C_{X_{22}} & 0 \\ C_{X_{13}}{}^* & 0 & C_{X_{33}} \end{pmatrix} =
$$

$$
\mathbf{U}^{-1} \begin{pmatrix} |B_h + B_v|^2 & (B_h + B_v)(B_h{}^* - B_v{}^*)\mathrm{sinc}(2\beta) & 0 \\ (B_h{}^* + B_v{}^*)(B_h - B_v)\mathrm{sinc}(2\beta) & \frac{1}{2}|B_h - B_v|^2(1 + \mathrm{sinc}(4\beta)) & 0 \\ 0 & 0 & \frac{1}{2}|B_h - B_v|^2(1 - \mathrm{sinc}(4\beta)) \end{pmatrix} \mathbf{U} \, , \tag{2}
$$

where the subscript "X" stands for X-Bragg model and **U** is a con-similarity linear transformation given by [38]:

$$
\mathbf{U} = \frac{1}{\sqrt{2}} \begin{pmatrix} 1 & 0 & 1 \\ 1 & 0 & -1 \\ 0 & \sqrt{2} & 0 \end{pmatrix} . \tag{3}
$$

In Equation (2), β is the surface local tilting angle, while B_h and B_v are the Fresnel complex coefficients for Bragg scattering under horizontal and vertical polarizations, respectively, which are given by [41]:

$$
\begin{cases} B_h = \dfrac{\cos(\mathrm{AOI}) - \sqrt{\epsilon - \sin^2(\mathrm{AOI})}}{\cos(\mathrm{AOI}) + \sqrt{\epsilon - \sin^2(\mathrm{AOI})}} \\[4ex] B_v = \dfrac{(\epsilon - 1)\left(\sin^2(\mathrm{AOI}) - \epsilon\left(1 + \sin^2(\mathrm{AOI})\right) \right)}{\left(\epsilon\cos(\mathrm{AOI}) + \sqrt{\epsilon - \sin^2(\mathrm{AOI})} \right)^2} \end{cases} . \tag{4}
$$

In the X-Bragg sea surface scattering model, β is related to the amount of surface roughness and it rules both the cross-polarized backscattering and the HH-VV coherence, while the Bragg scattering coefficients are related to the incidence angle and to the frequency-dependent relative electric permittivity of seawater, ϵ [41]. β was assumed to be uniformly distributed in the range $[-90°, +90°]$, where $\beta \approx 0°$ describes an almost flat sea surface (negligible cross-polarized backscattering and HH-VV coherence close to 1), while $\beta \approx \pm90°$ characterizes an extremely rough sea surface (significant cross-polarized backscattering and HH-VV coherence that tends to 0) [41]. In [15,40], it is found that reliable β values fall within the range $[-30°, 30°]$.

Once **C** is predicted according to the X-Bragg scattering model, the co-polarized phase difference can be obtained:

$$
\angle C_{X_{13}} = \angle \left(\frac{1}{2}|B_h + B_v|^2 - \frac{1}{4}|B_h - B_v|^2\left(1 - \mathrm{sinc}(4\beta)\right) - j\Im(B_h{}^* + B_v{}^*)(B_h - B_v)\mathrm{sinc}(2\beta) \right) , \tag{5}
$$

where $\angle(\cdot)$ and $\Im(\cdot)$ mean phase and imaginary part, respectively. Note that Equation (5) allows explicitly pointing out the relationship between the co-polarized phase difference and SAR acquisition parameters (through the Bragg scattering coefficients, i.e., AOI and frequency) and geometric/dielectric properties of sea surface (through β and ϵ). Hence, it is a starting point to understand the effects of such parameters on the co-polarized phase difference distribution.

3. Datasets

In this section, the SAR dataset and the ancillary data are briefly described.

3.1. SAR Dataset

The L-, C- and X-band polarimetric SAR dataset consists of 14 scenes acquired between June 2006 and July 2013. The two L-band Single-look complex (SLC) full-polarimetric Uninhabited Aerial Vehicle Synthetic Aperture Radar (UAVSAR) scenes were acquired with an AOI ranging from about 22° to

65° in the northern part of the Gulf of Mexico. The UAVSAR system operates at center frequency of 1.26 GHz and is characterized by a NESZ of about −53 dB at mid-swath [42].

The four L-band SLC full-polarimetric ALOS PALSAR-1 (AP) scenes were acquired between June 2006 and April 2009 with an AOI from about 22° to 25° in the northern part of the Gulf of Mexico and off the northeastern coast of Brazil. The Japanese AP system operates at center frequency of 1.27 GHz with a NESZ of −29 dB at near range (21.5°) [43].

The four C-band SLC full-polarimetric RADARSAT-2 (RS) scenes were collected between January 2009 and May 2010 with AOI ranging from about 22° to 41° in the northern part of the Gulf of Mexico and off the California coast. The Canadian RS system operates at center frequency of 5.40 GHz with a NESZ of −35 dB [44].

The four X-band SLC dual co-polarimetric HH-VV TSX scenes were collected between December 2011 and July 2012 with an AOI range of 25° to about 35° in the northern part of the Gulf of Mexico. The German TSX system operates at center frequency of 9.60 GHz with a mean NESZ of −22 dB [45].

Figure 1 shows excerpts (1000 × 1000 pixels) of the VV-polarized intensity images of the whole SAR dataset representing homogeneous sea surface areas. Figure 1 is organized in a matrix format, where rows refer to the different SAR sensors (UAVSAR, AP, RS and TSX from the top to the bottom) and columns are arranged according to increasing AOI (from left to right). Note that, even though the same decibel (dB) scale is adopted, i.e., [0 -40], the sea surface patterns and, therefore, the backscattering change significantly according to the SAR imaging parameters and meteo-marine conditions. An overview of the SAR dataset is provided in Tables 1 and 2. Note that the whole SAR dataset is partitioned according to three AOI ranges: low (22–27°), intermediate (31–35°) and high (38–42°) (see Table 2).

Table 1. General information on SAR imagery.

SAR Sensor	UAVSAR	AP	RS	TSX
Frequency (GHz)	1.26	1.27	5.40	9.60
Imaging mode	Full-polarimetric	Full-polarimetric	Full-polarimetric	Dual co-polarimetric
Slant range x azimuth resolution (m)	1.7 x 1.0	9.4 x 3.6	4.7 x 5.1	1.2 x 6.6
Number of scenes	2	4	4	4
Nominal NESZ (dB)	−53	−29	−35	−22

Table 2. Overview of the dataset.

	SAR Data			Ancillary Wind Info	
SAR Sensor	Scene ID	Acquisition Date	AOI Range (°)	Speed (m/s)	Direction (°)
UAVSAR	1	25/07/2013	22.0–27.0 31.0–35.0 38.0–42.0	4.2	115.4
	2	22/06/2010	22.0–26.0 31.0–35.0 38.0–42.0	7.1	321.0
AP	3	02/04/2009	22.7–25.0	3.1	232.0
	4	20/06/2006	22.7–25.0	4.9	298.6
	5	14/03/2007	22.7–25.0	7.4	332.2
	6	19/03/2009	22.7–25.0	8.0	243.3
RS	7	31/01/2009	22.6–24.2	9.0	215.1
	8	04/05/2010	23.4–25.3	4.3	129.3
	9	26/09/2009	31.3–33.0	4.7	142.9
	10	01/05/2010	39.3–40.7	12.0	331.3
TSX	11	01/07/2012	25.0–26.7	4.1	1.3
	12	05/12/2011	25.0–26.7	9.2	5.1
	13	17/05/2012	33.0–34.5	3.8	168.6
	14	19/06/2012	33.0–34.5	8.4	305.0

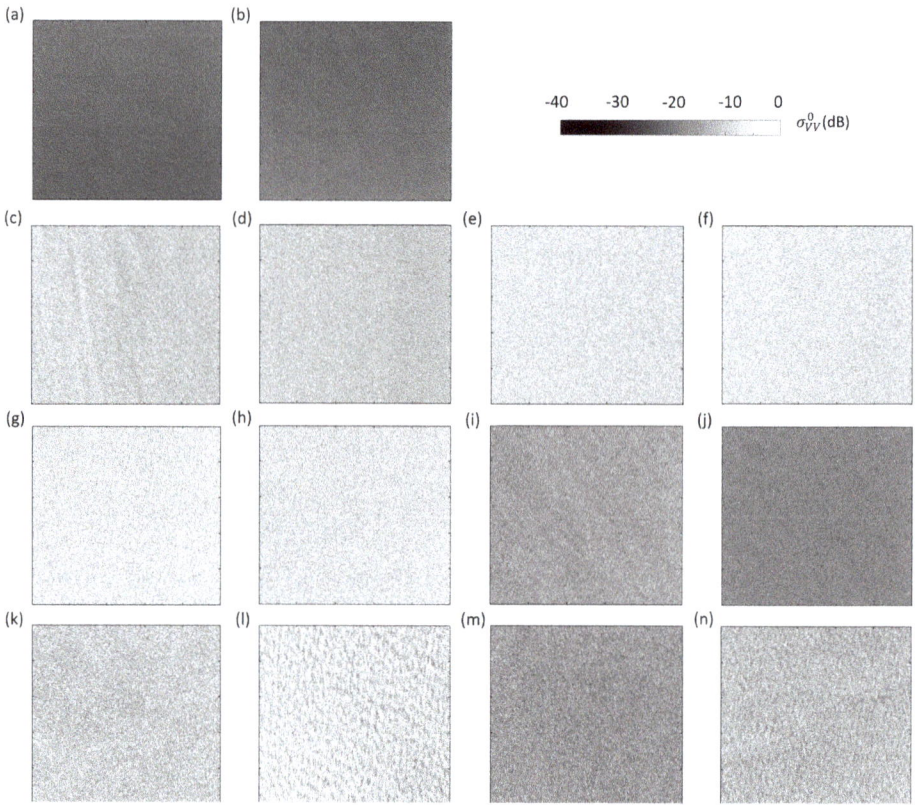

Figure 1. SAR dataset. VV-polarized intensity images (excerpts' size: 1000 × 1000), in dB scale, over homogeneous sea surface area. Rows refer to the different SAR sensors: (**a,b**) UAVSAR; (**c–f**) AP; (**g–j**) RS; and (**k–n**) TSX. Columns are organized according to increasing AOI: (**a,c,g,k**) low, i.e., 22–27°; (**d,e,h,i,l,m**) intermediate, i.e., 31–35°; and (**b,f,j,n**) high, i.e., 38–42°.

3.2. Ancillary Wind Field Information

The wind information was provided by the Physical Oceanography Distributed Active Archive Center (PO.DAAC) that makes available science data to a wide user community [46–48]. The wind information was collected from two different sources depending on the SAR imagery acquisition period. The wind field co-located to SAR scenes acquired before 2012 was collected from the Cross-Calibrated Multi-Platform (CCMP) project, while the wind field related to the SAR scenes acquired since 2012 was collected from the Advanced Scatterometer (ASCAT) instrument on MetOp-A (Meteorological Operational Satellites) satellite.

The CCMP product is characterized by a spatial resolution of about 25 km and combines cross-calibrated satellite winds from RMS Remote Sensing Systems derived from microwave radiometers and from scatterometers, in situ measurements and reanalysis data from the European Center for Medium-Range Weather Forecasts (ECMWF) ERA-40 Reanalysis [49]. The wind products obtained from ASCAT are characterized by an effective resolution of 50 km and are delivered by the European Organization for the Exploitation of Meteorological Satellites (EUMETSAT) Ocean and Sea Ice Satellite Application Facility (OSI SAF) through the Royal Netherlands Meteorological Institute (KNMI) [50].

The SAR scenes were all collected under low-to-moderate wind conditions, i.e., from about 3 m/s to 12 m/s. The wind information is summarized in Table 2.

4. Model-Based Experimental Results

In this section, the behavior of $\sigma_{\varphi C}$ against AOI and β is discussed using both the theoretical model presented in Section 2 and actual polarimetric SAR measurements collected by UAVSAR (see Tables 1 and 2). The latter, on the one side, is characterized by a very low NESZ that guarantees high-quality measurements even at larger AOIs, while, on the other side, UAVSAR scenes cover a broad range of AOIs (spanning from about 20° to 60°) and, therefore, they make possible comparing theoretical predictions with actual measurements over the whole range of Bragg AOIs.

To predict realistic $\sigma_{\varphi C}$ values, we used β values estimated from the UAVSAR scenes using the formula proposed in [51]:

$$\text{sinc}(4\beta) = \frac{Tr(\mathbf{C}) - 2C_{22} - 0.5\Re(C_{13})}{Tr(\mathbf{C}) - 0.5\Re(C_{13})} \quad , \tag{6}$$

where $Tr(\cdot)$ is the trace operator and $\Re(\cdot)$ means real part. Hence, for each SAR scene, three equal-sized homogeneous sea surface regions of interest (ROIs) were excerpted that are characterized by low (22–27°), intermediate (31–35°) and high (38–42°) AOI, respectively. Accordingly, to better understand the behavior of β, normalized histograms were computed (see Figure 2). It can be noted that the empirical statistical distribution of β (see blue histograms) resembles a Gaussian bell for any AOI, as witnessed by the fitted distribution (see red curves) that satisfies the chi-square test with 0.05 confidence interval.

Once β was obtained, to predict $\sigma_{\varphi C}$, a Monte Carlo approach based on 1000 independent simulations for each AOI was implemented.

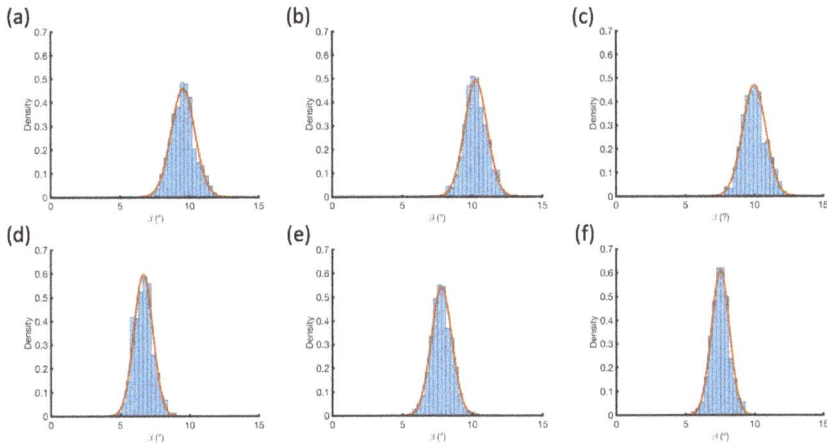

Figure 2. Normalized β histograms and fitted Gaussian distributions referring to: (**a–c**) UAVSAR scene ID 1 acquired at low (22–27°), intermediate (31–35°) and high (38–42°) AOIs, respectively; and (**d–f**) the same for UAVSAR scene ID 2.

For each simulation, β was randomly selected according to a Gaussian distribution whose mean and standard deviation values were obtained from the UAVSAR ROIs, i.e., 9.9° and 0.8° for the UAVSAR scene ID 1 and 7.3° and 0.7° for the UAVSAR scene ID 2 (see Table 3).

Simulation results are depicted in Figure 3a, where the continuous and dashed lines are related to β extracted from UAVSAR scenes ID 1 and ID 2, respectively. It can be noted that, in both cases, $\sigma_{\varphi C}$ increases with AOI, with $\sigma_{\varphi C}$ values ranging from about 0.01° at near range to about 0.15° at far range.

σ_{φ_C} values measured from actual UAVSAR scenes using a 9×9 moving window are depicted in Figure 3b, where continuous and dashed lines stand for UAVSAR scene ID 1 and ID 2, respectively. It can be noted that both measured and simulated σ_{φ_C} increase with the incidence angle. This witnesses that the X-Bragg model succeeds in predicting the actual behavior of σ_{φ_C} with respect to AOI. However, the simulated and measured σ_{φ_C} values are completely different, i.e., the simulated σ_{φ_C} values are about one order of magnitude smaller (measured σ_{φ_C} values range from about 0.5° at near range to about 4° at far range). Nonetheless, this is not a surprising result. The theoretical scattering model predicts a noise-free behavior referred to a low-depolarizing Bragg scattering surface, i.e., the co-polarized phase difference distribution should resemble, ideally, a Dirac delta function (i.e., $\sigma_{\varphi_C} = 0°$), while UAVSAR measurements, although very accurate, are noisy and refer to a real scattering surface, i.e., sensor and environmental parameters cause decorrelation between the co-polarized channels that results in a broader co-polarized phase difference distribution (i.e., larger σ_{φ_C} values). Note that a similar trend for σ_{φ_C} is experimentally observed in [42]. Notwithstanding that, the global σ_{φ_C} behavior with respect to AOI is well-described by the X-Bragg scattering model.

Figure 3 also shows that the model succeeds in predicting σ_{φ_C} values related to the UAVSAR scene ID 1, which are slightly larger than the corresponding ones related to the UAVSAR scene ID 2. This is because actual β values were used in the simulations.

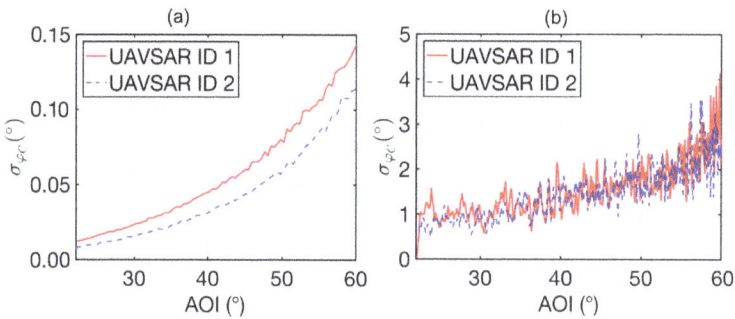

Figure 3. Simulated (**a**); and measured (**b**) behavior of σ_{φ_C} with respect to AOI relevant to UAVSAR scenes ID 1 (red line) and 2 (blue dashed line).

5. Experimental Results

The sensitivity of σ_{φ_C} with respect to acquisition parameters, i.e., AOI, NESZ and incident wavelength, and meteo-marine conditions, i.e., WS and β, was analyzed using actual polarimetric SAR data (Tables 1 and 2). When dealing with SAR imaging parameters, hardware and technical constraints suggest that there is a close relationship among NESZ, AOI and wavelength [45,52]. In addition, from a scattering viewpoint, when a Bragg surface backscattering is considered, the larger is the AOI, the lower is the signal-to-noise ratio (SNR), i.e., the mean ratio between co-polarized sea surface backscattering and the nominal NESZ [36,42]. When dealing with meteo-marine parameters, both β and near-surface wind speed are related to geometrical sea surface characteristics, i.e., sea surface roughness. Hence, a preliminary analysis devoted to understanding the relationship between β and WS is due.

To accomplish this task, the normalized β histograms were evaluated for each ROI belonging to AP and RS SAR scenes for low, intermediate and high AOIs (see Figure 4), where the mean WS values obtained from the external sources described in Section 3.2 were also annotated. First, it can be noted that β normalized histograms are broader (about 2–3 times) than the corresponding UAVSAR ones (see Figure 2). This is most likely due to the higher NESZ that characterizes AP and RS rather than rougher sea state conditions (see Table 3). In addition, no clear relationship between WS and β was observed. This suggests that, although both β and WS are related to the sea surface roughness, there is no clear link between them. Consequently, the sensitivity of $\sigma_{\varphi C}$ with respect to WS and β deserves to be analyzed separately.

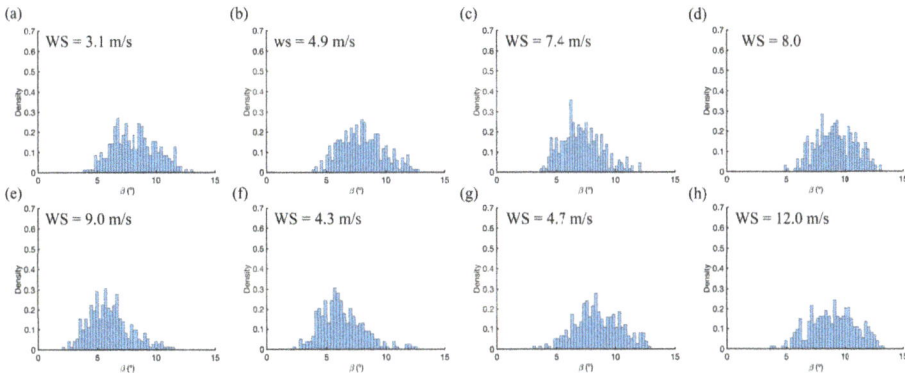

Figure 4. Normalized β histograms, evaluated over the homogeneous sea surface ROIs. The figure is organized in a matrix format. Rows refer to: AP (**a–d**); and RS (**e–h**) SAR sensors. Columns refer to the AOI: (**a,e**) low (22–27°); (**b,c,f,g**) intermediate (31–35°); and (**d,h**) high (38–42°). The corresponding mean WS values are also annotated.

Hence, the whole SAR dataset was processed according to the methodology described in Section 4 for the UAVSAR case. The only difference relies on the fact that AP, RS and TSX SAR scenes are characterized by a significantly narrower AOI range if compared to UAVSAR (less than 3° from near to far range, see Table 2) and, therefore, only one ROI was selected for each scene. $\sigma_{\varphi C}$ maps, obtained according to Migliaccio et al. [28], are shown in Figure 5, where it can be noted how $\sigma_{\varphi C}$ is affected by both SAR imaging parameters and environmental conditions. In fact, $\sigma_{\varphi C}$ values range from about 0° to more than 15° along the whole dataset. The lowest $\sigma_{\varphi C}$ values are shown in Figure 5a,b, corresponding to UAVSAR acquisitions (L-band, NESZ = −53 dB), while the largest values are related to TSX acquisitions (X-band, NESZ = −22 dB) (see Figure 5m,n). In addition, since columns in Figure 5 are organized according to increasing AOI (as in Figure 1), it can be noted that $\sigma_{\varphi C}$ increases with AOIs, as suggested by the model-based analysis presented in Section 4.

The visual interpretation was confirmed by the quantitative analysis undertaken on 1000 independent samples, randomly selected from the ROIs, whose $\sigma_{\varphi C}$ mean and standard deviation values were evaluated together with the average SNR (see Table 3). In Table 3, mean and standard deviation values of the β parameter estimated from the whole dataset according to Equation (6) are also listed. It must be noted that β values related to TSX SAR scenes are not available since full-polarimetric information is needed to estimate β. The following subsections focus on the quantitative analysis of the effects produced by sensor and scene parameters on sea surface $\sigma_{\varphi C}$, separately.

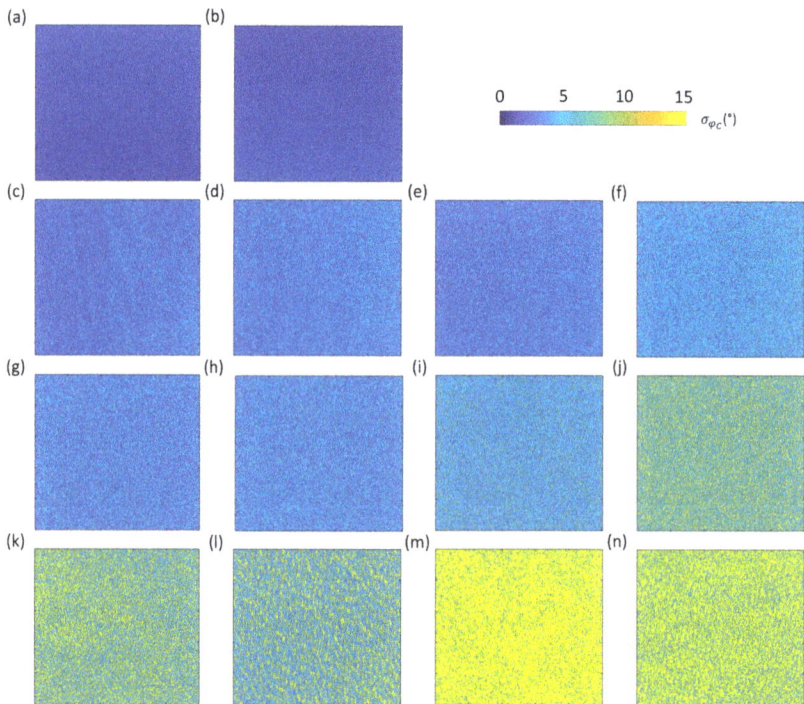

Figure 5. σ_{φ_C} (°) images evaluated over the ROIs shown in Figure 1. Rows refer to the different SAR sensors: (**a,b**) UAVSAR; (**c–f**) AP; (**g–j**) RS; and (**k–n**) TSX. Columns are organized according to increasing AOI: (**a,c,g,k**) low, i.e., 22–27; (**d,e,h,i,l,m**) intermediate, i.e., 31–35; and (**b,f,j,n**) high, i.e., 38–42.

5.1. Sensitivity Analysis: Meteo-Marine Parameters

The influence of WS and β on σ_{φ_C} was analyzed. When dealing with the effects of WS on σ_{φ_C}, according to Table 2, the AP SAR imagery were considered since the four SAR scenes were acquired under very similar SAR imaging parameters, i.e., the same AOI ($\approx 24°$) and incident wavelength (L-band), and by almost the same estimated mean SNR, i.e., ≈ 19–22 dB for both channels (see Table 3). Results listed in Table 3 clearly point out that, although the AP SAR scenes were collected under different wind conditions, i.e., ≈ 3 m/s and 8 m/s for SAR scene ID 3 and 6, respectively, they are characterized by very similar σ_{φ_C} mean values, i.e., $\approx 3.1°$. In addition, WS does not affect significantly the variability of σ_{φ_C}—see the standard deviation of σ_{φ_C} in Table 3, which is close to $\approx 1.3°$ for all AP SAR scenes. The same comments apply for SAR scene ID 4 and 5. Hence, it can be concluded that, under low-to-moderate wind conditions, σ_{φ_C} is almost unaffected by WS.

To discuss the effects of β on σ_{φ_C}, it is worth analyzing results that refer to AP and RS SAR imagery—see ID 5–8 in Table 3, since they refer to SAR scenes collected under similar acquisition parameters, i.e., L-band, AOI of $\approx 24°$ and average SNR of about 21 dB (C-band, AOI of $\approx 24°$ and average SNR of about 28 dB). It can be noted that, from the results in Table 3, SAR scenes ID 5 and 6 are characterized by almost the same σ_{φ_C} values, i.e., $3.1° \pm 1.3°$, although SAR scene ID 6 is characterized by a β value that is about 30% larger than the one related to the SAR scene ID 5. The same comments apply for SAR scene ID 7 and 8, which are characterized by almost the same σ_{φ_C} values, i.e., $4.0° \pm 2.0°$ and $3.7° \pm 1.7°$, respectively, although the β parameter estimated from SAR scene ID 7 is about 25% larger than the one evaluated from SAR scene ID 8.

Hence, this experimental analysis shows that there is no clear trend between β and $\sigma_{\varphi C}$. In particular, β does not play a dominant role in broadening/shrinking the co-polarized phase difference distribution.

Table 3. Values of $\sigma_{\varphi C}$, β and HH and VV SNR evaluated within ROIs selected over a homogeneous sea surface area. $\sigma_{\varphi C}$ and β are presented as mean ± standard deviation value.

SAR Sensor	ID	AOI (°)	$\sigma_{\varphi C}$ (°)	WS (m/s)	β (°)	SNR (dB) HH	VV
UAVSAR	1	22.0–27.0	1.0 ± 0.4		9.6 ± 0.8	39.0	38.7
		31.0–35.0	1.1 ± 0.5	4.2	10.2 ± 0.9	35.6	35.9
		38.0–42.0	1.4 ± 0.5		10.0 ± 0.7	30.1	32.1
	2	22.0–26.0	0.7 ± 0.3		6.6 ± 0.7	38	40.3
		31.0–35.0	1.0 ± 0.4	7.1	7.7 ± 0.7	35.7	38.7
		38.0–42.0	1.3 ± 0.5		7.5 ± 0.6	30.2	34.9
AP	3	22.7–25.0	3.0 ± 1.3	3.1	8.3 ± 2.0	18.7	20.3
	4	22.7–25.0	3.2 ± 1.3	4.9	8.2 ± 2.0	19.3	20.3
	5	22.7–25.0	3.2 ± 1.4	7.4	7.3 ± 1.7	21.2	22.0
	6	22.7–25.0	3.0 ± 1.2	8.0	9.5 ± 2.3	20.8	21.5
RS	7	22.6–24.2	4.0 ± 2.0	9.0	5.9 ± 1.7	27.8	28.2
	8	23.4–25.3	3.7 ± 1.7	4.3	6.7 ± 1.9	27.4	27.8
	9	31.3–33.0	5.0 ± 2.3	4.7	8.4 ± 2.6	19.2	21.4
	10	39.3–40.7	8.0 ± 4.0	12.0	9.6 ± 3.2	14.5	18.2
TSX	11	25.0–26.7	8.8 ± 6.4	4.1	–	9.2	9.8
	12	25.0–26.7	7.4 ± 5.7	9.2	–	10.2	10.4
	13	33.0–34.5	18.4 ± 16.5	3.8	–	2.7	3.8
	14	33.0–34.5	12.6 ± 10.0	8.4	–	5.8	7.9

5.2. Sensitivity Analysis: SAR Imaging Parameters

The influence of noise floor, incidence angle and wavelength on $\sigma_{\varphi C}$ was analyzed. As explicitly pointed out above, it is not straightforward to isolate their individual contribution since they are inter-connected.

To analyze the effects of NESZ on $\sigma_{\varphi C}$, SAR scenes collected by the L-band UAVSAR and AP platforms were considered since they are characterized by completely different NESZ values, i.e., -53 dB and -29 dB, respectively (see Table 1), while sharing the same operating frequency. In addition, the AOI range that characterizes AP SAR scenes was included into the one provided by UAVSAR. Hence, those datasets allow a fair analysis of the impact of noise floor in broadening/shrinking the co-polarized phase difference distribution. UAVSAR scenes ID 1 and ID 2 and AP SAR scenes ID 3–6 are characterized by a SNR equal to $\approx$39 dB and $\approx$21 dB, respectively (see Table 3). Their mean $\sigma_{\varphi C}$ values are 0.85° and 3.1°, respectively. Hence, one can note that, when halving the SNR, $\sigma_{\varphi C}$ increases of about four times. In addition, results listed in Table 3 show that a lower SNR corresponds to a larger $\sigma_{\varphi C}$ variability, whose standard deviation increases from 0.35° (UAVSAR scenes ID 1–2, low AOI) to 1.3° (AP SAR scenes ID 3–6), i.e., about four times.

When dealing with the effects of AOI on $\sigma_{\varphi C}$, first results presented and discussed in Section 4, obtained considering UAVSAR imagery, suggest that AOI significantly affects $\sigma_{\varphi C}$ (see Figure 3). It was found that UAVSAR $\sigma_{\varphi C}$ increases of about 23% (29%) when moving from low to intermediate (intermediate to high) AOI. Nonetheless, to further confirm those results, RS and TSX SAR datasets were analyzed according to Table 2. In particular, RS (TSX) SAR scene ID 9 (11) and SAR scene ID 10 (14) were considered for lower and higher AOI, respectively. It must be underlined that both couples of SAR images were acquired under the same incident wavelength and they are characterized by almost the same SNR. When dealing with RS SAR scenes, an increase of about 25% in AOI results in an increase of about 60% in the $\sigma_{\varphi C}$ mean value, i.e., from 5° to 8° when AOI moves from about 32° to about 40° (see Table 3). When dealing with TSX SAR scenes, the σ_{φ_C} mean value grows from about 9° to about 13° ($\approx$ 43%) when the AOI increases from about 26° to about 34° (31%) (see Table 3). The variability of σ_{φ_C} also increases when AOI increases.

When dealing with the effects of incident wavelength on σ_{φ_C}, according to Table 1, a fair analysis cannot be undertaken since each SAR platform is characterized by its NESZ. However, the L-band AP and C-band RS SAR imagery are considered since SAR scenes are characterized by the closest NESZ values within the dataset ($\approx$6 dB difference). In addition, they were observed under almost the same AOI (see SAR scene ID 3–8 in Table 3). It can be observed, in Table 3, that, when moving from L- to C-band SAR images, σ_{φ_C} values increase, on average, of $\approx$24% (from 3.1° to 3.85°), even though it must be pointed out that the mean SNR that characterizes L-band SAR measurements, i.e., 20.5 dB, is about 36% lower than the corresponding C-band one, i.e., 27.8 dB. Hence, although the larger mean SNR provided by C-band RS SAR measurements with respect to L-band ones would result in significantly lower σ_{φ_C} values (as discussed previously), an increasing trend of σ_{φ_C} values was observed when moving from L- to C-band SAR imagery, witnessing the key role played by the incident wavelength on the co-polarized phase difference distribution. In addition, the σ_{φ_C} variability, i.e., the standard deviation of σ_{φ_C}, also increases from L- to C-band of $\approx$42%. Results relevant to TSX SAR imagery listed in Table 3 also confirm the increasing trend of σ_{φ_C} when decreasing the incident wavelength. It can be observed how, under almost the same AOI, σ_{φ_C} values approximately doubled when moving from C- to X-band (see SAR scenes ID 7–8 and 11–12, and SAR scenes ID 9 and 14). Nonetheless, a completely different behavior applies for TSX SAR scene ID 13 that is characterized by the largest σ_{φ_C} mean and standard deviation values within the whole SAR dataset, i.e., 18.4° $\pm$ 16.5°. This is likely due to the fact that SAR scene ID 13 is severely corrupted by noise due to the high TSX NESZ (see Table 1): the average SNR lies very close to 3 dB, which is a threshold value usually adopted to judge the reliability of SAR measurements [24,42]).

6. Conclusions

A theoretical and experimental study aimed at investigating the sensitivity of σ_{φ_C} with respect to SAR acquisition parameters (NESZ, AOI, and incident wavelength) and meteo-marine conditions (WS and β) was performed. The X-Bragg polarimetric scattering model was adopted as a reference scattering framework to predict the behavior of σ_{φ_C} over sea surface, while experiments on actual measurements were accomplished considering a polarimetric SAR dataset that consists of 14 scenes collected over sea surface under different imaging configurations and environmental conditions. The main outcomes of this study are summarized as follows:

- The X-Bragg sea surface scattering model allows predicting the increasing trend of σ_{φ_C} with respect to AOI over sea surface along the whole range of Bragg scattering incidence angles, i.e., $\approx$20–60°.
- Under low-to-moderate sea state conditions, SAR imaging parameters have a stronger effect on σ_{φ_C} than meteo-marine parameters, which play a negligible role.
- Among SAR imaging parameters, incident wavelength and NESZ result in the most pronounced effect on sea surface σ_{φ_C}.

These outcomes can altogether support the design of polarimetric SAR architectures/algorithms that aim at enhancing the contrast between a given marine target of interest (ships, surfactants, icebergs, etc.) and sea clutter. Future works may include the extension of such sensitivity analysis on a larger SAR dataset (i.e., to include higher wind regimes) and the application of the proposed approach to find the most suitable SAR configuration to observe reference targets, e.g., ships.

Author Contributions: Conceptualization, A.B. and M.M.; Methodology, A.B. and F.N.; Software, A.B. and C.R.d.M.; Validation, C.R.d.M.; Formal Analysis, A.B. and F.N.; Investigation, A.B., C.R.d.M., F.N. and Domenico Velotto; Data Curation, C.R.d.M. and D.V.; Writing—Original Draft Preparation, A.B. and C.R.d.M.; Writing—Review and Editing, A.B. and F.N.; and Supervision, F.N., D.V. and M.M.

Funding: This study was partly funded by the Universitá degli Studi di Napoli Parthenope, project ID DING 202 and by the European Space Agency under the Dragon 4 project ID 32235.

Acknowledgments: The authors would like to thank the German Aerospace Center (DLR) that provided the TerraSAR-X SAR data under the AO OCE1045, NASA JPL that provided free of charge the UAVSAR SAR data and the wind field information under the PO.DAAC archive, the Japanese Space Agency (JAXA) that provided the ALOS PALSAR-1 SAR data under the RA-6 project ID 3064, and the Canadian Space Agency (CSA) that provided the RADARSAT-2 SAR data under the SOAR-EU project.

Conflicts of Interest: The authors declare no conflict of interest. The founding sponsors had no role in the design of the study; in the collection, analyses, or interpretation of data; in the writing of the manuscript, and in the decision to publish the results.

Abbreviations

The following abbreviations are used in this manuscript:

AOI	Angle of incidence
AP	ALOS PALSAR
ASCAT	Advanced Scatterometer
CCMP	Cross calibrated multi platform
dB	Decibel
ECMWF	European Center for Medium-Range Weather Forecasts
EUMETSAT	European Organization for the Exploitation of Meteorological Satellites
HH	Horizontal transmit-horizontal receive
KNMI	Royal Netherlands Meteorological Institute
ID	Identifier
MetOp	Meteorological operational satellite
NASA	National Aeronautics and Space Administration
NESZ	Noise-equivalent sigma zero
OSI SAF	Ocean and Sea Ice Satellite Application Facility
PO.DAAC	Physical Oceanography Distributed Active Archive Center
RMS	Remote Sensing Systems
ROI	Region of interest
RS	RADARSAT-2
SAR	Synthetic aperture radar
SLC	Single-look complex
SNR	Signal-to-noise ratio
TSX	TerraSAR-X
VV	Vertical transmit-vertical receive
WS	Wind speed

References

1. Costanza, R. The ecological, economic, and social importance of the oceans. *Ecol. Econ.* **1999**, *31*, 199–213. [CrossRef]
2. Visbeck, M. Ocean science research is key for a sustainable future. *Nat. Commun.* **2018**, *9*, 690. [CrossRef] [PubMed]
3. Fanning, L.; Mahon, R.; Baldwin, K.; Douglas, S. Transboundary waters assessment Programme (TWAP) assessment of governance arrangements for the ocean, Volume 1: Transboundary large marine ecosystems. *IOC Tech. Ser.* **2015**, *119*, 91.
4. European Commission. *Report on the Blue Growth Strategy Towards More Sustainable Growth and Jobs in the Blue Economy*; European Commission: Brussels, Belgium, 2017.
5. Nations, U. *The Sustainable Development Goals Report 2016*; United Nations: New York, NY, USA, 2016.
6. National Academies of Sciences, Engineering, and Medicine and others. Active earth remote sensing for ocean applications. In *A Strategy for Active Remote Sensing Amid Increased Demand for Radio Spectrum*; National Academies Press: Washington, DC, USA, 2015.

7.   Ding, X.; Nunziata, F.; Li, X.; Migliaccio, M. Performance analysis and validation of waterline extraction approaches using single-and dual-polarimetric SAR data. *IEEE J. Sel. Top. Appl. Earth Observ. Remote Sens.* **2015**, *8*, 1019–1027. [CrossRef]

8.   Nunziata, F.; Buono, A.; Migliaccio, M.; Benassai, G. Dual-polarimetric C-and X-band SAR data for coastline extraction. *IEEE J. Sel. Top. Appl. Earth Observ. Remote Sens.* **2016**, *9*, 4921–4928. [CrossRef]

9.   Yu, Y.; Acton, S.T. Automated delineation of coastline from polarimetric SAR imagery. *Int. J. Remote Sens.* **2004**, *25*, 3423–3438. [CrossRef]

10.   Velotto, D.; Nunziata, F.; Migliaccio, M.; Lehner, S. Dual-polarimetric TerraSAR-X SAR data for target at sea observation. *IEEE Geosci. Remote Sens. Lett.* **2013**, *10*, 1114–1118. [CrossRef]

11.   Nunziata, F.; Migliaccio, M.; Brown, C.E. Reflection symmetry for polarimetric observation of man-made metallic targets at sea. *IEEE J. Ocean. Eng.* **2012**, *37*, 384–394. [CrossRef]

12.   Li, H.; Perrie, W.; He, Y.; Lehner, S.; Brusch, S. Target detection on the ocean with the relative phase of compact polarimetry SAR. *IEEE Trans. Geosci. Remote Sens.* **2013**, *51*, 3299–3305. [CrossRef]

13.   Migliaccio, M.; Nunziata, F. On the exploitation of polarimetric SAR data to map damping properties of the Deepwater Horizon oil spill. *Int. J. Remote Sens.* **2014**, *35*, 3499–3519. [CrossRef]

14.   Migliaccio, M.; Nunziata, F.; Buono, A. SAR polarimetry for sea oil slick observation. *Int. J. Remote Sens.* **2015**, *36*, 3243–3273. [CrossRef]

15.   Buono, A.; Nunziata, F.; Migliaccio, M.; Li, X. Polarimetric analysis of compact-polarimetry SAR architectures for sea oil slick observation. *IEEE Trans. Geosci. Remote Sens.* **2016**, *54*, 5862–5874. [CrossRef]

16.   Velotto, D.; Migliaccio, M.; Nunziata, F.; Lehner, S. Dual-polarized TerraSAR-X data for oil-spill observation. *IEEE Trans. Geosci. Remote Sens.* **2011**, *49*, 4751–4762. [CrossRef]

17.   Marino, A.; Dierking, W.; Wesche, C. A depolarization ratio anomaly detector to identify icebergs in sea ice using dual-polarization SAR images. *IEEE Trans. Geosci. Remote Sens.* **2016**, *54*, 5602–5615. [CrossRef]

18.   Dabboor, M.; Geldsetzer, T. Towards sea ice classification using simulated RADARSAT Constellation Mission compact polarimetric SAR imagery. *Remote Sens. Environ.* **2014**, *140*, 189–195. [CrossRef]

19.   Gill, J.P.; Yackel, J.J. Evaluation of C-band SAR polarimetric parameters for discrimination of first-year sea ice types. *Can. J. Remote Sens.* **2012**, *38*, 306–323. [CrossRef]

20.   Touzi, R.; Charbonneau, F.; Hawkins, R.; Vachon, P. Ship detection and characterization using polarimetric SAR. *Can. J. Remote Sens.* **2004**, *30*, 552–559. [CrossRef]

21.   Nunziata, F.; Buono, A.; Migliaccio, M. COSMO-SkyMed Synthetic Aperture Radar Data to Observe the Deepwater Horizon Oil Spill. *Sustainability* **2018**, *10*, 3599. [CrossRef]

22.   Kudryavtsev, V.N.; Chapron, B.; Myasoedov, A.G.; Collard, F.; Johannessen, J.A. On dual co-polarized SAR measurements of the ocean surface. *IEEE Geosci. Remote Sens. Lett.* **2013**, *10*, 761–765. [CrossRef]

23.   Liu, C.; Vachon, P.; Geling, G. Improved ship detection with airborne polarimetric SAR data. *Can. J. Remote Sens.* **2005**, *31*, 122–131. [CrossRef]

24.   Minchew, B. Determining the mixing of oil and sea water using polarimetric synthetic aperture radar. *Geophys. Res. Lett.* **2012**, *39*. [CrossRef]

25.   Marino, A.; Velotto, D.; Nunziata, F. Offshore Metallic Platforms Observation Using Dual-Polarimetric TS-X/TD-X Satellite Imagery: A Case Study in the Gulf of Mexico. *IEEE J. Sel. Top. Appl. Earth Obs. Remote Sens.* **2017**, *10*, 4376–4386. [CrossRef]

26.   Ryu, E.; Yang, C.S.; Ouchi, K. Comparison of Ship Detection Accuracy Based on Image Contrast with Different Combinations HH-and VV-polarization TerraSAR-X SAR Images. *IEICE Tech. Rep.* **2017**, *117*, 13–16.

27.   Salberg, A.B.; Rudjord, Ø.; Solberg, A.H.S. Oil spill detection in hybrid-polarimetric SAR images. *IEEE Trans. Geosci. Remote Sens.* **2014**, *52*, 6521–6533. [CrossRef]

28.   Migliaccio, M.; Nunziata, F.; Gambardella, A. On the co-polarized phase difference for oil spill observation. *Int. J. Remote Sens.* **2009**, *30*, 1587–1602. [CrossRef]

29.   Dierking, W.; Wesche, C. C-Band radar polarimetry—Useful for detection of icebergs in sea ice? *IEEE Trans. Geosci. Remote Sens.* **2014**, *52*, 25–37. [CrossRef]

30.   Moen, M.A.; Doulgeris, A.P.; Anfinsen, S.N.; Renner, A.H.; Hughes, N.; Gerland, S.; Eltoft, T. Comparison of feature based segmentation of full polarimetric SAR satellite sea ice images with manually drawn ice charts. *Cryosphere* **2013**, *7*, 1693–1705. [CrossRef]

31.   Joughin, I.R.; Winebrenner, D.P.; Percival, D.B. Probability density functions for multilook polarimetric signatures. *IEEE Trans. Geosci. Remote Sens.* **1994**, *32*, 562–574. [CrossRef]

32. Schuler, D.L.; Lee, J.S.; Hoppel, K.W. Polarimetric SAR image signatures of the ocean and Gulf Stream features. *IEEE Trans. Geosci. Remote Sens.* **1993**, *31*, 1210–1221. [CrossRef]

33. Lee, J.S.; Hoppel, K.W.; Mango, S.A.; Miller, A.R. Intensity and phase statistics of multilook polarimetric and interferometric SAR imagery. *IEEE Trans. Geosci. Remote Sens.* **1994**, *32*, 1017–1028.

34. Guissard, A. Phase calibration of polarimetric radars from slightly rough surfaces. *IEEE Trans. Geosci. Remote Sens.* **1994**, *32*, 712–715. [CrossRef]

35. Nunziata, F.; Gambardella, A.; Migliaccio, M. A unitary Mueller-based view of polarimetric SAR oil slick observation. *Int. J. Remote Sens.* **2012**, *33*, 6403–6425. [CrossRef]

36. Buono, A.; Nunziata, F.; de Macedo, C.R.; Velotto, D.; Migliaccio, M. A sensitivity analysis of the standard deviation of the co-polarized phase difference for sea oil slick observation. *IEEE Trans. Geosci. Remote Sens.* **2018**, doi:10.1109/TGRS.2018.2870738. [CrossRef]

37. Nunziata, F.; de Macedo, C.R.; Buono, A.; Velotto, D.; Migliaccio, M. On the analysis of a time series of X-band TerraSAR-X SAR imagery over oil seepages. *Int. J. Remote Sens.* **2018**, in print. [CrossRef]

38. Lee, J.S.; Pottier, E. *Polarimetric Radar Imaging: From Basics to Applications*; CRC Press: Boca Raton, FL, USA, 2009.

39. Yin, J.; Yang, J.; Zhou, Z.S.; Song, J. The extended Bragg scattering model-based method for ship and oil-spill observation using compact polarimetric SAR. *IEEE J. Sel. Top. Appl. Earth Observ. Remote Sens.* **2015**, *8*, 3760–3772. [CrossRef]

40. Buono, A.; Nunziata, F.; Migliaccio, M. Analysis of Full and Compact Polarimetric SAR Features over the Sea Surface. *IEEE Geosci. Remote Sens. Lett.* **2016**, *13*, 1527–1531. [CrossRef]

41. Hajnsek, I.; Pottier, E.; Cloude, S.R. Inversion of surface parameters from polarimetric SAR. *IEEE Trans. Geosci. Remote Sens.* **2003**, *41*, 727–744. [CrossRef]

42. Minchew, B.; Jones, C.E.; Holt, B. Polarimetric analysis of backscatter from the Deepwater Horizon oil spill using L-band synthetic aperture radar. *IEEE Trans. Geosci. Remote Sens.* **2012**, *50*, 3812–3830. [CrossRef]

43. EORC JAXA: ALOS Data Users Handbook Rev. C. 2008. Available online: https://www.eorc.jaxa.jp/ALOS/en/doc/fdata/ALOS_HB_RevC_EN.pdf (accessed on 5 October 2018).

44. Slade, B. *RADARSAT-2 Product Description*; Tech. Rep. RN-SP-S2-1238; MDA Ltd.: Richmond, BC, Canada, 2009.

45. Eineder, M.; Fritz, T. *TerraSAR-X Basic Product Specification Document*; Technical Report, TX-GS-DD-3302; 2010. Available online: https://www.dlr.de/dlr/Portaldata/1/Resources/documents/TX-GS-DD-3302_Basic-Product-Specification-Document_1_7.pdf (accessed on 18 October 2018).

46. Carvalho, D.; Rocha, A.; Gómez-Gesteira, M.; Alvarez, I.; Santos, C.S. Comparison between CCMP, QuikSCAT and buoy winds along the Iberian Peninsula coast. *Remote Sens. Environ.* **2013**, *137*, 173–183. [CrossRef]

47. Gao, G.; Wang, X.; Niu, M. Statistical modeling of the reflection symmetry metric for sea clutter in dual-polarimetric SAR data. *IEEE J. Ocean. Eng.* **2016**, *41*, 339–345.

48. Migliaccio, M.; Nunziata, F.; Montuori, A.; Paes, R.L. Single-look complex COSMO-SkyMed SAR data to observe metallic targets at sea. *IEEE J. Sel. Top. Appl. Earth Observ. Remote Sens.* **2012**, *5*, 893–901. [CrossRef]

49. Atlas, R.; Hoffman, R.N.; Ardizzone, J.; Leidner, S.M.; Jusem, J.C.; Smith, D.K.; Gombos, D. A cross-calibrated, multiplatform ocean surface wind velocity product for meteorological and oceanographic applications. *Bull. Am. Met. Soc.* **2011**, *92*, 157–174. [CrossRef]

50. Verhoef, A.; Stoffelen, A. ASCAT Wind Product User Manual Version 1.15. Document external project: SAF/OSI/CDOP/KNMI/TEC/MA/126; 2018, EUMETSAT. Available online: http://projects.knmi.nl/publications/fulltexts/ss3_pm_ascat_1.15.pdf (accessed on 18 October 2018).

51. Mattia, F.; Le Toan, T.; Souyris, J.C.; De Carolis, C.; Floury, N.; Posa, F.; Pasquariello, N. The effect of surface roughness on multifrequency polarimetric SAR data. *IEEE Trans. Geosci. Remote Sens.* **1997**, *35*, 954–966. [CrossRef]

52. Younis, M.; Huber, S.; Patyuchenko, A.; Bordoni, F.; Krieger, G. Performance Comparison of Reflector- and Planar-Antenna Based Digital Beam-Forming SAR. *IEEE J. Antennas Propag.* **2009**, *2009*, 614931. [CrossRef]

remote sensing

MDPI

Article

Extension of Ship Wake Detectability Model for Non-Linear Influences of Parameters Using Satellite Based X-Band Synthetic Aperture Radar

Björn Tings *, Andrey Pleskachevsky, Domenico Velotto and Sven Jacobsen

German Aerospace Center, Henrich-Focke-Str. 4, 28199 Bremen, Germany; andrey.pleskachevsky@dlr.de (A.P.); domenico.velotto@dlr.de (D.V.); sven.jacobsen@dlr.de (S.J.)
* Correspondence: bjoern.tings@dlr.de

Received: 29 January 2019; Accepted: 25 February 2019; Published: 7 March 2019

Abstract: The physics of the imaging mechanism underlying the emergence of ship wakes in Synthetic Aperture Radar (SAR) images has been studied in the past by many researchers providing a well-understood theory. Therefore, many publications describe how well ship wakes are detectable on SAR under the influence of different environmental conditions like sea state or local wind, ship properties like ship speed or ship heading, and image acquisition parameters like incidence angle or satellite heading. The increased imaging capabilities of current satellite SAR missions facilitate the collection of large datasets of moving vessels. Such a large dataset of high resolution TerraSAR-X acquisitions now enables the quantitative analysis of the previously formulated theory about the detectability of ship wakes using real data. In this paper we propose an extension of our wake detectability model by using a non-linear basis which allows consideration of all the influencing parameters simultaneously. Such an approach provides new insights and a better understanding of the non-linear influence of parameters on the wake detectability and their interdependencies can now be represented. The results show that the non-linear, interdependent influence of the different influencing parameters on the detectability of wakes matches well to the oceanographic expectations published in the past. Also possible applications of the model for the extraction of missing parameters and automatic for wake detection systems are demonstrated.

Keywords: Wake detection; Synthetic Aperture Radar; support vector machines; detectability model

1. Introduction

The detection of ships on space-borne Synthetic Aperture Radar (SAR) imagery is hardly possible, when the ship's construction material is non-conductive and in turn the ship's SAR-signatures are not or badly represented in the SAR images [1]. Instead of searching for the signatures of ships directly, their presence can be determined indirectly through the ship's wakes [2]. Since the automatic detection of ships on SAR has become of interest in earth observation, also the automatic detection of wakes is taken into account [3]. However, the maturity of automatic ship detection methods is further developed than of automatic wake detection methods, what is also reflected by the diverging amount of publications proposing different methods for the detection approaches [4–9].

A similar context can also be observed in the existence of approaches for modeling the detectability of ship or wake signatures in dependency to parameters influencing their detectability in SAR imagery. In the following these parameters will be called influencing parameters. While different approaches for modeling ship detectability have been published in past years and tuned for various SAR missions [10–13], a model for the detectability of wake signatures using real data has only been published recently [14]. However, theoretical assumptions about the dependency of influencing parameters with the detectability of certain components of wakes using simulated data and physical contemplations exist since decades [15]. The well-known effects of tilt modulation, hydrodynamic modulation, and velocity bunching are the basis for general theories about the emergence of ocean surface waves and their visibility on SAR imagery [16–18].

SAR signatures of ship wakes are categorized into the four main wake components: turbulent wake, narrow V-wakes, ship-generated internal waves, and the Kelvin wake pattern [19]. The turbulent wake consists of a rough sea region (white water) up to two ship beams aft the ship induced by the propeller and a calm sea region beginning after it and persisting up to tens of kilometers caused by the attenuation of ambient short surface waves by ascending bubbles and surfactants. The rough sea region is responsible of high radar backscatter due to the strong turbulence and breaking waves while the calm sea region is responsible of low radar backscatter due to the smoothed surface [20–22]. By turbulent wake in this study we only refer to the long and smooth region, as the white water region is observable almost independently from the influencing parameters and easily confused with the ship signature itself. The Kelvin wake pattern consists of transverse, divergent, and cusp waves [23]. The cusp waves lie on the Kelvin wake arms and constitute the dominant backscattering responsible for the detectability of the V-shaped Kelvin envelope [15,19]. In this study only high resolution SAR data acquired from the TerraSAR-X satellite at a center frequency of 9.65 GHz (X-band) is used. Therefore, the narrow V-wakes cannot be taken into account as their half angle on X-Band is so small that they are expected being consumed by the turbulent wake [19,24,25]. According to [26] the visibility of Bragg-based scattering generated by both narrow V-wakes and ship-generated internal waves depends on variety of mechanisms making them also difficult to discriminate. Indeed, Bragg waves can be modulated by ship-generated internal waves on the ocean surface making them also visible on X-Band [27], but their emergence depends among other things on the water depth and the presence of either surface films or interactions with surface currents [19,28]. These kinds of influences can neither be derived from SAR automatically nor be provided by oceanographic models in adequate resolution. The requirement for a large dataset in this study only permits the consideration of influencing parameters which are automatically derivable. This means internal waves cannot be regarded appropriately and the scope of this study is restricted to the detectability of only two main wake components: turbulent wake and Kelvin wake.

Literature is rich in publications describing the dependency of influencing parameters with the detectability of turbulent wake and Kelvin wake. Most of the conclusions are in good agreement with each other. The following paragraphs are dedicated to reviewing previous studies about detectability of ship wakes in dependency to the influencing parameters investigated in this study.

Lyden et al. [19] state that the whole Kelvin envelope is best detectable when observed in alignment with the azimuth direction, and the individual cusp waves are better visible when traveling into range direction. This statement is also supported by [15] and [23], while in [15] it is pointed out that cusp waves propagating towards the radar-looking direction show an increase in backscatter relative to the surrounding and cusp waves propagating away from the radar-looking direction show a decrease in relative backscatter. Lyden et al. [19] further suggest that for turbulent wakes the relative-looking direction to the ship track is less influencing, but a relative looking direction perpendicular to the ship track produces the best results.

In [19] the authors pointed out that turbulent wakes and Kelvin wakes are best visible under moderate wind conditions, i.e. wind speed in the value range 2.5–10 m/s [22]. Hence, the minimum wind speed considered in this study is 2 m/s being also the minimum wind speed retrievable from SAR imagery [29], and the maximum wind speed is 10 m/s. Hennings et al. [15] describe that the Kelvin wake arms are better visible with lower wind speeds, as the contrast between cusp waves and background sea surface roughness decreases under the influence of higher wind speeds. Additionally, an outcome of the analysis conducted in [15] is the low dependency of detectability of Kelvin wake arms to relative wind direction.

In [30] the author suggests that signatures of turbulent wakes are visible also under moderate sea state conditions due the destruction of the ambient wind-generated waves by the wake's currents, although their extent is larger in low sea state conditions. This is confirmed by Tunaley at al. [24] at least for large vessels. Additionally, wakes generated by large vessels with slow velocities would further be visible under low sea state conditions and azimuthal wind direction. Reference [24] further stated that the Kelvin wakes are less visible in high sea state conditions and proposed that velocity bunching produced by transverse waves is responsible for this. The worse detectability of turbulent and Kelvin wakes in high sea state conditions is also confirmed by Zilman et al. [18].

Kelvin wakes are expected being better visible in HH-polarized images compared to VV-polarized images [15,23]. For the turbulent wakes no definite conclusion about the difference regarding HH and VV polarization was found in the literature, but the SAR imaging of water surfaces smoothed by the turbulent wakes is similar compared to water surfaces smoothed by oceanic surface films. For oceanic surface films, Gade et al. [31] found that there is no significant difference between detectability on HH- or VV-polarized images. A slightly better total detectability on HH images compared to VV images was also found by [14,21]. As the difference in dependency of detectability to influencing parameters is insignificant, HH and VV are not distinguished in this study.

The influence of the radar's incidence angle on wake detectability is similar for turbulent and Kelvin wakes. While [15] states that the detectability of Kelvin wake arms decreases, when the incidence angle increases, according to [32] the detectability of smoothed ocean surfaces, such as turbulent wakes, also decreases with increasing incidence angle.

The model we proposed in [14] only takes three influencing parameters into account at a time and assumes a linear dependency between the influencing parameters and the detectability of wakes. Especially for influencing parameters with units measured by degree, a linear model basis is assumed insufficient. Further, all influencing parameters must be considered simultaneously in one model in order to obtain comparable probability of detection estimates. This paper presents results achieved using a model with higher complexity and able to take into account all the influencing parameters introduced in [14] together. Besides these influencing parameters, additional influencing parameters are included and evaluated. Finally, possible applications of the extended wake detectability model are demonstrated.

2. Materials and Methods

The flow-chart in Figure 1 displays the overall procedure divided in three main steps: Extraction of labelled wake samples (Figure 1A); extraction of the influencing parameters from these samples (Figure 1B); and building of the detectability model and its visualization (Figure 1C). The first two steps are fully described in the remainder of this section while the third step deserves a dedicated section which follows-on.

The data used in this study is based on a dataset of 791 high resolution TerraSAR-X scenes acquired between the years 2013 to 2017 in North Sea, Baltic Sea and Mediterranean Sea. The scenes were acquired in Stripmap or Spotlight mode mostly with HH-polarization (in detail: 530 HH-polarized, 81 VV-polarized and 180 dual-co-polarized images). For all images, at least one ship's self-reporting message via the Automatic Identification System (AIS) was available. AIS provides information about ship properties like speed over ground or vessel location, which were used as influencing parameters as well as to identify possible locations of wake signatures. For the latter, an automatic intersection of AIS with the SAR images was executed to assign AIS messages to image regions. A manual correction of these colocations was performed to let the unreliable AIS data fulfill ground truth requirements, which means colocations have been discarded in case of large amounts of artifacts like ambiguities or marine objects being present. Then on the basis of these two datasets co-located in space and time a manual search for moving vessels was conducted (Figure 1(A.1)). During the search the background of the moving vessels was checked for unambiguous visibility or non-visibility of wake signatures. By doing so to each wake sample either the class label "detected" or "not detected" was assigned (Figure 1(A.2)). Detailed information about the manual inspection procedure can be found in [14].

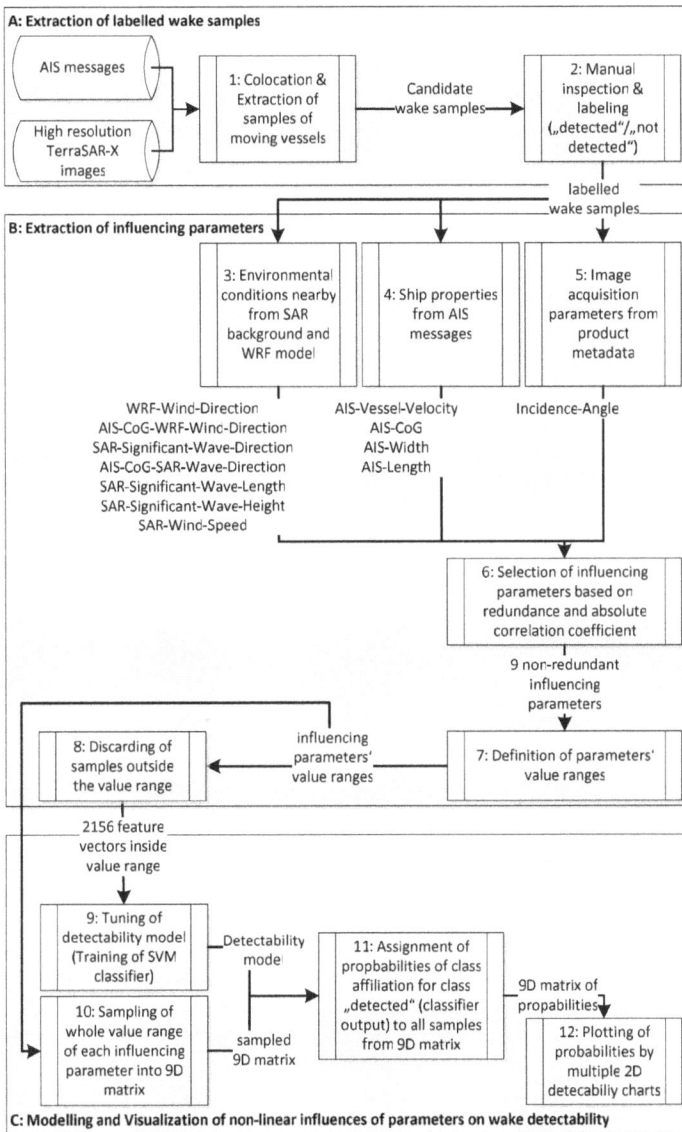

Figure 1. Flow-chart describing the overall process from data preparation (**A**) over retrieval of relevant parameter influencing the detectability (**B**) up to building of the detectability model and its visualization (**C**). AIS: Automatic Identification System; SAR: Synthetic Aperture Radar.

The detectability of wake signatures is affected by a number of influencing parameters (Figure 1(B.3),(B.4),(B.5)). A comparison of Pearson product–moment correlation coefficients was conducted in order to select influencing parameters with distinct physical background. The correlation coefficients are calculated for each influencing parameter between the parameter's magnitudes of all wake samples and their respective class labels, where "detected" was quantified as 1 and "not detected" as 0. Figure 2 shows a graph with all twelve compared influencing parameters and their absolute correlation scores. From influencing parameters with redundant physical background the respective influencing parameters with lower absolute correlation score were discarded (Figure 1(B.6)). The discard applies to the following three redundant influencing parameters: WRF-Wind-Direction, which describes the relative wind direction from the Weather Research and Forecasting Model (WRF) towards the radar beam, SAR-Significant-Wave-Direction, which describes the relative wave direction towards the radar beam, and AIS-Width, which defines the width of the ship producing the wake.

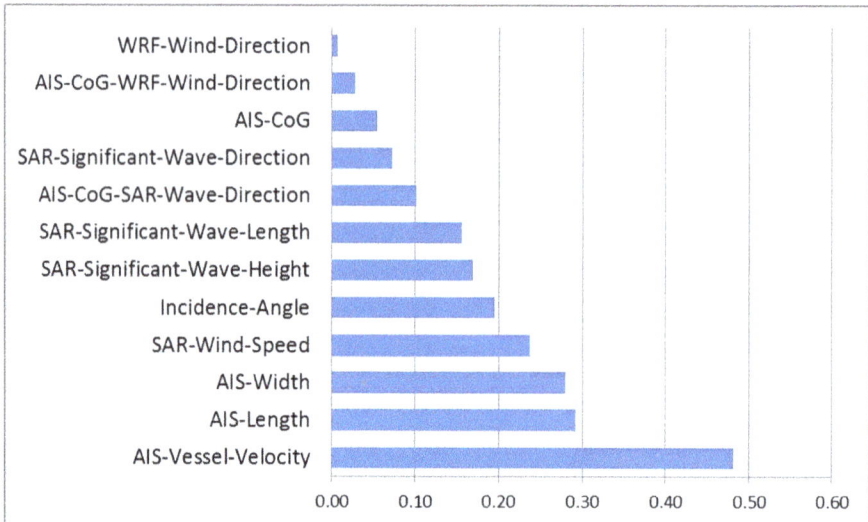

Figure 2. Plot of absolute correlation coefficients (Pearson product-moment correlation coefficients) of each influencing parameter with the wake visibility used for selection of parameters with distinct physical background. WRF: Weather Research and Forecasting Model.

A description of the remaining nine influencing parameters can be found in Table 1. It was decided to not apply any further dimensionality reduction technique as a meaningful, distinct physical background of the influencing parameters is supposed to be more important than expressive or independent parameters. Extreme characteristics of the influencing parameters only rarely occur in reality, e.g., small ships can hardy travel at high wind speeds and sea state conditions, or large ships cannot exceed their hull speed. Thus, the value range of the nine selected influencing parameters was restricted in order to obtain a nine dimensional space non-sparsely filled with wake samples, reducing the risk of curse of dimensionality as well (Figure 1(B.7)). Additionally, the value range of influencing parameters measured in degrees was projected down from 0°–360° to 0°–90°. The performed projection is displayed in Figure 3. In this way the complexity of the detectability model is reduced significantly, because only one detectability peak must be modeled, instead of two with reversed direction as in [15]. After discarding all wake samples with influencing parameters outside the defined value range, a training dataset consisting of 2156 labelled wake samples was concluded (Figure 1(B.8))

Table 1. List of the nine influencing parameters considered in the detectability model along with a description and the value range, in which all samples in the training dataset lie; also a default parameter setting used for the plots in Section 3 is provided.

Influencing Parameter Name	Description	Value Range *(Default Setting)*
AIS-Vessel-Velocity	Velocity of the vessel derived from AIS messages interpolated to the image acquisition time	0 m/s to 12 m/s *(6 m/s)*
AIS-Length	Length of the corresponding vessel based on AIS information	10 m to 390 m *(100 m)*
SAR-Wind-Speed	Wind speed estimated from the SAR background around the vessel using the XMOD-2 geophysical model function [29,33]	2 m/s to 10 m/s *(6 m/s)*
Incidence-Angle	Incidence angle of the radar cropped to TerraSAR-X's full performance value range	20° to 45° *(30°)*
SAR-Significant-Wave-Height	Significant wave height estimated from the SAR background around the vessel using the XWAVE_C empirical model function [34]	0 m to 3 m *(0.5 m)*
SAR-Significant-Wave-Length	Wave length estimated from the SAR background around the vessel using the XWAVE_C empirical model function [34]	75 m to 350 m *(150 m)*
AIS-CoG-SAR-Wave-Direction	Absolute angular difference between AIS-CoG and wave direction estimated from the SAR background around the vessel using the XWAVE_C empirical model function [34]. The 0°–360° value range has been projected to 0°–90° as displayed in Figure 3.	0° to 90° *(45°)*
AIS-CoG	The course over ground based on AIS information relative to the radar looking direction (0° means parallel to range and 90° mean parallel to Azimuth). The 0°–360° value range has been projected to 0°–90° as displayed in Figure 3.	0° to 90° *(45°)*
AIS-CoG-WRF-Wind-Direction	Absolute angular difference between AIS-CoG and wind direction estimated by the Weather Research and Forecasting Model (WRF) [35] nearby the vessel. The 0°–360° value range has been projected to 0°–90° as displayed in Figure 3.	0° to 90° *(45°)*

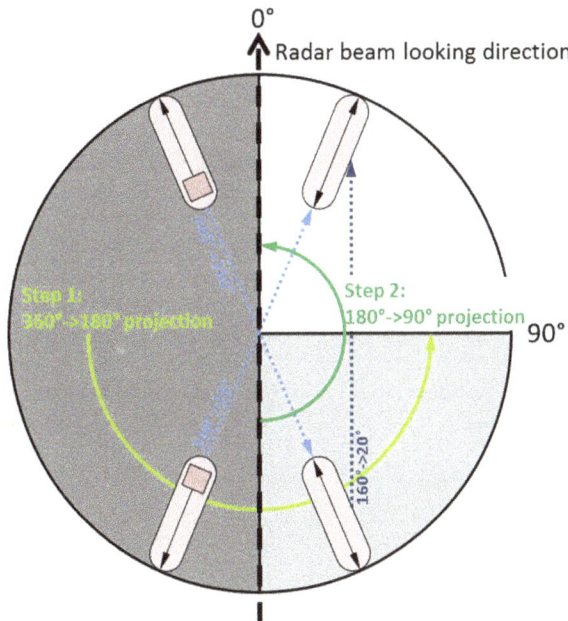

Figure 3. Example for a projection of a ship's heading from 0°–360° down to 0°–90°. The same projection has been applied to all influencing parameters measured in degrees.

Remote Sens. **2019**, *11*, 563

3. Results

Figure 1 part C displays, where the procedures and results explained in this section are integrated in the overall workflow executed for this study. In this section the detectability model is detailed and the selection of the best-performing hyperparameters for adjustment of the model's complexity is explained. Further, is explained how the model can be used to visualize the detectability with respect to the influencing parameters. Finally, the model results are displayed and described.

3.1. Tuning of the 9D SVM Detectability Model

In [14] a linear logistic regression classifier was used for binary classification of wake samples with the class labels "detected" or "not detected" based on various combinations of three influencing parameters. The probability of class affiliation to the class "detected" was used to express the probability of detection. The same approach is adopted for this study, but a Support Vector Machine (SVM) classifier is trained on all nine influencing parameters together (Figure 1(C.9)). SVM classifiers have the advantage among other classifiers that they can be easily tuned in their complexity [36]. Actually SVMs are not capable of providing probability estimates of class affiliation, but this drawback is overcome by training the probability estimates after classifier training as proposed in [37]. By only considering the probability estimates of class affiliation to the class "detected", which is calculated from the nine-dimensional input to the SVM, the model can be expressed by the following simplified formula:

$$PoD = f(x_1, x_2, x_3, x_4, x_5, x_6, x_7, x_8, x_9), \tag{1}$$

where $x_i \ \forall i \in \{i \in \mathbb{N} | 1 \le i \le 9\}$ denotes one of the nine influencing parameters listed in Table 1 using the subscript i as index, f is the SVM detectability model and $PoD \in [0,1]$ the derived probability of detection.

Using a linear kernel in the SVM is similar to the linear logistic regression classifier approach used in [14]. Multiple settings for hyperparameters of the SVM model were tested for this study. Most of the settings allowed a representation of non-linear influences of the nine selected parameters on the detectability and their interdependencies. It turned out that a polynomial kernel with a degree of two can outperform the linear model. First, a much higher complexity, induced by higher degree of the polynomial kernel or due to radial-basis or sigmoid-kernel, leads to overfitting [36]. Second, for all the nine selected influencing parameters only one detectability peak is expected, therefore a polynomial kernel with a degree of two is sufficient to model this one peaked maximum. Different cost-parameter values ranging from 0.01 to 100 at steps of multiples of 10 were tried and finally a low cost-parameter was set, which allows the SVM model to adopt a large margin and account for the noisy dataset. The gamma-parameter controls the magnitude of curvature of the separating hyperplane. Its tuning is dependent on the cost-parameter, as a narrower margin requires a stronger bending (i.e., a lower gamma-parameter) in case of non-perfectly separable classes and vice-versa. Gamma-parameter values ranging from 0.001 to 0.1 at steps of multiples of 10 were tried, and the best performance was achieved when gamma is set to 0.01 and cost to 0.1. The tuning of the coef0-parameter was performed over the value range of 0 to 1000 at steps of multiples of 10. However the effect of this parameter on the performance turned out negligible. This can be explained by the fact that a polynomial kernel with a degree of two requires less trading off between the first-order and second-order terms of the polynomial, compared to polynomial kernels with higher degrees. More detailed information about tuning of SVM's hyperparameters can be found in [38]. 10-fold-cross validation was applied to quantify the model's performance [39]. The best performing parameterization of the SVM model is given in Table 2, with which a classification accuracy of ≈87% is achieved.

Table 2. Settings of the Support Vector Machine (SVM)'s hyperparameters achieving highest 10-fold-cross validation accuracy on the training dataset.

Hyperparameter Name	Value
Kernel type	polynomial
Kernel degree	2
Cost	0.1
Gamma	0.01
Coef0	100

3.2. Visualization of 9D Detectability Model

As the full visualization of a model with nine dimensions is not feasible by two dimensional figures, only extracted views into the model can be presented here. The views are constructed in a way that they visualize the direct dependencies of influencing parameters with the probability of detection and the interdependencies between the influencing parameters also with regard to detectability. This means one view into the model can only display the value range of two influencing parameters (one on the x-axis and one on the y-axis) at a time, when the probability of detection itself is represented by a color-coded scale (quantifying in a restricted manner values on a z-axis). Such views into the model are in the flowing denoted as 2D detectability charts or 2D charts. By plotting multiples of these 2D detectability charts, each with a different fixed value setting for the influencing parameters not displayed in the chart, the various effects of the influencing parameters on the detectability can be analyzed. This way of visualizing the models was already proposed in [13] and [14] for the data-driven 3D and 4D detectability models. While for the 3D model only one and for the 4D model only two influencing parameters would have to be set to a fixed value in order to obtain one 2D detectability chart, for the demonstrated 9D model seven influencing parameters must be set to a fixed value. Therefore, the investigation of the 9D space required repeated combination of settings for the influencing parameters and repeated plotting and chart interpretation.

The fastest way of producing each required 2D chart was to first sample the whole value range of the 9D dimensional feature space into a 9D matrix (Figure 1(C.10)). Then each 9D sample from the matrix was fed into the SVM model and the probability of class affiliation for the class label "detected" was assigned to the respective sample, expressing the wake detectability for the respective setting (Figure 1(C.11)). Finally, only the 9D matrix of probability estimates needed to be read out by accessing the probability values according to the desired settings of influencing parameters for which the 2D charts were required (Figure 1(C.12)).

3.3. Characteristics of Influences on Wake Detectability

The characteristics of how an influencing parameter affects the detectability can be categorized into four types. The influences of parameters can also vary in dependency to other influencing parameters. Such dependencies between parameters are here called interdependencies. In this section the parameter with index c for which the characteristics are described is denoted as x_c and its value range as I_c. All the respective other parameters x_o with indices $o \in \{i \in \mathbb{N} | 1 \leq i \leq 9 \wedge i \neq c\}$ are in the set $X_{o \neq c}$ and their respective value ranges are denoted I_o.

3.3.1. Influencing Parameters with No Influence on Detectability

When no significant variation of probability of detection is observed for all magnitudes of the characterized influencing parameter over its whole value range in combination with various magnitude settings of respective other influencing parameters, then the characterized parameter is defined as having no influence on the detectability. This means:

$$f'(x_c, X_{o \neq c}) = \frac{\partial f}{\partial x_c} = 0, \forall x_c \in I_c, \forall x_o \in I_o \tag{2}$$

3.3.2. Influencing Parameters with Independent Monotonic Influence on Detectability

Detectability models with a linear basis like the ones presented in [13,14] are only capable of representing independent monotonic influences of parameters on the detectability. Such an independent monotonic influence on detectability can still be observed for the presented non-linear SVM detectability model. The monotonic influence of such parameters is independent from the magnitudes of any other influencing parameters. Therefore, the parameters reach relatively high absolute correlation coefficients. However, the gradient of detectability's variation for these parameters can change with different magnitudes of other influencing parameters, while for a linear model the gradients are constant. This means:

$$f'\left(x_c, X_{o \neq c}\right) = \frac{\partial f}{\partial x_c} \lessgtr 0, \forall x_c \in I_c, \forall x_o \in I_o \tag{3}$$

3.3.3. Influencing Parameters with a One-peaked Maximum Influence on Detectability

The benefit of training a polynomial model is best demonstrated by parameters characterized with a one-peaked maximum influence on detectability. For these the probability of detection reaches one maximum at $x_{c,max}$, which is located inside the value range of the influencing parameter's magnitudes. The gradient of detectability's variation switches its sign at this maximum. On either side of the maximum the influence on detectability is monotonic. This means:

$$f'\left(x_{c,max}, X_{o \neq c}\right) = \frac{\partial f}{\partial x_c} = 0, \exists x_{c,max} \in I_c, \forall x_o \in I_o \tag{4}$$

and

$$f''\left(x_{c,max}, X_{o \neq c}\right) = \frac{\partial f'}{\partial x_c} \neq 0, \exists x_{c,max} \in I_c, \forall x_o \in I_o \tag{5}$$

3.3.4. Influencing Parameters with Interdependent Monotonic Influence on Detectability

The presented more complex model is also capable of representing monotonic influences on the detectability, which are not independent from the other influencing parameters. Such interdependent monotonic influences mean that the influencing parameter's gradient of detectability's variation can switch its sign, when the magnitude combination of other influencing parameters reaches a certain setting $I_{o,turn}$. On either side of this turning point the influence remains monotonic:

$$f'\left(x_c, X_{o \neq c}\right) = \frac{\partial f}{\partial x_c} \lessgtr 0, \forall x_c \in I_c, \forall x_o \in I_o \cap I_{o,turn} \tag{6}$$

Such a turning point in the setting of the interdependent other influencing parameters is characterized by either the influence of the characterized parameter showing no effect on detectability (7) or the influence showing an insignificant one-peaked maximum at $x_{c,max}$ over the parameter's value range (8) and (9):

$$f'\left(x_c, X_{o \neq c}\right) = \frac{\partial f}{\partial x_c} = 0, \forall x_c \in I_c, \exists x_o \in I_{o,turn} \tag{7}$$

or

$$f'\left(x_{c,max}, X_{o \neq c}\right) = \frac{\partial f}{\partial x_c} = 0, \exists x_{c,max} \in I_c, \exists x_o \in I_{o,turn} \tag{8}$$

and

$$f''\left(x_{c,max}, X_{o \neq c}\right) = \frac{\partial f'}{\partial x_c} \neq 0, \exists x_{c,max} \in I_c, \exists x_o \in I_{o,turn} \tag{9}$$

3.4. Categorization of Influencing Parameters by Characteristics of Influences

In the following subsections an extract of 2D detectability charts out of all investigated charts is presented. These charts were selected in a representative way so that they provide an insight into the parameter's influences on detectability and the interdependencies between them. The default setting of fixed values for the influencing parameters is given in brackets in the right column of Table 1.

3.4.1. AIS-CoG-WRF-Wind-Direction

AIS-CoG-WRF-Wind-Direction has no effect on the detectability of ship wakes. Figure 4 shows that the probability of detection remains constant for any magnitude of AIS-CoG-WRF-Wind-Direction for each setting of SAR-Wind-Speed and SAR-Significant-Wave-Height.

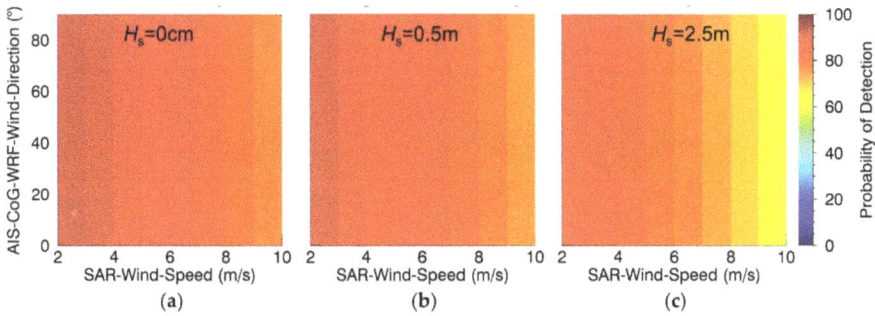

Figure 4. 2D detectability charts based on SAR-Wind-Speed, AIS-CoG-WRF-Wind-direction and from left to right SAR-Significant-Wave-Height with (**a**) 0 m, (**b**) 0.5 m, and (**c**) 2.5 m.

3.4.2. AIS-Vessel-Velocity

AIS-Vessel-Velocity has an independent monotonic influence on detectability. This characteristic can be observed in the Figure 5, but also in the Figures 8 and 9, which are presented later in this section, when AIS-Vessel-Velocity is contrasted with the other two influencing parameters, Incidence-Angle and SAR-Significant-Wave-Length, respectively. Already the relatively high absolute correlation coefficient of this parameter indicates that it is also the parameter with most influence compared to the other influencing parameters. With increasing magnitude of AIS-Vessel-Velocity, the detectability increases.

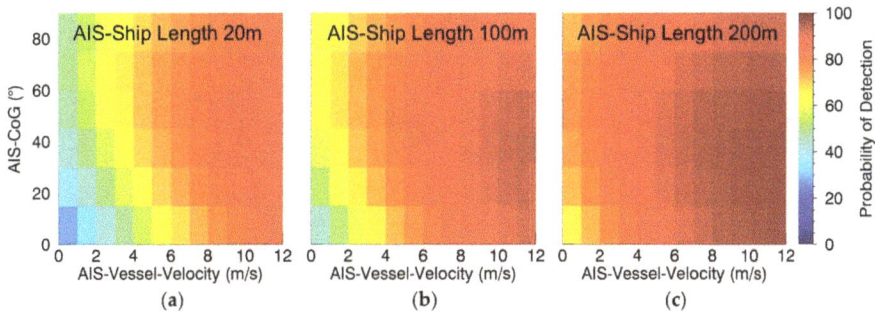

Figure 5. 2D detectability charts based on AIS-Vessel-Velocity, AIS-CoG and from left to right AIS-Length with: (**a**) 20 m, (**b**) 100 m, and (**c**) 200 m.

3.4.3. AIS-Length

Figure 5 also shows that an independent monotonic influence is present for the AIS-Length influencing parameter. With increasing magnitude of AIS-Length, the detectability increases.

3.4.4. SAR-Wind-Speed

The independent monotonic influence of SAR-Wind-Speed on detectability is observable in the Figure 4. With increasing magnitude of SAR-Wind-Speed, the detectability decreases.

3.4.5. SAR-Significant-Wave-Height

The fourth influencing parameter with independent monotonic influence is SAR-Significant-Wave-Height. Its characteristics can be retrieved from Figure 4. With increasing magnitude of SAR-Significant-Wave-Height, the detectability decreases.

3.4.6. AIS-CoG

From Figure 5 a one-peaked maximum influence of AIS-CoG on detectability can be derived. It is located between magnitudes from ≈30° to ≈70°, interdependently from AIS-Vessel-Velocity. The maximum is located around 30° for low AIS-Vessel-Velocity and shifts towards 70° with increasing AIS-Vessel-Velocity.

3.4.7. AIS-CoG-SAR-Wave-Direction

For AIS-CoG-SAR-Wave-Direction the detectability reaches its one-peaked maximum around magnitudes from ≈60° to ≈70°, which can be observed in the Figures 6 and 7. The interdependency to other influencing parameters is negligible.

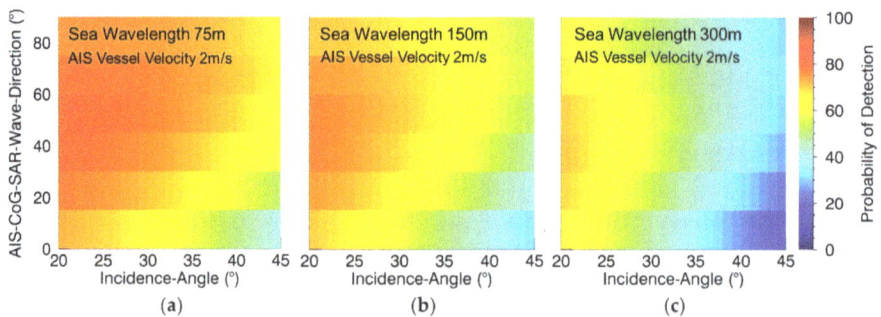

Figure 6. 2D detectability charts based on Incidence-Angle, AIS-CoG-SAR-Wave-direction and from left to right SAR-Significant-Wave-Length with: (**a**) 75 m, (**b**) 150 m, and (**c**) 300 m, deviating from the default setting for this chart AIS-Vessel-Velocity was set to 2 m/s.

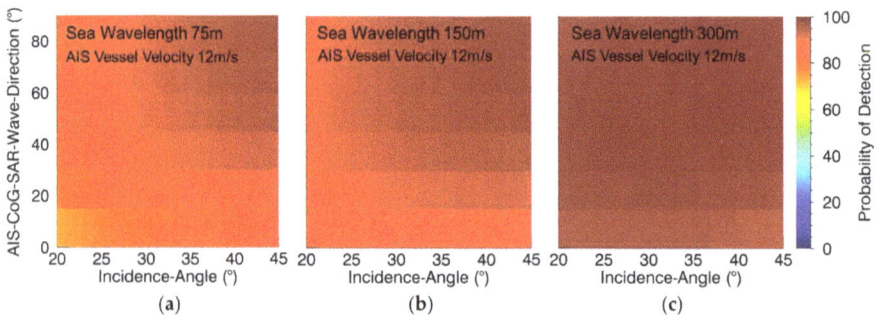

Figure 7. 2D detectability charts based on Incidence-Angle, AIS-CoG-SAR-Wave-direction and from left to right SAR-Significant-Wave-Length with: (**a**) 75 m, (**b**) 150 m, and (**c**) 300 m, deviating from the default setting for this chart AIS-Vessel-Velocity was set to 12 m/s.

3.4.8. Incidence-Angle

Incidence-Angle has an interdependent monotonic influence on wake detectability. Its monotonic influence is interdependent to AIS-Vessel-Velocity, SAR-Wind-Speed and SAR-Significant-Wave-Height. In the 2D detectability charts in Figure 8 the general sea surface roughness characterized by SAR-Wind-Speed and SAR-Significant-Wave-Height is expressed by the Beaufort-Scale number (abbreviated as bft) [40]. In the Figure 6 to Figure 8 the interdependency to AIS-Vessel-Velocity is observable. When describing the influence of Incidence-Angle on detectability, four different combinations with different parameter magnitudes must be considered

- For smooth ocean surface the turning point is located around 9 m/s of AIS-Vessel-Velocity:

  o Below 9 m/s with increasing magnitude of Incidence-Angle, the detectability decreases by few percentage points close to 9 m/s up to ~35 percentage points close to 0 m/s
  o Above 9 m/s no influence of Incidence-Angle on the detectability is observed

- For rough ocean surface the turning point is located around 6 m/s of AIS-Vessel-Velocity:

  o Below 5 m/s with increasing magnitude of Incidence-Angle, the detectability decreases by few percentage points close to 6 m/s up to ~20 percentage points close to 0 m/s
  o Above 5 m/s with increasing magnitude of Incidence-Angle, the detectability increases by few percentage points close to 6 m/s up to ~20 percentage points close to 12 m/s

- This means, the turning point, at which the gradient of detectability's variation of Incidence-Angle switches its sign, decreases from 9 m/s to 6 m/s when the ocean surface gets rougher.

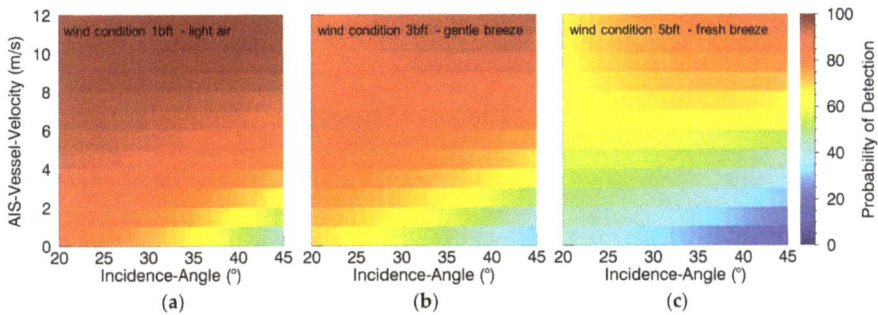

Figure 8. 2D detectability charts based on Incidence-Angle, AIS-Vessel-Velocity and from left to right Beaufort numbers with: (**a**) 1 bft, (**b**) 3 bft, and (**c**) 5 bft.

3.4.9. SAR-Significant-Wave-Length

Also SAR-Significant-Wave-Length has a monotonic influence, which is interdependent from AIS-Vessel-Velocity, SAR-Wind-Speed, and SAR-Significant-Wave-Height. All interdependent monotonic influences are shown in Figure 9. When describing the influence of the SAR-Significant-Wave-Length on detectability, again four different combinations with different parameter magnitudes must be considered

- For smooth ocean surface the turning point is located around 3 m/s of AIS-Vessel-Velocity:

  o Below 3 m/s with increasing magnitude of SAR-Significant-Wave-Length, the detectability decreases by few percentage points close to 3 m/s up to ~10 percentage points close to 0 m/s o   Above 3 m/s with increasing magnitude of SAR-Significant-Wave-Length, the detectability increases by few percentage points close to 3 m/s up to ~5 percentage points close to 12 m/s

- For rough ocean surface the turning point is located around 6 m/s of AIS-Vessel-Velocity:

    o   Below 6 m/s with increasing magnitude of SAR-Significant-Wave-Length, the detectability decreases by ~5 percentage points close to 6 m/s up to ~25 percentage points close to 0 m/s

    o   Above 6 m/s with increasing magnitude of SAR-Significant-Wave-Length, the detectability increases by few percentage points close to 6 m/s up to ~20 percentage points close to 12 m/s

- This means, the turning point, at which the gradient of detectability's variation of SAR-Significant-Wave-Length switches its sign, increases from 3 m/s to 6 m/s when the ocean surface gets rougher

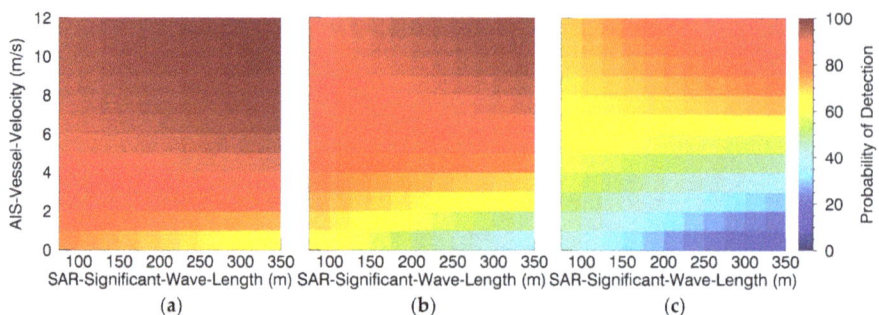

Figure 9. 2D detectability charts based on SAR-Significant-Wave-Length, AIS-Vessel-Velocity and from left to right Beaufort numbers with: (**a**) 1 bft, (**b**) 3 bft, and (**c**) 5 bft.

4. Discussion

The results obtained by the presented detectability model are partially based on influencing parameters retrieved using other models, i.e., the XWAVE_C model for sea state retrieval, the XMOD-2 model for wind speed estimation and the WRF model for estimation of wind direction. Models possess only limited capabilities of representing reality and different measures for accuracy apply for each of them. Therefore, an interesting result of this study is that the detectability model trained on the basis of these imperfect models and real data is able to reproduce many oceanographic expectations stated by other researchers in the past. The accuracies of XWAVE_C and XMOD-2 were also considered in the following discussion.

4.1. AIS-CoG-WRF-Wind-Direction

Hennings et al. [15] found out that the wind direction has a slight influence on the detectability of the Kelvin wake arms. By using the influencing parameter AIS-CoG-WRF-Wind-Direction in the detectability model proposed in this study, this behavior could not be reproduced. First, using the WRF model as a substitute for the actually required high resolution wind direction is insufficient in terms of local wind field variability. Higher resolution wind direction models or an automatic extraction of wind direction from the SAR image is required. Second, the real influence is low and the presented detectability model may also not be sensitive enough to represent this influence or the real influence is interdependent on influencing parameters, which are not considered in this study.

4.2. AIS-Vessel-Velocity

The most pronounced influencing parameter is AIS-Vessel-Velocity. Figure 10 shows wakes for the three different vessel velocity classes: slow with AIS-Vessel-Velocity up to 4 m/s

(Figure 10a), medium with AIS-Vessel-Velocity between 4 m/s and 9 m/s (Figure 10b) and fast with AIS-Vessel-Velocity above 9 m/s (Figure 10c). The differences between these classes are connected with forcing waves of different amplitude, period and propagation. For example, for low speed vessels the Kelvin waves and their SAR signature are weaker than the SAR signatures of turbulent wakes.

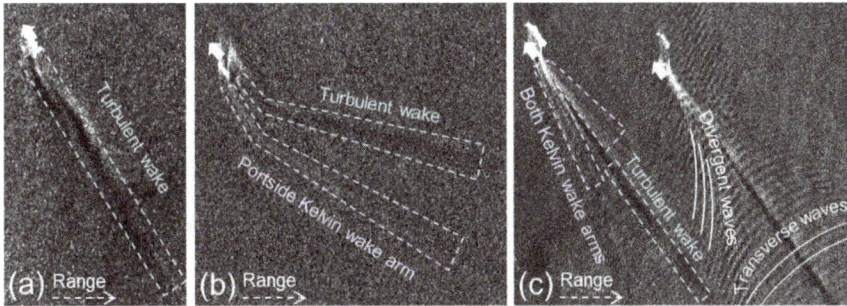

Figure 10. Slow vessel (**a**), middle speed vessel (**b**) and high speed vessel (**c**); for the slow vessel only the turbulent wake is visible, for the middle speed the portside Kelvin wake arm produces a stronger signal, for the high speed vessel the Kelvin wave is steeper producing bright backscatter. All other influencing parameters have comparable setting for the above wake patches. The transvers waves and divergent waves producing the Kelvin wakes arms are clearly observable behind the fast right vessel (**c**).

Briefly, faster vessels are better detectable as depicted by Figure 5 due to two reasons.

- First, a larger velocity results in a more extensive area of the ocean surface being affected in a shorter time and larger wake signatures are better recognizable.
- Second, the compressed divergent waves are especially well imaged on SAR, because of the steep waves with dense wave crests and high amplitudes forming shapes similar to corner reflectors and also the resulting wave breaking [41,42].

In the following, a detailed explanation is given, why wakes from faster vessels are better detectable. The propagation of wakes is based on the wake's Froude number [43,44], a non-dimensional measure for the wave drag behind the ship calculated using length and velocity of a ship by:

$$Fr = V / \sqrt{gL}, \tag{10}$$

where V is the AIS-Vessel-Velociy, L the AIS-Length and g the gravitational acceleration. The amplitude of the transversal waves decreases for larger Froude numbers and the V-shaped wave pattern becomes narrower due to compressing of the divergent waves when the ship exceeds its hull speed [43,44]. Hence, the radar backscatter resulting from these waves is higher for larger incidence angles and so wakes from fast vessels are better detectable. This is indicated by Figures 7 and 8.

Figures 6 and 8 on the other hand show that for slow vessels lower incidence angles are better for detection, what can be explained as follows. The V-shaped Kelvin envelope with its delimiting constant angle of ≈19.47° as defined by [45] is visible on SAR due to constructive interference between the wave crests of divergent and transversal waves. The resulting higher amplitude and sometimes wave breaking are only present for small Froude numbers, as the transversal waves vanish for larger Froude numbers so that less constructive interference is present. The backscattering of this effect is low, when compared to the effects occurring with large Froude numbers. Hardly affected by the ship velocity is the turbulent wake, which is represented by a smooth ocean surface originating from the ship's propeller.

Similar to the detection of oil spills, the detection of the ship's turbulent wake is easier with lower incidence angles as the contrast between the smooth wake and the rougher surrounding ocean surface is more distinct [32]. With higher ship velocity only the extent of the smoothed ocean surface increases, not the smoothness itself. In general the detection of smooth ocean surface areas is more difficult than of the compressed divergent waves. All these different effects finally constitute the reason why an increase of AIS-Vessel-Velocity generally implies a better detectability of wakes. Figure 8 indicates additionally that a distinct change of the wake signature due to larger Froude numbers in most cases is observable with an AIS-Vessel-Velocity between 5 m/s and 9 m/s.

4.3. AIS-Length

Beside the AIS-Vessel-Velocity also the AIS-Length is required for calculation of Froude number. The Froude number is inversely proportional to AIS-Length, but the ascending slope of the Froude number is higher for small magnitudes of AIS-Length, while for large magnitudes the gradient is much smaller. This means small ships have in general larger Froude numbers than large ships. The increased detectability of ship wakes with larger Froud numbers, as explained in Section 4.2, compensates to some degree the fact that large ships produce higher waves, what makes them better detectable. Still, the better detectability of large ships can be observed in Figure 2. When taking into account that an increase if AIS-Length by a factor of 10 from 20 m to 200 m only leads to an increase of detectability of only around 10 percentage points, then the effect of this influencing parameter on detectability is almost negligible, what was already stated in [14].

4.4. SAR-Wind-Speed and SAR-Significant-Wave-Height

The influencing parameters SAR-Wind-Speed and SAR-Significant-Wave-Height are in practice proportionally connected to each other as they both describe the roughness of the ocean surface as a result of striking winds. Wake samples with high SAR-Wind-Speed and low SAR-Significant-Wave-Height (storm formation) or low SAR-Wind-Speed and high SAR-Significant-Wave-Height (only swell) occur rarely in the dataset, especially in the study area North Sea, Baltic Sea and Mediterranean Sea. Thus, in the Figures 8 and 9 both influencing parameters are combined using the Beaufort scale [40]. From Figure 4 a significant decrease of detectability can be observed with increasing SAR-Wind-Speed and also a decrease of detectability with increasing SAR-Significant-Wave-Height is pronounced. First, the rougher ocean conditions interfere with the formation and propagation of all wake components in reality, what means wake signatures can only occur less distinct and smaller. Second, the bright and inhomogeneous appearance of the ocean surface, which is surrounding and superimposing the wake signatures, impedes the unambiguous perception of the respective structures in the images [15].

4.5. AIS-CoG

The propagation direction of cusp waves can be approximated by propagation direction of divergent waves [15]. The detectability of cusp waves forming the Kelvin wave pattern is expected to be sensitive to AIS-CoG as the high amplitudes and breaking waves deploy their best backscattering properties, when exposed perpendicularly by the radar beam contrary to the wave's running direction [15,19,23]. Actually the maximum detectability should then be approached for magnitudes of AIS-CoG around 70.53°. Indeed, two Kelvin arms exist for each wake, but after projection of the ship's heading onto the 0° to 90° value range both arms are projected onto 19.47°, where 70.53° is the projection of the perpendicular direction. Thus, also the information about the respective Kelvin arm's running direction is dropped during the projection of the ship's heading. Figure 5 shows that generally for slow ships and also for fast and small ships the maximum is indeed located around 70°, but for fast and large ships the maximum is around 30°. As explained above, for fast ships the Kelvin wake arms represent the more distinct feature and therefore actually the maximum should be located around 70° especially for fast ships. The maximum around 30° is contradicting here. However, the small gradients around the maximum illustrate that the general dependency of detectability to AIS-CoG is so marginal that inaccuracies in the AIS dataset could lead to false estimations for this condition, as only few samples of fast and large ships are presents in the data. The maximum around 30° for fast and large ships should therefore be neglected.

4.6. AIS-CoG-SAR-Wave-Direction

The maximum detectability of the influencing parameter AIS-CoG-SAR-Wave-Direction is shifted from ≈70° towards ≈60° for different magnitudes of SAR-Significant-Wave-Length as depicted in Figures 6 and 7. The wavelength is directly estimated from SAR subscenes using a 2D fast Fourier transform (FFT) and a consecutive search for the peak wavelength. Cases of non-pronounced imaged wave signatures are not discarded from the dataset and this is recognized to lead to some inaccuracies in the estimation of wavelength. Thus, this shift is also connected to these inaccuracies in wavelength estimation and should be neglected. Similar to AIS-CoG also for the AIS-CoG-SAR-Wave-Direction parameter the projection of the angles leads to a perpendicular angle of ocean waves towards Kelvin wake arms in the magnitude of 70.53°. A peak around 70.53° means that the waves in the two Kelvin wake arms are colliding with the in parallel running ocean surface waves. Already the resulting constructive interference and the in turn resulting heightened wave amplitude and wave breaking increase the backscattering and therefore the wake's detectability. In cases where the in-parallel running waves collide with the Kelvin wake arm's waves in the opposite running directions, this effect is even more intense. Thus, the maximum around 70.53° matches oceanographic expectations.

4.7. Incidence-Angle

According to [15] and [32] Kelvin wake arms and turbulent wake are both less visible under high incidence angle conditions, which is also observable in the Figure 8 for slow AIS-Vessel-Velocity. However, according to Figure 8 for higher AIS-Vessel-Velocity the dependency is reversed, which is an unexpected result regarding oceanographic expectations. This reversed dependency for high AIS-Vessel-Velocity is not documented in the literature, but it is explained in Section 4.1 by the high Froude numbers of fast vessels, which produce high amplitude waves and wave breaking at the Kelvin wake arms.

4.8. SAR-Significant-Wave-Length

Generally, the interaction of wind waves with short wavelengths and steep crests with the ship's wake principally differs from the interaction of waves produced by swell with longer wavelength and smooth crests. The complex dependency of detectability from different magnitudes of SAR-Significant-Wave-Length is best indicated by Figure 9. In general, for short SAR-Significant-Wave-Length the Kelvin wave pattern is less visible. The short waves mean the dominant local steep wind waves, which interfere with the Kelvin wake arms in a destructive manner. Further, the more inhomogeneous, brighter ocean background makes the recognition of the Kelvin wave pattern more difficult. This is also observable in the Figures 6 and 7, where the gradients around the maximum detectability regarding AIS-CoG-SAR-Wave-Direction are more pronounced for short SAR-Significant-Wave-Length compared to long SAR-Significant-Wave-Length, what means less influence of AIS-CoG-SAR-Wave-Direction on detectability for this condition.

In case of swell sea state conditions less or no collisions of ocean waves with the Kelvin wake arms occur, as the swell waves only increase and decrease the general sea surface height without interference [16]. On the contrary, small magnitudes of SAR-Significant-Wave-Length are slightly better for recognizing turbulent wakes, what is indicated by Figure 9 for slow AIS-Vessel-Velocity. The reason is that the darker turbulent wakes have a better contrast to the larger amount of surrounding bright wave crests, which are flattened by the ship's propeller in the wake.

5. Applications

Given the case that a wake has been detected, it is possible to reverse the detectability model in order to derive information about the four missing parameters with independent monotonic influence on detectability. Three of these, namely AIS-Length, SAR-Wind-Speed, and SAR-Significant-Wave-Height, are available in most of the cases or can directly be derived from the SAR image. Therefore, the model reversion can be used to provide rough estimations of the underlying AIS-Vessel-Velocity, which a moving object must possess as a minimum to produce a detectable wake signature. For the plots in Figure 11, a probability of detection threshold of 80% is defined and the minimum AIS-Vessel-Velocity is derived, for which the model provides a probability of detection above the threshold. Figure 11 also provides a redundant view on the complex dependency of Incidence-Angle as well as SAR-Significant-Wave-Length on AIS-Vessel-Velocity, SAR-Wind-Speed, and SAR-Significant-Wave-Height with regard to detectability.

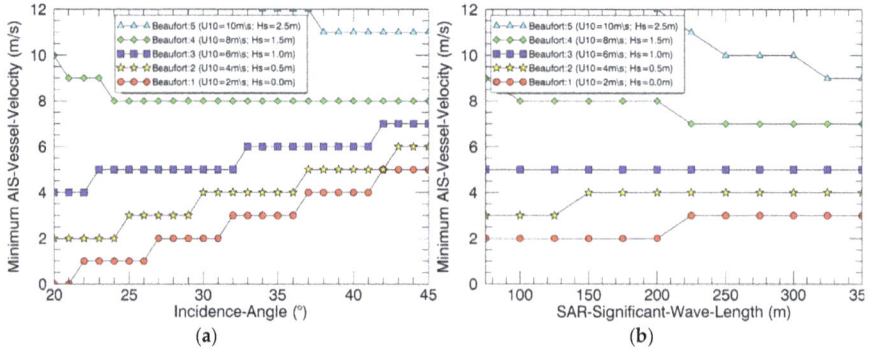

Figure 11. Minimum vessel velocity calculated for the default parameters and changing Incidence-Angle (**a**) and SAR-Significant-Wave-Length (**b**) respectively. Results obtained by varying SAR-Wind-Speed and SAR-Significant-Wave-Height are displayed by different line markers. The minimum probability of detection was set to 80%.

Beside the estimation of minimum values for missing parameters, the detectability model can be used to control an automatic wake detection process. Automatic wake detection based on Radon Transform as developed by [5] could be accelerated by limiting the search space in the Radon domain to wake headings, for which a certain level of probability of detection is reached. Also the search for specific wake components may be skipped in case these components are not detectable under certain characteristics of influencing parameters.

6. Conclusions

The linear wake detectability model presented in [14] has been extended by a non-linear basis using a Support Vector Machine classifier with a polynomial kernel of second grade. The model classifies the input data between the classes "detected" and "not detected", where the probability of class affiliation to the class "detected" is taken as measure for the probability of detection of ship wakes in SAR imagery. Nine influencing parameters, which are affecting the detectability, are considered simultaneously in a single model. Thus, the model can represent not only the dependency of detectability from the influencing parameters, but also depict interdependencies between them. The influencing parameters describe different environmental conditions (i.e., wind speed, wind direction, sea state height, sea state direction and sea state wave length), ship properties (i.e., size, heading, and velocity) and image acquisition settings (i.e., incidence angle, beam looking direction).

Most of the statements about the influencing parameters are theoretically expected, but in this publication they are quantitatively proven using real data. The main outcomes are:

- The higher the vessel velocity the higher the detectability
- The radar beam looking direction and the ocean waves' traveling direction should be perpendicular to the angle of Kelvin wake arms for higher detectability

- Rough, inhomogeneous ocean surface conditions worsen the detectability
- Slow ships are better detectable with lower incidence angles or shorter wavelengths of ocean surface waves and fast ships are better detectable with higher incidence angles and longer wavelengths of ocean surface waves

Beside the statements about the interdependencies of the different influencing parameters, the presented detectability model can also be applied to control an automatic wake detection system. Another possible application of the model is the estimation of minimum vessel velocities, which must be present in order to make the ship produce a detectable wake signature, by inverting the model and setting a fixed level for probability of detection.

Author Contributions: "conceptualization, B.T.; methodology, B.T.; software, B.T.; validation, B.T.; formal analysis, B.T., A.P. and S.J.; investigation, B.T.; resources, B.T.; data curation, B.T. and D.V.; writing—original draft preparation, B.T.; writing—review and editing, D.V.; visualization, B.T.; supervision, S.J. and D.V.; project administration, S.J.; funding acquisition, S.J."

Funding: This research received no external funding.

Acknowledgments: TerraSAR-X data has been acquired using the science proposals OCE2203, OCE3207 and OCE3596. The study includes copyrighted material of vesseltracker.com GmbH. The study includes copyrighted material of JAKOTA Cruise Systems GmbH.

Conflicts of Interest: Authors declare no conflict of interest.

References

1. Tings, B.; Bentes, C.; Lehner, S. Dynamically adapted ship parameter estimation using TerraSAR-X images. *Int. J. Remote Sens.* **2016**, *37*, 1990–2015. [CrossRef]
2. Copeland, A.C.; Ravichandran, G.; Trivedi, M.M. Localized Radon Transform-Based Detection of Ship Wakes in SAR Images. *IEEE Trans. Geosci. Remote Sens.* **1995**, *33*, 35–45. [CrossRef]
3. Eldhuset, K. An Automatic Ship and Ship Wake Detection System for Spaceborne SAR Images in Coastal Regions. *IEEE Trans. Geosci. Remote Sens.* **1996**, *34*, 1010–1019. [CrossRef]
4. Crisp, D.J. *The State-of-the-Art in Ship Detection in Synthetic Aperture Radar Imagery*; DSTO Information Sciences Laboratory: Edinburgh, Scotland, 2004.
5. Graziano, M.D.; D'Errico, M.; Rufino, G. Wake Component Detection in X-Band SAR Images for Ship Heading and Velocity Estimation. *Remote Sens.* **2016**, *6*, 498. [CrossRef]
6. Biondi, F. Low-Rank Plus Sparse Decomposition and Localized Radon Transform for Ship Wake Detection in Synthetic Aperture Radar Images. *IEEE Geosci. Remote Sens. Lett.* **2017**, *15*, 117–121. [CrossRef]
7. Biondi, F. (L+ S)-RT-CCD for Terrain Paths Monitoring. *IEEE Geosci. Remote Sens. Lett.* **2018**, *15*, 1209–1213. [CrossRef]
8. Biondi, F. A Polarimetric Extension of Low-Rank Plus Sparse Decomposition and Radon Transform for Ship Wake Detection in Synthetic Aperture Radar Images. *IEEE Geosci. Remote Sens. Lett.* **2018**, *16*, 75–79. [CrossRef]
9. Schurmann, S.R. Radar characterization of ship wake signatures and ambient ocean clutter features. *IEEE Aerosp. Electron. Syst. Mag.* **1989**, *4*, 182–187. [CrossRef]
10. Vachon, P.; Campbell, J.; Bjerkelund, C.; Dobson, F.; Rey, M. Ship Detection by the RADARSAT SAR: Validation of Detection Model Predictions. *Can. J. Remote Sens.* **1997**, *23*, 48–59. [CrossRef]
11. Vachon, P.; Wolfe, J.; Greidanus, H. Analysis of Sentinel-1 marine applications potential. In Proceedings of the 2012 IEEE International Geoscience and Remote Sensing Symposium, Munich, Germany, 22–27 July 2012.
12. Vachon, P.; English, R.; Sandirasegaram, N.; Wolfe, J. *Development of an X-Band SAR Ship Detectability Model: Analysis of TerraSAR-X Ocean Imagery*; Defence R&D Canada: Ottawa, ON, Canada, 2013.
13. Tings, B.; Bentes, C.; Velotto, D.; Voinov, S. Modelling Ship Detectability Depending On TerraSAR-X-derived Metocean Parameters. *CEAS Space J.* **2018**, 1–14. [CrossRef]
14. Tings, B.; Velotto, D. Comparison of ship wake detectability on C-band and X-band SAR. *Int. J. Remote Sens.* **2018**, *39*, 4451–4468. [CrossRef]
15. Hennings, I.; Romeiser, R.; Alpers, W.; Viola, A. Radar imaging of Kelvin arms of ship wakes. *Int. J. Remote Sens.* **1999**, *20*, 2519–2543. [CrossRef]

16. Alpers, W.R.; Ross, D.B.; Rufenach, C.L. On the detectability of ocean surface waves by real and synthetic aperture radar. *J. Geophys. Res.* **1981**, *86*, 6481–6498. [CrossRef]
17. Panico, A.; Graziano, M.D.; Renga, A. SAR-Based Vessel Velocity Estimation From Partially Imaged Kelvin Pattern. *IEEE Geosci. Remote Sens. Lett.* **2017**, *14*, 2067–2071. [CrossRef]
18. Zilman, G.; Zapolski, A.; Marom, M. On Detectability of a Ship's Kelvin Wake in Simulated SAR Images of Rough Sea Surface. *IEEE Trans. Geosci. Remote Sens.* **2015**, *53*, 609–619. [CrossRef]
19. Lyden, J.D.; Hammond, R.R.; Lyzenga, D.R.; Shuchman, R. Synthetic Aperture Radar Imaging of Surface Ship Wakes. *J. Geophys. Res.* **1988**, *93*, 12293–12303. [CrossRef]
20. Reed, A.M.; Milgram, J.H. Ship Wakes and Their Radar Images. *Annu. Rev. Fluid Mech.* **2002**, *34*, 469–502. [CrossRef]
21. Soloviev, A.; Gilman, M.; Young, K.; Brusch, S.; Lehner, S. Sonar Measurements in Ship Wakes Simultaneous with TerraSAR-X Overpasses. *IEEE Trans. Geosci. Remote Sens.* **2010**, *48*, 841–851. [CrossRef]
22. Milgram, J.H.; Peltzer, R.D.; Griffin, O.M. Supression of Short Sea Waves in Ship Wakes: Measurements and Observations. *J. Geophys. Res.* **1993**, *98*, 7103–7144. [CrossRef]
23. Alpers, W.; Romeiser, R.; Hennings, I. On the radar imaging mechanism of Kelvin arms of ship wakes. In Proceedings of the IEEE International Geoscience and Remote Sensing, Symposium Proceedings, Seattle, WA, USA, 6–10 July 1998.
24. Tunaley, J.K.E.; Buller, E.H.; Wu, K.H.; Rey, M.T. The Simulation of the SAR Image of a Ship Wake. *IEEE Trans. Geosci. Remote Sens.* **1991**, *29*, 149–156. [CrossRef]
25. Gu, D.; Phillips, O. On narrow V-like ship wakes. *J. Fluid Mech.* **1994**, *275*, 301–321. [CrossRef]
26. Stapleton, N.R. Ship wakes in radar imagery. *Int. J. Remote Sens.* **1997**, *18*, 1381–1386. [CrossRef]
27. Thompson, D.R.; Gasparovic, R.F. Intensity modulation in SAR images of internal waves. *Nature* **1986**, *320*, 345–348. [CrossRef]
28. Alpers, W. Theory of radar imaging of internal waves. *Nature* **1985**, *314*, 245–247. [CrossRef]
29. Li, X.-M.; Lehner, S. Algorithm for Sea Surface Wind Retrieval from TerraSAR-X and TanDEM-X Data. *IEEE Trans. Geosci. Remote Sens.* **2014**, *52*, 2928–2939. [CrossRef]
30. Shemdin, O.H. Synthetic Aperture Radar Imaging of Ship Wakes in the Gulf of Alaska. *J. Geophys. Res.* **1990**, *95*, 16319–16338. [CrossRef]
31. Gade, M.; Alpers, W.; Hühnerfuss, H.; Wismann, V.R.; Lange, P.A. On the Reduction of the Radar Backscatter by Oceanic Surface Films: Scatterometer Measurements and Their Theoretical Interpretation. *Remote Sens. Environ.* **1998**, *66*, 52–70. [CrossRef]
32. Minchew, B.; Jones, C.E.; Holt, B. Polarimetric Analysis of Backscatter from the Deepwater Horizon Oil Spill Using L-Band Synthetic Aperture Radar. *IEEE Trans. Geosci. Remote Sens.* **2012**, *50*, 3812–3830. [CrossRef]
33. Jacobsen, S.; Lehner, S.; Hieronimus, J.; Schneemann, J.; Kühn, M. Joint Offshore Wind Field Monitoring with Spaceborne SAR and Platform-Based Doppler LiDAR Measurements. *Int. Arch. Photogramm. Remote Sens. Spat. Inf. Sci.* **2015**, 959–966. [CrossRef]
34. Pleskachevsky, A.; Rosenthal, W.; Lehner, S. Meteo-marine parameters for highly variable environment in coastal regions from satellite radar images. *ISPRS J. Photogramm. Remote Sens.* **2016**, *119*, 464–484. [CrossRef]
35. Skamarock, W.C.; Klemp, J.B.; Dudhia, J.; Gill, D.O.; Barker, D.M.; Duda, M.G.; Huang, X.-Y.; Wang, W.; Powers, J.G. *A Description of the Advanced Research WRF Version 3*; NCAR Technical Notes; National Center for Atmospheric Research: Boulder, CO, USA, 2008.
36. Berthold, M.; Hand, D.J. *Intelligent Data Analysis—An Introduction*; Springer: Heidelberg, Germany, 2003.
37. Platt, J.C. Probabilistic Outputs for Support Vector Machines and Comparisons to Regularized Likelihood Methods. *Adv. Large Margin Classif.* **2000**, *10*, 61–74.
38. Ben-Hur, A.; Weston, J. A User's Guide to Support Vector Machines. In *Data Mining Techniques for the Life Sciences*; Humana Press: Totowa, NJ, USA, 2010; pp. 223–239.
39. Kohavi, R. A study of cross-validation and bootstrap for accuracy estimation and model selection. In Proceedings of the Fourteenth International Joint Conference on Artificial Intelligence, Montreal, QC, Canada, 20–25 August 1995; Volume 2.
40. Office, M. Beaufort Wind Force Scale. Met Office. 3 March 2016. Available online: https://www.metoffice.gov.uk/guide/weather/marine/beaufort-scale (accessed on 20 January 2019).
41. Wackerman, C.; Clemente-Colón, P. Wave Refraction, Breaking and Other Near-Shore Processes. In *Synthetic Aperture Radar*; NOAA NESDIS Office of Research and Applications: Washington, DC, USA, 2000; pp. 171–189.

42. Lehner, S.; Pleskachevsky, A.; Velotto, D.; Jacobsen, S. Meteo-Marine Parameters and Their Variability Observed by High Resolution Satellite Radar Images. *J. Oceanogr.* **2013**, *26*, 80–91. [CrossRef]
43. Rabaud, M.; Moisy, F. Ship wakes: Kelvin or Mach angle? *Phys. Rev. Lett.* **2013**, *110*, 21. [CrossRef] [PubMed]
44. Darmon, A.; Benzaquen, M.; Raphaël, E. Kelvin wake pattern at large Froude numbers. *J. Fluid Mech.* **2014**, *738*. [CrossRef]
45. Kelvin, L. On the waves produced by a single impulse in water of any depth. *Proc. R. Soc. Lond. Ser. A* **1887**, *42*, 80–83.

remote sensing

MDPI

Article

Capabilities of Chinese Gaofen-3 Synthetic Aperture Radar in Selected Topics for Coastal and Ocean Observations

Xiao-Ming Li [1,2,3,*], Tianyu Zhang [1,4], Bingqing Huang [1] and Tong Jia [1,2]

[1] Key Laboratory of Digital Earth Science, Institute of Remote Sensing and Digital Earth,
    Chinese Academy of Sciences, Beijing 100094, China; zhangty@radi.ac.cn (T.Z.); huangbq@radi.ac.cn (B.H.);
    jiatong@radi.ac.cn (T.J.)
[2] Laboratory for Regional Oceanography and Numerical Modeling, Qingdao National Laboratory for Marine
    Science and Technology, Qingdao 266235, China
[3] Hainan Key Laboratory of Earth Observation, Sanya 572029, China
[4] University of Chinese Academy of Sciences, Beijing 101408, China
* Correspondence: lixm@radi.ac.cn; Tel.: +86-10-8217-8168

Received: 1 October 2018; Accepted: 28 November 2018; Published: 30 November 2018

Abstract: Gaofen-3 (GF-3), the first Chinese spaceborne synthetic aperture radar (SAR) in C-band for civil applications, was launched on August 2016. Some studies have examined the use of GF-3 SAR data for ocean and coastal observations, but these studies generally focus on one particular application. As GF-3 has been in operation over two years, it is essential to evaluate its performance in ocean observation, a primary goal of the GF-3 launch. In this paper, we offer an overview demonstrating the capabilities of GF-3 SAR in ocean and coastal observations by presenting several representative cases, i.e., the monitoring of intertidal flats, offshore tidal turbulent wakes and oceanic internal waves, to highlight the GF-3's full polarimetry, high spatial resolution and wide-swath imaging advantages. Moreover, we also present a detailed analysis of the use of GF-3 quad-polarization data for sea surface wind retrievals and wave mode data for sea surface wave retrievals. The case studies and statistical analysis suggest that GF-3 has good ocean and coastal monitoring capabilities, though further improvements are possible, particularly in radiometric calibration and stable image quality.

Keywords: synthetic aperture radar; GF-3; coast and ocean observation; sea surface roughness

---

1. Introduction

A major reason that synthetic aperture radar (SAR) is favored for many applications in ocean observations is its high spatial resolution as an imaging radar. Simply put, SAR images allow us to observe the fine structures of many interesting oceanic and atmospheric phenomena and processes. Furthermore, as an active radar, SAR has the capability to work independent of sunlight and the ability to penetrate cloud cover and, to some extent, rain. Despite the short 106-day lifetime of the first civil ocean SAR onboard the SEASAT mission, it showed great potential for spaceborne SAR ocean observations [1]. The successful launch by the European Space Agency (ESA) of ERS-1 in 1991 and ERS-2 in 1995, both of which had onboard SAR sensors, enabled the operational acquisition of data for a long period of twenty years. In the 1990s, together with the ERS-1 and 2 SARs, another C-band SAR, Radarsat-1 (launched in 1995) and an L-band SAR, JERS-1 provided a large and diverse body for earth observations and significantly advanced our knowledge of the ocean, coastal zones, and polar regions.

The ocean is vast, with high spatial variability, therefore, it is preferable for spaceborne SAR to capture images with a large swath in addition to good spatial resolution. During the last century, the Radarsat-1 and Advanced SAR (ASAR) have acquired images with a swath width over a few hundred kilometers. Wide-swath or ScanSAR images generally have a spatial resolution of tens of meters and, more importantly, can map a large area of the open sea and coast, which makes them particularly suitable for studying meso-scale oceanic and atmospheric processes, e.g., by mapping the distribution of internal ocean waves [2], observing atmospheric solitons [3], estimating the wind speed of tropical cyclones [4], and measuring sea surface velocity [5].

The year 2007 marks an important advance in the development of spaceborne SAR; two X-band spaceborne SAR, the TerraSAR-X (TSX) and Cosmo-SkyMed and the C-band SAR, Radarsat-2 (R2), were launched. Compared with previous spaceborne SAR missions, the new generation of SAR sensors has several advantages. One advantage is that the new generation can acquire images with a high spatial resolution of up to 1 m in spotlight mode [6–8]. This offers a unique opportunity to detect targets in the ocean and coast, e.g., for ship detection [9–11]. The other advantage is that these SAR sensors have polarimetric capabilities of acquiring data in different polarization combinations of VV, HH, VH and HV. These SAR polarimetric data are widely exploited for oil spill detection or classification [12–14], analysis of objects scattering or their classification in coastal intertidal flats [15–17], and sea ice detection and classification [18–20]. In addition to the general advantages of the aforementioned high spatial resolution and polarimetry, advanced SARs have constellation configuration design. The Cosmo-SkyMed, TSX/TanDEM-X (TDX) and Sentinel-1A/1B missions, as well as the forthcoming Radarsat Constellation Mission (RCM), all operate in constellations, which significantly reduces the temporal intervals of SAR data acquisition and therefore enhances the capture of dynamic sea surface information [21–23]. In particular, TSX can cooperate with its twin, TDX, to achieve along-track interferometry and to retrieve sea surface currents in high spatial resolution from space [24].

During the development of spaceborne SAR, the wave mode of the ESA's SAR missions has played an interesting role in ocean observations. These missions acquired "imagettes" (small size images, approximately 5–10 by 5–10 km) continuously over the global oceans and they are dedicated to ocean wave measurements, as the name indicates. The wave mode began during the ERS/SAR mission [25–28] and became operationally available for public users with the delivery of standard Level-1b (single-look-complex data) and Level-2 (swell spectrum) products. Along with the ASAR wave mode data, available from October 2002 to April 2012, statistical analysis of global ocean waves using these data provides additional insight into, e.g., ocean swell propagation and crossing in global oceans [29]. The current Sentinel-1A/1B SAR missions are continuing to acquire wave mode data with a larger image size of 20 km by 20 km and alternative incidence angles of 23° and 36° [30]. Since the Sentinel-1 wave mode image size is comparable to standard stripmap images, and these images are acquired globally and consecutively, we can expect wider applications in ocean observations based on these data. Besides the wave mode data of Sentinel-1A/1B acquired in global ocean, the Interferometry Wide (IW) swath (250 km) and Extra-Wide (EW) swath (400 km) modes data are acquired intensively in European water for ocean monitoring, as well as in polar regions for sea ice monitoring. Both the image modes employ the TOPSAR (Terrain Observation with Progressive Scans SAR, [31]) technique to avoid scalloping [32] and generate homogenous SAR images in large coverage. This technique can lead to better sea surface wind retrievals and the Sentinel-1A/1B sea surface wind products (one of the OCN products) become operationally available.

In August 2016, the Chinese first civil spaceborne SAR, named Gaofen-3 (a phonetic rendition of the Chinese word for "high spatial resolution"), joined the list of spaceborne SAR missions in orbit. Several studies have used GF-3 for ocean monitoring, mainly focusing on sea surface wind and wave retrievals. In [33], 56 data pairs of GF-3 collocations with buoy measurements were used to preliminarily assess the quality of sea surface wind retrieval; the results indicated a root-mean-square-error (RMSE) of 2.46 m/s. Ren et al. [34] conducted a more detailed analysis of sea surface wind retrieval from GF-3 Quad-Polarization Stripmap (QPS) data, not only in the VV but also HH and HV polarizations. An empirical algorithm was proposed in [35] to derive significant wave height (SWH) from GF-3 wave mode data. In total, 12 coefficients of the empirical algorithm were tuned using the collocations of the GF-3 wave mode data with the WaveWatch III mode results.

However, it seems that only cut-off information derived from the GF-3 data [36] may also yield a reasonable SWH compared with the rather complicated empirical algorithm in [35].

After over two years of operation, GF-3 has acquired a large amount of global data; therefore, an overall assessment of its data quality and potential applications is essential. In this paper, we provide an overview of GF-3's capabilities in ocean and coastal observations, focusing on presenting representative cases over a few "super" test sites where similar studies have been conducted using other spaceborne SAR data, to evaluate the full polarimetry, high spatial resolution and wide-swath imaging capabilities of GF-3. We also address the quantitative retrieval of sea surface wind and wave information from GF-3 data. Although some studies on wind and wave retrieval from GF-3 data have been reported, as previously mentioned, it is important to conduct an intensive investigation on how accurately we can derive sea surface wind and wave parameters from GF-3 data.

This paper is organized as follows. Section 1 presents a brief introduction of GF-3. Section 2 provides some representative examples of the use of GF-3 for ocean and coastal observations. In Section 3, we focus on evaluating data for sea surface wind and wave retrievals. Finally, conclusions and an outlook are given in Section 4.

2. Brief Introduction of GF-3

GF-3, which was launched by the "Long March 4C" rocket on 10 August 2016, operates in C-band (5.3 GHz) at an altitude of 755 km in a polar sun-synchronous orbit. The repeat cycle of the orbits is 29 days. Currently, four ground stations in China are receiving the GF-3 SAR data, the Miyun station (in the Beijing suburb), Kashi station (in Xinjiang, western China), Sanya station (in Hainan, southern China) and Mudanjiang station (in the northeastern China), as well as an overseas station in Kiruna, Sweden.

GF-3 has flexible imaging modes. The five general modes are spotlight, stripmap, ScanSAR and wave. GF-3 also has several subclasses for the various general imaging modes. For example, the stripmap mode has standard stripmap, quad-polarization stripmap and fine stripmap modes. In addition, GF-3 can acquire data operationally in full polarization of VV (Vertical-Vertical), HH (Horizontal-Horizontal), VH and HV (Horizontal-Vertical) with various swath widths (up to 50 km) and spatial resolutions (up to 8 m). Table 1 lists the details of the available imaging modes of GF-3 and their technical specifications.

Table 1. Available GF-3 imaging modes and the corresponding technical specifications.

No.	Imaging Mode	Incidence Angle (°)	Nominal Resolution (m)	Swath Width (km)
1	Spotlight Mode	20–50	1	10 × 10
	Stripmap Mode			
	Superfine	20–50	3	30
	Fine	19–50	5	50
2	Wide Fine	19–50	10	100
	Standard	17–50	25	130
	Quad-pol. 1	20–41	8	30
	Quad-pol. 2	20–38	25	40
	ScanSAR Mode			
3	Narrow	17–50	50	300
	Wide	17–50	100	500
	Global	17–53	500	650
4	Wave Mode	20–41	10	5 × 5

The standard GF-3 products include Level-1a single-look-complex data, Level-1b intensity data and so-called Level-2 projected and georeferenced intensity data. The data are stored in TIFF format for each polarization channel, and the ancillary information (i.e., the metadata) is stored in XML files, similar to the standard products of other present spaceborne SAR missions. Along with the rational polynomial coefficients (RPC) file, one can derive geolocation information for individual pixels of each GF-3 SAR image.

3. Uses of GF-3 for Coastal and Open Ocean Observations

In this section, representative cases including coastal observations of an intertidal flat and offshore wind farm turbulent wakes and open ocean observations of internal waves, are presented to demonstrate GF-3's polarimetry, high spatial resolution and wide-swath imaging capabilities. Some of these cases are similar to previous studies using other spaceborne SAR data, such as TSX, Radarsat-2 and ENVISAT/ASAR.

3.1. Determination of the Scattering Characteristics of an Intertidal Flat in the Subei Shoal with GF-3 Full Polarimetric Data

To demonstrate the polarimetric capabilities of GF-3, we use an example from the Subei Shoal, which has unique radiation characteristics. Surface objects in the region include a complex mixture of mud flats, tidal current channels, aquaculture rafts and offshore wind farm turbines. Along with tidal variations, the radar backscatter characteristics of the objects in this area show high spatial and temporal variations in spaceborne SAR images, particularly in different polarization channels, as they have different scattering mechanisms. Therefore, the Subei Shoal is an appropriate site to test the polarimetric capability of SAR.

Figure 1a shows the radiometrically calibrated HV polarization from GF-3 Quad-Polarization Stripmap (QPS) data acquired on 5 October 2017. In the HV-polarized GF-3 SAR image, the distinct bright and ordered strip features are rafts composed of bamboo, ropes and nets for *Porphyra* aquaculture. Figure 1b is a photo of a single raft, while the photo in Figure 1c shows an array of numbers of rafts seen from sky. The unique structures of these rafts can induce strong volume scattering and therefore they are presented as bright patterns in the HV-polarized SAR image. The radar backscatter features of mud flats in the HV-polarized image are complicated, showing both bright and dark patterns. Thanks to high spatial resolution of the image, we can see the veering black lines, which are water channels in the mud flats.

In a previous study [16], by exploiting dual-polarization TSX data (HH and VV polarizations) and quad-polarization R2 data (VV, HH, VH and HV polarizations), we conducted a detailed analysis of the polarimetric characteristics of different objects in this area. Here, we apply the same method, i.e., the four-component scattering power decomposition [37] to GF-3 quad-polarization data to show various scattering characteristics of objects in the intertidal flat area. Figure 1d is false-color composite image from the GF-3 QPS data, using the four-component decomposition method where the red, blue and green channels represent the double bounce, surface and volume scattering, respectively. A brief description of the four-component scattering power decomposition method follows.

Equation (1) derives the Pauli vector of the polarimetric SAR data, and the coherency matrix is given in Equation (2):

$$\vec{k}_p = \frac{1}{\sqrt{2}} \begin{bmatrix} S_{HH} + S_{VV} \\ S_{HH} - S_{VV} \\ 2S_{HV} \end{bmatrix} \tag{1}$$

$$\langle [T] \rangle = \langle \vec{k}_p \cdot \vec{k}_p^* \rangle = \begin{bmatrix} T_{11} & T_{12} & T_{13} \\ T_{21} & T_{22} & T_{23} \\ T_{31} & T_{32} & T_{33} \end{bmatrix} \tag{2}$$

where S_{HH}, S_{VV}, S_{HV} are the scattering matrix elements, and it is assumed that $S_{HV} = S_{VH}$ satisfies the reciprocity condition. The symbol $\langle \cdot \rangle$ denotes the ensemble average in an imaging window, and the superscript ∗ denotes the complex conjugation. The coherency rotation after a rotation by angle θ is obtained using Equation (3), as follows:

$$[T(\theta)] = [R(\theta)]\langle[T]\rangle[R(\theta)]^*$$ (3)

$$\text{where } [R(\theta)] = \begin{bmatrix} 1 & 0 & 0 \\ 0 & \cos 2\theta & \sin 2\theta \\ 0 & -\sin 2\theta & \cos 2\theta \end{bmatrix}$$ (4)

The rotated coherency matrix $T(\theta)$ is further decomposed into four scattering components corresponding to the surface, double bounce, volume, and helix scattering mechanisms, as follows:

$$\langle[T(\theta)]\rangle = f_s\langle|T|\rangle_{surface} + f_d\langle|T|\rangle_{double} + f_s\langle|T|\rangle_{volume} + f_h\langle|T|\rangle_{helix} = P_s + P_d + P_v + P_h$$ (5)

where f_s, f_d, f_v and f_h are the contributing coefficients, and P_s, P_d, P_v, P_h and are the decomposition powers for the surface, double bounce, volume and helix scattering mechanisms, respectively. To determine the dominant scattering mechanism from the decomposition results, each scattering component is normalized using Equation (6).

$$N_i = \frac{P_i}{P_s + P_d + P_v + P_h}(i = s, d, v, h)$$ (6)

The decomposition result suggests that the polarimetric characteristics of the objects in this area are complex. The aquaculture rafts show highly variable polarimetric characteristics in different areas. Near the sea, they generally appear green in the false-color composite image, which indicates that volume scattering is dominant. This result is likely induced by the dense *Porphyra* attached to the rafts (see Figure 1b), as October–November is the high season for *Porphyra* aquaculture in the area [38]. Away from the sea, the water level decreases, and some rafts are exposed to the air; therefore, the dihedral angle between the bamboo grid of the rafts and the sea surface can induce double bounce scattering. Thus, they appear red and yellow (indicating a mixture of double bounce and volume scattering) in the false-color image. Some rafts even appear magenta, which suggests a mixture of double bounce and surface scattering. This phenomenon probably indicates few *Porphyra* attached on the rafts, so no volume scattering is induced.

An interesting feature is the yellow area in the left-hand part of the false color composition map, which indicates a mixture of double bounce and volume scattering in the mud flat, whereas the corresponding HV-polarized signal is very weak, as well as in other polarization channels. We do not have a plausible explanation for this feature. We conducted an in situ experiment in our previous study using TSX and R2 and found that visual inspection is very helpful for interpretation of polarimetric decomposition results in this area. However, in this case, we have only the SAR data and therefore, our analysis is mainly based on previous experience. The Subei Shoal is a unique area where both natural and man-made objects can show variable polarimetric characteristics due to changes in the tidal level and human activities. Further experiments to analyze the GF-3 polarimetric capabilities should focus on acquiring time-series data, as well as with essential field work to better understand the different polarimetric characteristic of various objects.

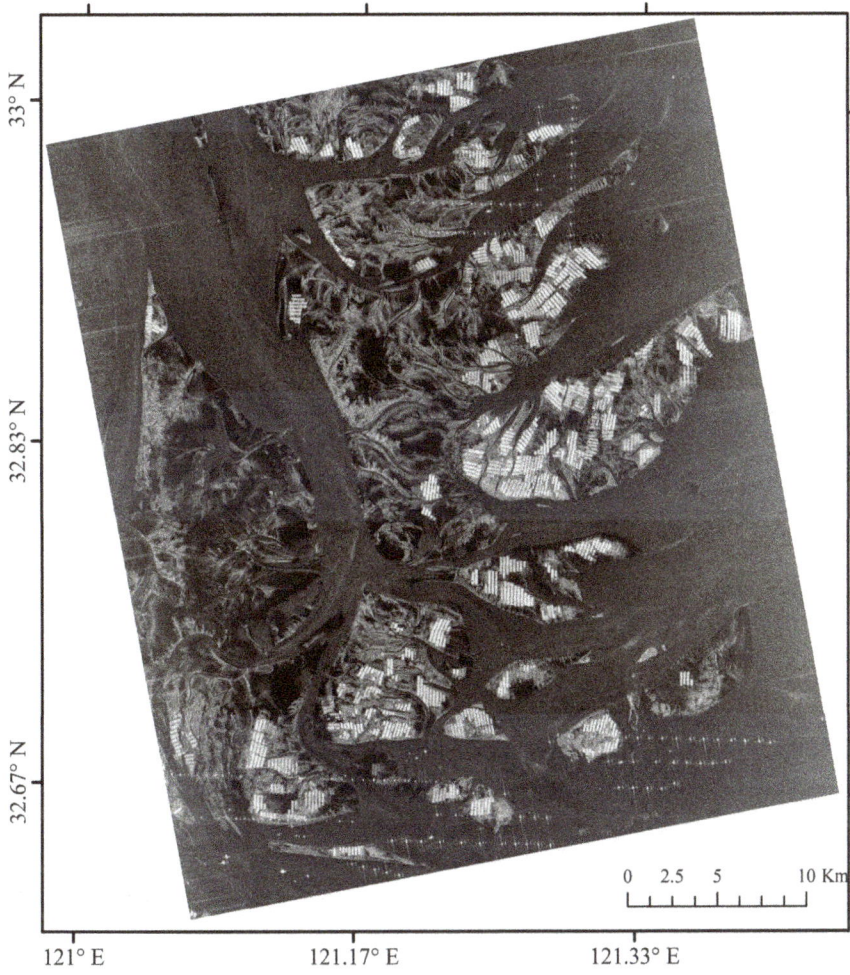

(a)

(b)

(c)

Figure 1. *Cont.*

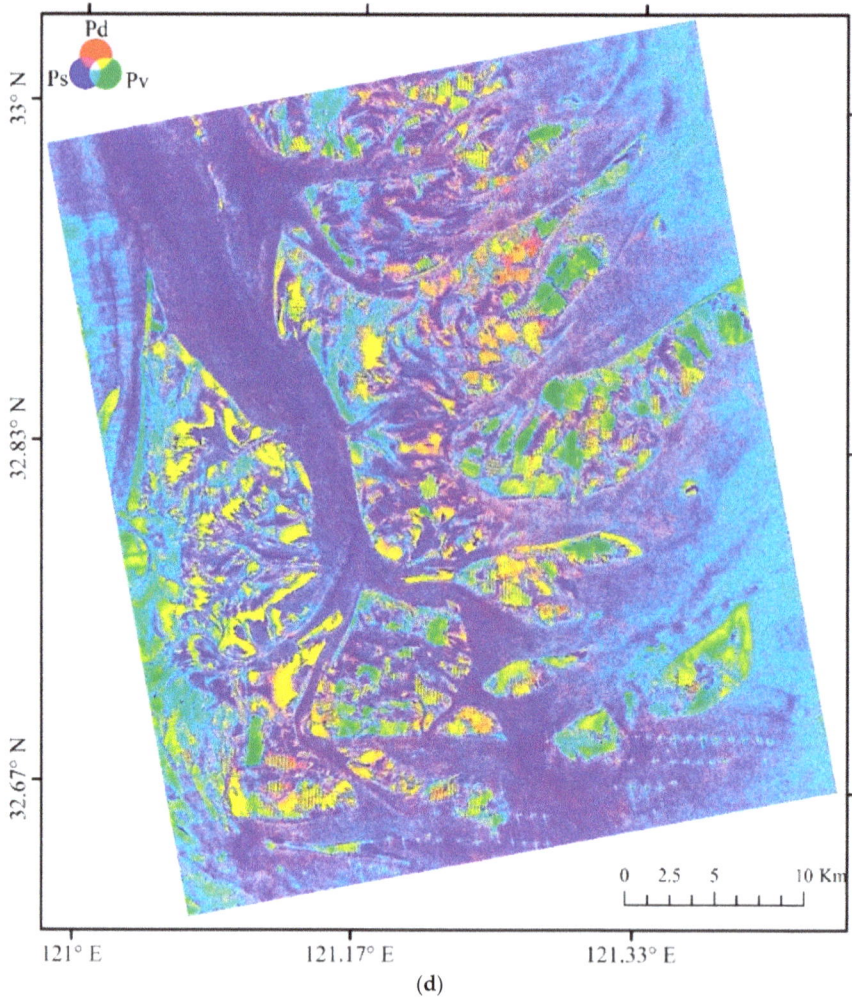

(d)

Figure 1. (**a**) Radiometrically calibrated HV polarization channel of the GF-3 QPS data acquired on 5 October 2017, in an ascending orbit. (**b**) Photo of a single raft in the study area, adapted from [39]. (**c**) Top-view photo of aquaculture rafts taken in the GF-3 imaged area. (**d**) False-color composite image generated using the normalized four-component decomposition of the GF-3 QPS data.

3.2. Observations of Offshore Wind Turbine Tidal Current Wakes

Numerous offshore wind farms have been constructed worldwide. The offshore wind turbine wake phenomenon is an important factor that needs to be considered for wind farm construction and operation. These wakes are generated because the turbine height, which often exceeds 90 m above the sea surface and the turbine's rotation induces spirals downstream in the air. When the helix vortices "touch" the sea surface and modulate Bragg waves, the sea surface roughness is consequently changed; therefore, when these areas are imaged by SAR, they often appear dark [40]. Spaceborne SAR, due to its high spatial resolution as an image radar, demonstrates unique advantages in monitoring offshore wind turbine wakes in terms of determining wake length, deficit velocity and wake meandering [41,42].

Interestingly, the wakes in offshore wind farms observed by SAR are not always induced by the rotating turbine turbulence in the air. In a previous study [43], a TSX image acquired over the East

China Sea offshore wind farm (near Shanghai) shows a distinct wake pattern downstream from each wind turbine. Based on multiple satellite observations and a numerical simulation, it was concluded that these patterns were generated by turbulence induced by interactions between the offshore wind farm foundation and a strong tidal current in Hangzhou Bay. Compared with the offshore wind turbine wakes generated by turbulence in the air, these wakes have a small spatial scale of approximately 500–1000 m in length, which varies with tidal current intensity and turbine foundation size. Therefore, these relatively small wakes are usually observed in SAR images with a high spatial resolution, e.g., in the TSX stripmap image with a spatial resolution of 3 m.

GF-3 has a fine stripmap image (FSI) mode, with a nominal spatial resolution of 5 m in both the azimuth and range directions. Figure 2a shows a GF-3 FSI image (HH polarization) acquired at 9:45 UTC on 15 February 2017, over the offshore wind farm in the East China Sea. The upper left part shows the urban area of Shanghai, and the upper right shows the Changjiang river estuary, where the visible linear features are induced by ebb tide currents. The wind farm is located to both the east and west of the Donghai Bridge (the veering line in the middle of the image). The sub-image in Figure 2b shows better visualization of sea surface features over the offshore wind farm area. In this sub-image, one can see linear patterns downstream of each offshore wind turbine with an approximately west-east orientation, whose lengths vary between 500 and 2000 m. One example of variations of the Normalized Radar Cross Section (NRCS) in the turbulent wake is shown in Figure 2c, which suggests that the NRCS gradually recovers downstream to a state comparable with that upstream of the turbine; therefore, the wake length is approximately 1000 m, as marked by the red dashed line in the figure. These patterns, which were induced by the water turbulence generated by the interaction between tidal currents and offshore wind turbine piles, are the same as those observed in a TSX image [41]. The piles are rounded, with diameters of approximately 15 m. In the southern part of the wind farm, there is a longer and more prominent wake pattern at the bridge induced by water turbulence from the interaction between tidal currents and the bridge piers. In the figure, we can also observe much longer bands, particularly apparent in the northwest of Xiaoji Hill, which are wind wakes. The European Centre for Medium-Range Weather Forecasts (ECMWF) ERA-interim model suggests the sea surface wind direction was approximately 320° at 9:00 UTC on 15 February 2017, which is consistent with northwest-southeast orientations of these wide and long bands, i.e., suggesting they are wind wakes.

Interestingly, we note that the sea surface wind direction in both the GF-3 and the TSX cases are cross with the tidal current direction. As previously discussed in the TSX case [43], wind stress plays an important role in the manifestation of the tidal current wake on SAR images. When the sea surface wind has a perpendicular component in the turbulent wake direction, it further enhances convergence and divergence of the wakes. This phenomenon represents one reason why the tidal current wakes in this case are not as distinct as those in the TSX case, because the sea surface wind speed of this case is approximately 6.0 m/s (ERA-Interim model result at 9:00 UTC) versus above 9.0 m/s in the TSX case. Second, the TSX case occurred closer to the spring tide and would have had a stronger tidal current than the GF-3 case. Finally, the TSX image has a steeper incidence angle of 19.7°–23.2° than that of the GF-3 of 23.9°–27.7° and therefore, one can generally expect strong radar backscattered signal. Moreover, the X-band SAR is more sensitive to the short scale Bragg waves than the C-band SAR, according to the Bragg resonant mechanism.

Compared with the wind wakes apparent in high resolution SAR images, the tidal current wakes have smaller spatial scales. While the former wakes often have a length greater than several kilometers (and up to tens of kilometers), the latter wakes generally have a length less than a few kilometers. The high spatial resolution capability of spaceborne SAR allows us to identify distinct tidal current wakes induced by man-made objects in shallow water. We note that these distinct wind farm turbulent wakes have also been identified in the North Sea offshore wind farm parks [44]. The promotion of clean offshore wind energy must not neglect the changes in local hydrodynamics [45] caused by wind farms and possible associated environmental issues.

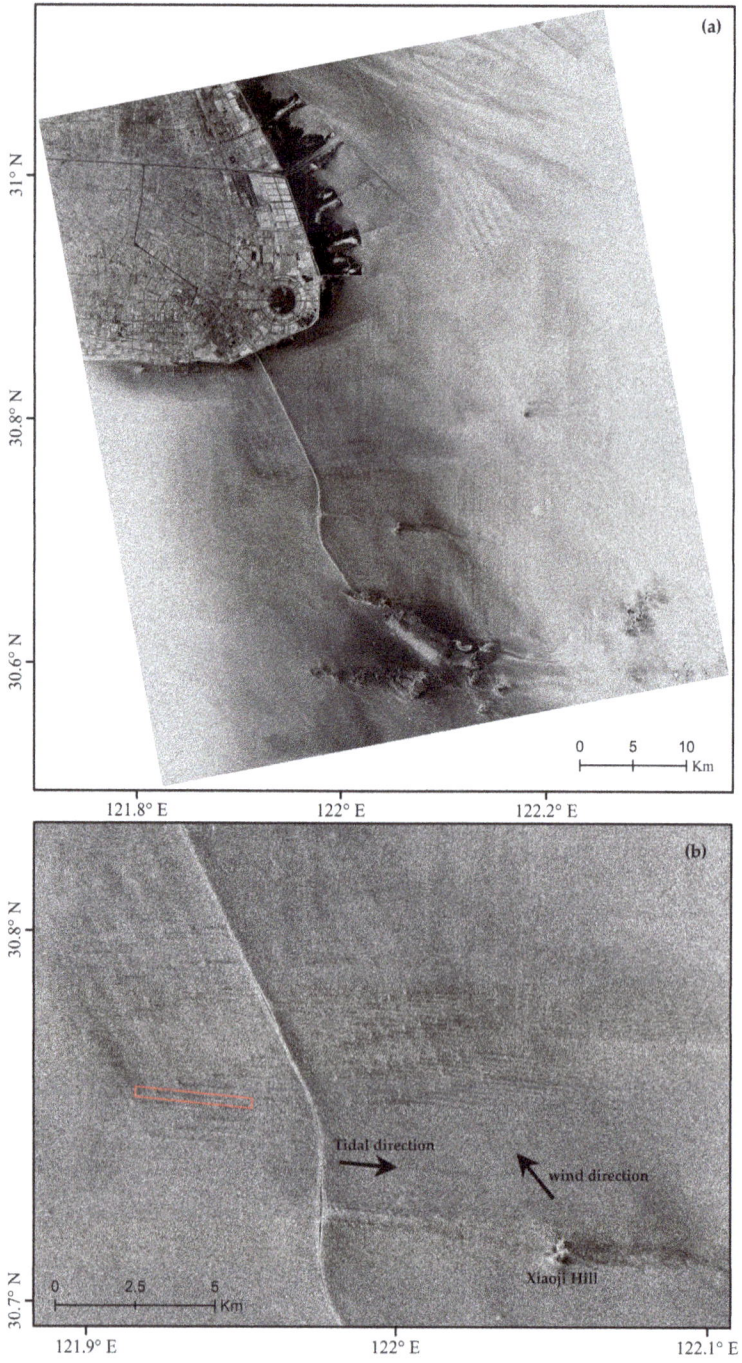

Figure 2. *Cont.*

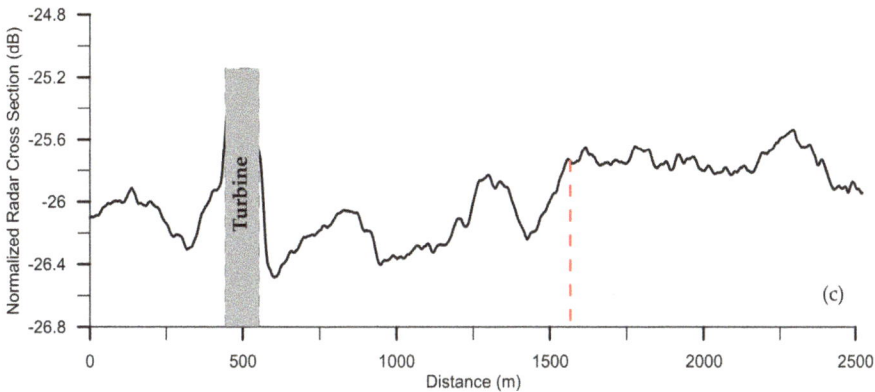

Figure 2. (a) A GF-3 FSI image in HH polarization acquired at 9:45 UTC on 15 February 2014, over the eastern Hangzhou Bay and Changjiang River estuary, in an ascending orbit. (b) The sub-image of (a) over the offshore wind farm, showing distinct tidal current wake patterns. (c) Variation of Normalized Radar Cross Section (NRCS) along a transect (marked by the red rectangle in (b)) through a tidal turbulent wake. The dashed line indicates recovery of the turbulent wake at approximately 1000 m downstream of the turbine.

3.3. Observation of Internal Waves in the South China Sea

It is generally understood that the internal waves (IWs) in the northeastern South China Sea (SCS) are primarily generated by interactions between barotropic tides and sills in the Luzon Strait; these IWs then propagate westward. Dongsha Atoll is in the pathway of these IWs in the northeastern SCS. The complicated bathymetry and various oceanic stratification patterns lead to significant spatial variations in IW at Dongsha Atoll.

The IW refraction, diffraction, reconnection and even reflection dynamics in Dongsha Atoll are all recorded by spaceborne SAR [46,47]. As there are often a few IW packets arriving at Dongsha Atoll with varying distances, and these IWs generally have long crests of a few hundred kilometers, it is preferable to use wide swath SAR images to clearly visualize IW propagation in this area. Therefore, Dongsha Atoll is a good site to test the capability of ocean surface imaging in wide swath by spaceborne SAR for observation of IW dynamics.

Figure 3 shows two GF-3 SAR images acquired in the narrow ScanSAR mode (three beams) with a swath width of approximately 300 km over Dongsha Atoll on 21 October and 24 October 2017, in ascending and descending orbits, respectively. The two images have almost identical spatial coverages. The clear arc-shape signatures of the internal solitary waves (ISWs) in the two SAR images reveal that the waves experienced significant spatial variations, i.e., wave refraction and diffraction, as they passed through Dongsha Atoll (as shown by the clearly visible round shape in the GF-3 image on 24 October). On the right side of Figure 3, the upper panels show two sub-images covering part of the ISW southern arms over Dongsha Atoll. The lower panels show the gray value variations in the GF-3 images across the transects through the ISW packets, indicated by the red lines in the two sub-images. Because IWs can induce convergence and divergence of sea surface flow, the sea surface roughness and radar backscatter are consequently changed because the Bragg waves are modulated by the sea surface flow [48]. Therefore, the significant changes in the image gray values suggests there are three IW packets (labeled S1, S2 and S3) in the 21 October 2017 image and two wave packets (labeled S1 and S2) in the 24 October 2017 image. The distance between neighboring ISW crests in one packet appears to monotonically decrease from front to rear, suggesting that these ISWs were propagating westward. As waves are assumed to propagate in the direction normal to the wave crests, it is estimated that the ISWs were traveling toward 272°–295° (clockwise relative to north). Variations in the gray values also suggest that the front section of the westward ISW is bright and the rear section is dark, indicating that the ISWs are depression waves. In the two SAR images, the distance between the leading wave crests P1 and P2 is approximately 107 km, close to the distance between S1 and S2, whereas the distance between P2 and P3 is 53 km. Li et al. noted that tidal daily inequality can lead to different inter-packet distances over Dongsha Atoll [49].

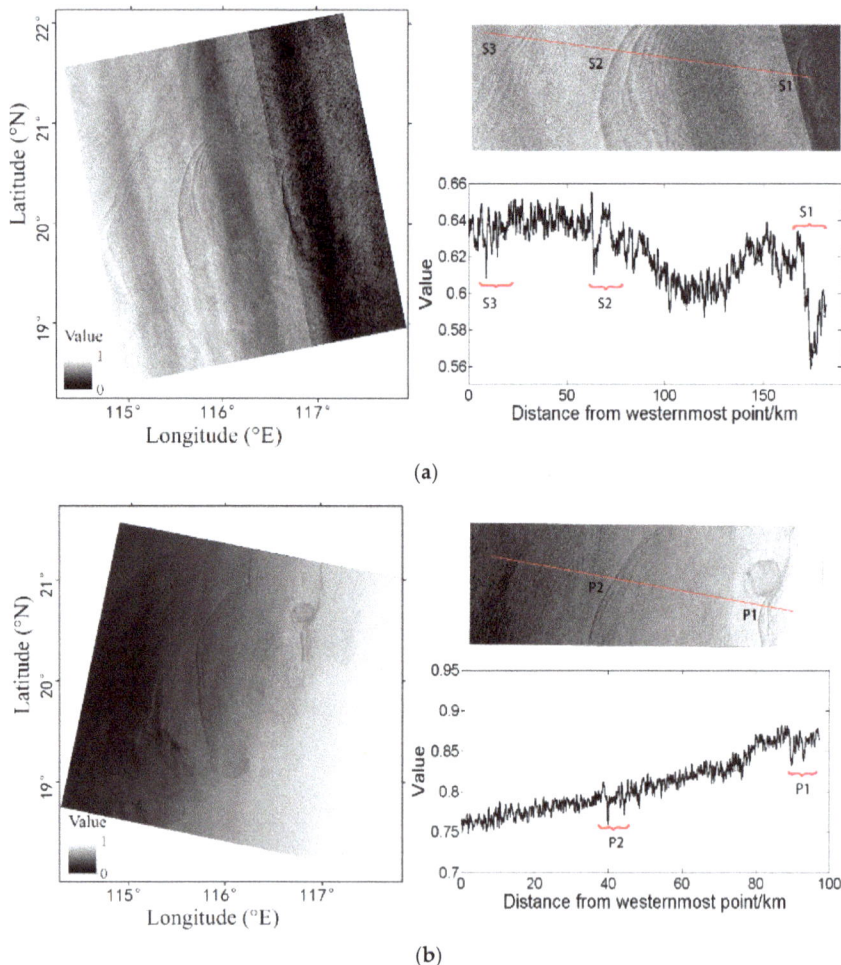

(a)

(b)

Figure 3. (**a**) A GF-3 narrow ScanSAR image acquired on 21 October 2017, in ascending orbit, over Dongsha Atoll showing ISW signatures (left panel), the sub-image encompassing part of the ISWs in the southern Dongsha Atoll (upper right panel) and variations in the gray values of the transect corresponding to the red line in the sub-image (lower-right panel). (**b**) The same as (**a**) but for the image acquired on 24 October 2017, in a descending orbit.

The ISWs in the northeastern SCS often have long crests up to a few hundred kilometers, appear in successive packets, and have spatial extents of a few hundred kilometers. Therefore, ScanSAR images are favorable for the observation of IWs in the SCS. The two images in Figure 3 yield a broad view of the dynamic ISW refraction and reconnection processes in Dongsha Atoll, highlighting the wide swath imaging capability of GF-3. However, some problems with the GF-3 ScanSAR image remain to be addressed. In the image acquired on 21 October, the rightmost beam presents a noticeable grayscale inhomogeneity compared with the two neighboring beams. In contrast, the image acquired on 24 October shows a homogeneous gray level transition from the SAR near to far range. Note that the two images were both radiometrically calibrated, although the normalized values are used for the current presentation. We were not able to eliminate the distinct sea surface radar backscatter inhomogeneity from beam to beam. While the image is suitable for qualitative analysis of IW dynamics, such inhomogeneities can lead to significant bias in quantitative retrieval of marine-meteorological parameters, such as sea surface wind speed, where absolute radar backscatter values are used.

4. Retrieval of Sea Surface Wind and Wave

In the previous sections, we present three cases demonstrating the full polarimetry, high spatial resolution and wide-swath imaging capabilities of GF-3. In this section, we examine GF-3's performance in the quantitative retrieval of sea surface wind and wave.

4.1. Sea Surface Wind Retrieval Using QPS Mode Data

The QPS mode is a promising GF-3 SAR imaging mode that obtains surface radar backscatter in four polarization channels of VV, HH, VH and HV. In the previous section, we presented an analysis of surface object radar backscatter characteristics based on polarimetric decomposition using QPS mode data. SAR VV polarization data is preferable for sea surface wind retrieval, as the sea surface generally has stronger radar backscatter in VV than that in other polarization channels. However, the VV polarized signal becomes insensitive to sea surface wind speeds above 25 m/s. Recent studies have suggested that SAR cross-polarization signals (VH or HV) increase linearly with increasing sea surface wind speeds [50,51] and are less dependent on wind direction and incidence angle than the VV polarization data [52].

Therefore, we started with the QPS mode data for sea surface wind retrieval. Notably, QPS is among a few imaging modes with the largest volume of data acquired by GF-3 over the ocean. However, when we attempted to compare the sea surface wind speed derived from QPS VV polarized data with those derived from wind models or other satellite measurements, we found distinct discrepancies. As the retrieval of sea surface wind information using C-band SAR data in VV polarization is a mature method, i.e., a method based on the geophysical model function (GMF), which relates the radar backscatter cross section σ_0 with the sea surface wind speed and direction and the radar incidence angles, we deduce that the σ_0 values of the original GF-3 QPS mode data probably have some biases. Any SAR data used for quantitively deriving marine-meteorological parameters must be radiometrically calibrated well. Equation (7) is the general radiometric calibration procedure of SAR data given in the unit dB. The DN value is the digital number recorded by the instrument, and the external calibration constant ($k_{cali\_constant}$, positive values in Equation (7)) is provided in the SAR data annotation file.

$$\sigma_0(db) = 10 \log_{10}\left(DN^2\right) - K_{cali\_constant} \tag{7}$$

To verify the accuracy of the GF-3 QPS mode data in VV polarization, we conducted a simulation experiment. As the C-band GMF (CMODs), can approximately reflect the sea surface radar backscatter σ_0 of C-band SAR in VV polarization given the sea surface wind speed and direction, we used CMOD5.N to simulate the GF-3 QPS data σ_0 for comparison. A total of 2841 QPS GF-3 SAR images acquired from September 2016 to November 2017 were collected. Each QPS image was divided into a few 5 km by 5 km subscenes, which were further collocated with the ECMWF ERA-Interim sea surface wind field (available every 3 h in a grid of 0.125°) (available at: http://apps.ecmwf.int/datasets/data/interim-full-daily/) using a temporal window less than 0.5 h and spatial distance less than 12.5 km. Next, the simulated σ_{0_sim} was achieved by inputting the collocated ERA-Interim sea surface wind speed, direction and radar incidence angle of each scene into CMOD5.N.

The GF-3 SAR radar backscatter without radiometric calibration is denoted σ_{0_raw}, which is equal to $10\log_{10}(DN^2)$. Then, the difference between σ_{0_sim} and σ_{0_raw}, $\Delta\sigma$, was treated as the external calibration constant value, assuming the simulated σ_{0_sim} is close to the truth of the normalized radar cross section and neglecting other factors that may affect the GF-3 radiometric calibration accuracy:

$$\Delta\sigma = \sigma_{0\_sim} - \sigma_{0\_raw} = \sigma_{0sim} - 10\log_{10}\left(DN^2\right) = K_{cali\_rvd} \tag{8}$$

Figure 4 shows the diagram of variations in $\Delta\sigma$ with incidence angles. The gray dots are the $\Delta\sigma$ values of each collocated data pair, on which the whiskers are overlaid. The results show that the $\Delta\sigma$ values are scattered and the acquired data irregularly distributed at different incidence angles. According to Equation (8), in an ideal situation, the value of $\Delta\sigma$ should be equal to k_{cali_rvd}; however, because the $\Delta\sigma$ values for each incidence angle are highly variable, we used the median values of the whiskers (red solid lines within the boxes) as k_{cali_rvd}. In the diagram, the overlaid blue triangles indicate the "old" calibration constants with respect to the "new" ones released in May 2018, which are marked by green triangles. The new calibration constants are generally higher than the old ones by an average of approximately 1.4 dB over different incidence angles. The recently updated calibration constants are close to the revised ones, i.e., k_{cali_rvd}, particularly for incidence angles ranging from 35° to 40°, where also the QPS mode data amount are the largest. The overall difference between the old calibration constants and k_{cali_rvd}, is 2.56 dB, whereas the difference between the new calibration constants and k_{cali_rvd}, decreased to 1.68 dB.

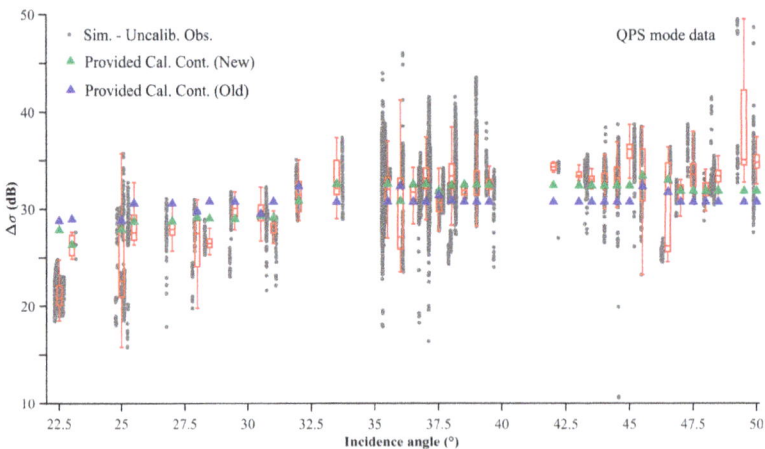

Figure 4. Distribution of $\Delta\sigma$ for different incidence angles of the GF-3 QPS mode data acquired from September 2016 to November 2017. The whiskers (red) are calculated from the samples of $\Delta\sigma$. The upper extreme of each whisker is equal to Q3 + 1.5*IQR, where the IQR is the interquartile range of the samples (i.e., Q3–Q1) and the lower extreme is equal to Q1 − 1.5 × IQR.

To further verify our assumption, the sea surface wind speeds were retrieved using the three groups of different calibration constants, which were collocated with WindSat sea surface wind speeds (available from http://www.remss.com/), treated as an independent dataset because the ERA-Interim data were used to derive k_{cali_rvd}. Figure 5 shows the comparison. When the old calibration constants were used, the comparison yields a large bias of 0.74 m/s and an RMSE of 2.38 m/s. When the new calibration constants were used, the bias and RMSE decrease to −0.15 m/s and 1.72 m/s, respectively; these results are similar to those achieved using the revised calibration constant for sea surface wind speed retrieval from GF-3, which has bias and RMSE values of −0.21 m/s and 1.74 m/s, respectively. The new calibration constants yield better retrievals, with results similar to those derived using the revised calibration constants. This suggests that continuous efforts to improve the accuracy of radiometric calibration are strongly recommended to achieve high quality sea surface wind retrieval results from GF-3 data.

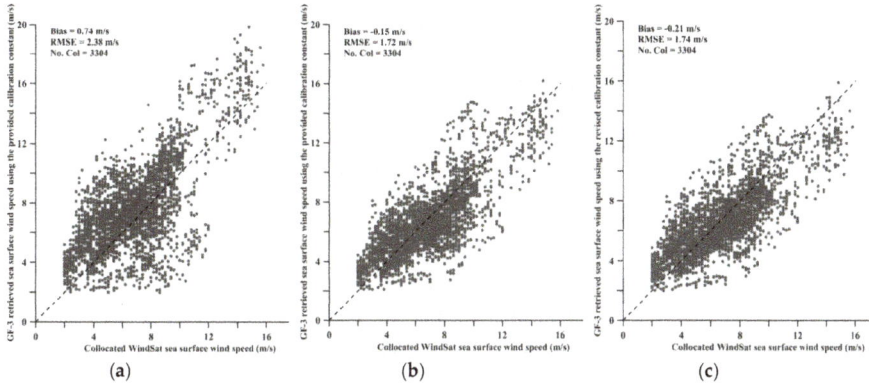

Figure 5. Retrieval of sea surface wind speed from the GF-3 QPS mode data in VV polarization using (**a**) the old calibration constant, (**b**) the new calibration constant released in May 2018, and (**c**) the revised calibration constants derived in this study.

Following the analysis of sea surface wind speed retrieval from QPS VV polarization data, we further examined the possibility of deriving sea surface wind speeds from VH polarization data. In view of the analysis above, the newly released calibration constants of the VV polarization data yielded better retrieval results than the old calibration constants, which are close to the retrievals using our revised calibration constants. This suggests that the radiometric calibration accuracy is improved. Therefore, we used the new calibration constants to derive the normalized radar cross section σ^0_{VH} of the QPS cross-polarization data. Next, the mean σ^0_{VH} of the VH polarized data subscenes were collocated with the ECMWF ERA-Interim sea surface wind field, as shown in Figure 6a and the boxes and whiskers were overlaid on the scatter plot. The plot suggests some important information about the GF-3 QPS VH data radar backscatter. First, the lower extremes of the whiskers suggest that the noise sigma equivalent zero (NESZ) is approximately −38 dB, comparable with the R2 and Sentinel-1 values of −35 dB [50,53]. Furthermore, the median values suggest a linear increase in σ^0_{VH} with sea surface wind speed, as indicated by the red dotted line. Thus, we fitted a linear relationship between σ^0_{VH} and the sea surface wind speed, U_{10}, as follows:

$$\sigma^0_{VH} = 0.6476 \times U_{10} - 37.1879 \tag{9}$$

We then used the relationship to derive sea surface wind speed from the VH polarization data and compared them with the independent wind measurements from WindSat, as shown in Figure 6b. The comparison yielded a bias and an RMSE of −0.33 m/s and 1.83 m/s, respectively, which are similar to the values from comparison with the VV retrieved sea surface wind speed (a bias of −0.15 m/s and an RMSE of 1.72 m/s; Figure 5b). We also compared the sea surface wind speed derived from both the VV and VH polarization data of GF-3, as shown in Figure 7a. Overall, sea surface wind speeds retrieved from both channels are in good agreement, with a bias of 0.11 m/s and an RMSE of 1.87 m/s. For sea surface wind speed values lower than 6 m/s, the retrieved sea surface wind speeds from the VV polarized data are generally higher than those from the VH polarized data. For sea surface wind speed values above 8 m/s, the opposite is true and the difference trend tends to increase with wind speed, which suggests that the GF-3 QPS cross-polarization data can yield a better retrieval of sea surface wind speed for high winds.

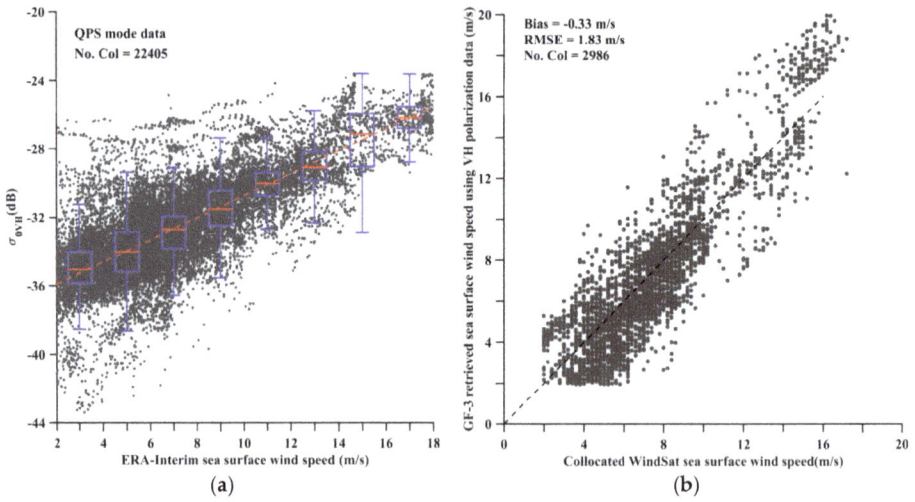

Figure 6. (**a**) Scatter plot of the GF-3 QPS σ_{VH}^0 values and the collocated ERA-Interim SSW. Boxes and whiskers, as well as linear fitting line are overlaid on the plot. (**b**) Comparison of the sea surface wind speed retrieval from GF-3 quad-polarization VH polarized data using Equation (9) with the collocated WindSat SSW speed data.

Figure 7b shows an example of retrieved sea surface wind speed from the VV (left) and VH (right) polarized channels of GF-3 QPS data. The collocated WindSat wind vectors are also overlaid on the plots. The sea surface wind streaks are clearly visible in the maps and are consistent with the wind directions of the WindSat data. The sea surface wind speeds derived from both polarized channels are consistent but present discrepancies, particularly in the near range of the SAR geometry (which has a steeper incidence angle than the far range). The dependence of σ_{VH}^0 on the incidence angles should be further investigated as more data are collected.

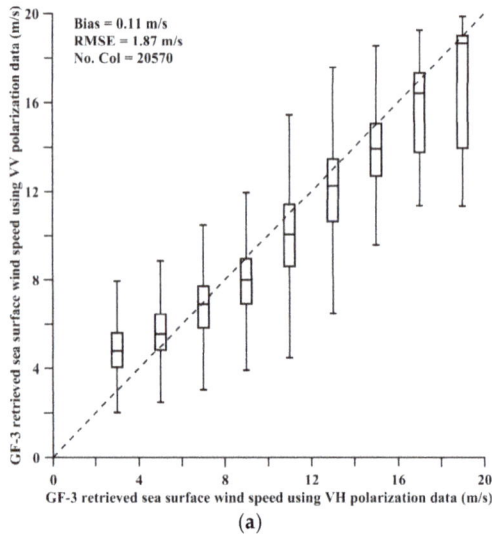

Figure 7. *Cont.*

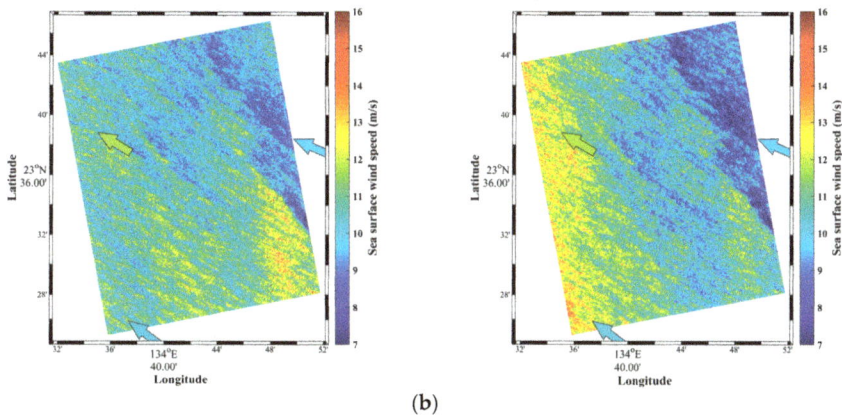

(b)

Figure 7. (**a**) Comparison of the sea surface wind speeds derived from the VV and VH polarized data of GF-3 QPS data. (**b**) Example sea surface wind speed maps derived from both the VV (left) and VH (right) polarized data of QPS data acquired on 21 August 2017. The overlaid wind vectors are from the collocated WindSat data.

4.2. Wave Mode for Ocean Wave Retrieval

As previously mentioned, wave mode is a powerful SAR imaging mode for global ocean wave measurements because this mode not only acquires two-dimensional ocean wave information but also regularly samples the global ocean. Correctly retrieving two-dimensional wave spectra from SAR has been a long effort, because it is generally thought that the imaging process of ocean waves by SAR is nonlinear, particularly for short waves or waves under a relatively rough sea state. Various approaches and methods have been attempted to derive full ocean wave spectra, swell spectra or integral wave parameters. Many of these algorithms are based on the nonlinear retrieval approach proposed in [54,55], which is called the Max-Planck Institute (MPI) approach. In this algorithm, a first guess spectrum is generally achieved by running a wave model such as Wave Model (WAM) to compensate for the lost (short) wave information and solve the 180° wave propagation ambiguity during the imaging process. In this study, we applied this classical algorithm to the GF-3 wave mode data to derive two-dimensional wave spectra. For a detailed description of this method, one can refer to the relevant literatures [54,55].

The ECMWF ERA-Interim reanalysis sea surface wind on a 0.125° grid was used to force the WAM model (cycle 4.5.1). The model outputs wave spectra every three hours at the same grid size as the input wind. The two-dimensional wave model spectrum has 25 bins in frequency and 12 bins in direction. These two-dimensional wave spectra are used as priori in the MPI scheme to retrieve the SAR wave spectra.

Figure 8a shows geolocations of GF-3 wave mode data acquired from two orbits on February 13, 2017 off the western coast of the United States. The locations of buoys 46004 and 51000 are marked with triangles. The SAR-retrieved SWH from these wave mode data were compared with the wave watch III (WW3) model results (available from http://polar.ncep.noaa.gov/waves/ensemble/download.shtml), as presented in Figure 8b, c for the two orbits. The sea state covering the area for the orbit 2701 was slightly low (less than 2.5 m), while that for 2698 was relatively rough (higher than 2.5 m). In general, the SAR retrievals are consistent with the WW3 model results, but they also pre-sent spatial sea state variations, particularly in the sea over which orbit 2701 passed.

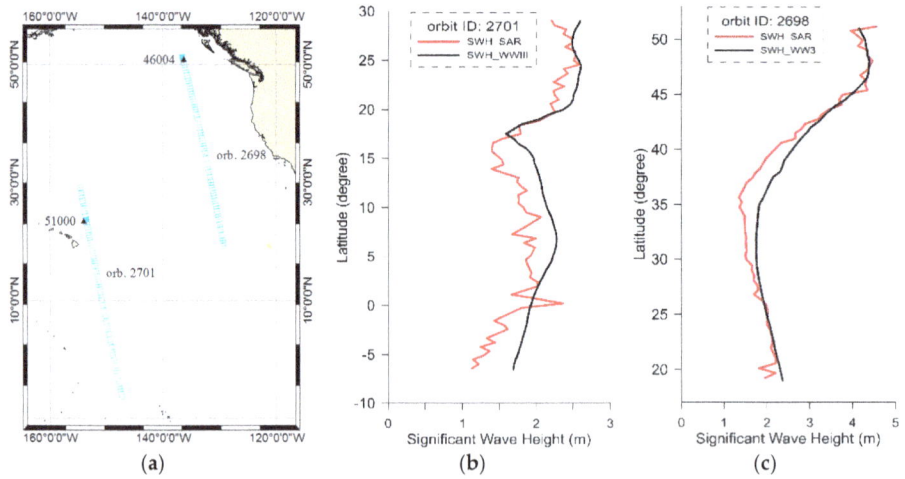

Figure 8. (**a**) Geolocation of GF-3 wave mode data in orbits 2701 and 2698 on 13 February 2017 (blue squares) and the locations of buoys 51000 and 46004 (black triangles). (**b**) SAR-retrieved SWH (red line) from the wave mode data of orbit 2701 and the collocated WW3 SWH (black line). (**c**) The same as (**b**) but for the wave mode data of the orbit 2698.

Figure 9 shows the two GF-3 wave mode imagettes near the two buoys and their corresponding retrievals. Both imagettes present clear swell patterns (the first row in Figure 9). The WAM model spectra (the second row) in the closest grid to the SAR acquisitions suggest that the swell propagated southeast and northwest, respectively, in the sea where the two images were acquired. The retrieval processes did not change much of the model spectra shapes, i.e., the first guess spectra, as found in the retrieved two-dimensional wave spectra (the third row). However, the retrieval did change the swell peak wave energy in the comparisons of one-dimensional wave spectra (the fourth row). Thus, the retrieved SWH is closer than the wave model to the buoy measurements. Interestingly, the two retrieved wave spectra indicate two different swell systems, although their peak wavelengths are almost identical. With respect to the retrieval of the GF-3 wave mode imagettes near buoy 51000, we deduce these swells came from North Pacific storms, whereas for the other image, which was nearer to the U.S. west coast, the swells likely came from the Southern Ocean, even though orbit 2698 is around two thousand kilometers east of orbit 2701.

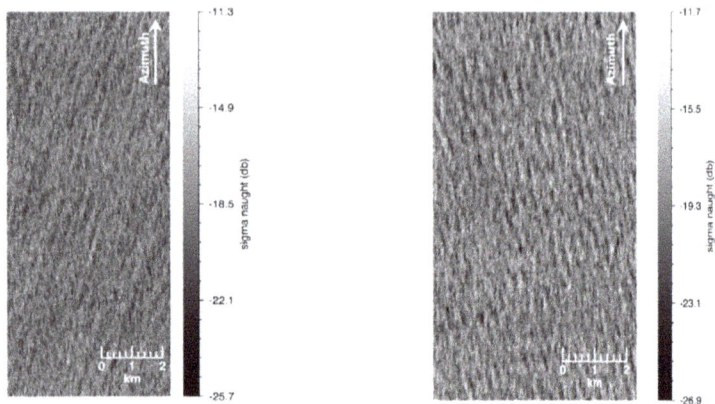

Figure 9. *Cont.*

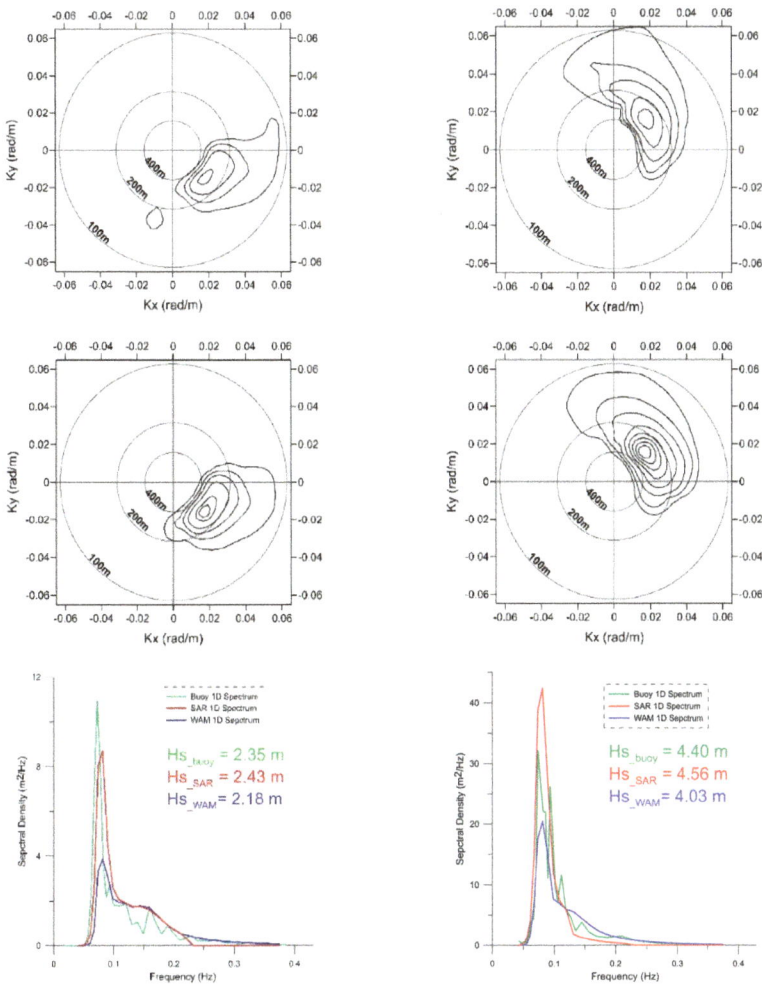

Figure 9. Two retrieval examples of the GF-3 wave mode data from orbit 2701 (left column) and orbit 2698 (right column). The SAR imagettes are in the first row. The collocated WAM spectra and the retrieved two-dimensional spectra are in the second and the third row, respectively. The one-dimensional spectra are compared in the last row.

5. Summary and Conclusions

In this study, we provided an overall assessment of the GF-3 SAR's capability for selected ocean and coastal observation. To study radar backscatter mechanisms of complicated objects, specifically seaweed aquaculture in the Subei Shoal, polarization decomposition based on GF-3 QPS data appropriately reflects the dominant scattering mechanisms of aquaculture areas. To demonstrate the high spatial resolution of GF-3, we investigated the Donghai Bridge offshore wind farm, which highlighted the capability of GF-3 data to observe the tidal current wake rather than the wind wakes generated by the rotating wind turbines. Tidal current wakes have a smaller spatial scale of hundreds of meters than wind turbine wakes of tens of kilometers, and we previously used a high spatial resolution TSX image to identify this type of fine feature. Notably, the GF-3 image of the Donghai Bridge area clearly showed similar tidal current wakes. Finally, IWs are mesoscale phenomena that often have long

Remote Sens. **2018**, *10*, 1929

wave crests exceeding a few hundred kilometers in the SCS, appearing in wave packets separated with varying distances. Therefore, it is preferable to use wide-swath SAR images for IWs. The case study of two GF-3 narrow ScanSAR images acquired in Dongsha Atoll demonstrated that the ScanSAR images could differentiate ISW refraction and diffraction around Dongsha Atoll and suggested variations in different wave packets arriving at Dongsha Atoll in one tidal cycle. However, the ScanSAR images manifested inhomogeneities in radar backscatter from beam to beam. Although this feature is rare, care should still be taken with the data processing system. We also noticed that the scalloping effect exists in some ScanSAR and Wide ScanSAR images of GF-3, particularly in the cross-polarization channels. De-scalloping should be undertaken during postprocessing (e.g., presented in [32]) for quantitative retrieval of marine-meteorological parameters.

Following these case studies, we investigated the capabilities of GF-3 QPS data for sea surface wind retrieval. A major conclusion of this investigation is that the recently provided calibration constants for the quad-polarization data significantly improve sea surface wind retrieval, with a bias of -0.15 m/s and an RMSE of 1.72 m/s, for wind speeds ranging from 2 m/s to 16 m/s. While the results were very close to the retrieved wind speed using the derived calibration constants in this study, they suggest that there is room to improve the GF-3 radiometric calibration accuracy. We also derived a linear function to derive sea surface wind speed from the GF-3 VH polarization data. The collection of more data will allow further investigation of the weak dependence of σ_0^{VH} on incidence angles, as well as its performance in retrieving high wind speeds. Nevertheless, with the current data, we can derive SSW from both VV and VH polarization data in a consistent manner.

In addition to the ESA's SAR mission, the Chinese GF-3 can also acquire wave mode data, albeit only in some regions thus far due to the limited coverage of the ground receiving stations. Nevertheless, our preliminary studies on the wave mode retrieval of two-dimensional wave spectra using nonlinear inversion demonstrate the usefulness of GF-3 for wave measurements. We expect that more wave mode data will be acquired and anticipate joint measurements with Sentinel-1A/1B wave mode data. In addition, the GF-3 wave mode data are acquired in full polarimetry and might provide a good opportunity to derive ocean wave information in a polarimetric manner [56].

Although we have presented a few informative examples, further detailed and dedicated efforts are needed to examine and improve data quality, considering that GF-3 has 12 available imaging modes and various polarization combinations. For instance, accurate radiometric calibration and noise estimation are particularly important for deriving the marine-meteorological parameters of sea surface wind and wave. Furthermore, stable performance is also important for an operational SAR data processing system. As more GF-3 data are acquired and analyzed, some abnormalities have been identified. The reasons underlying the occurrence of these cases should be investigated, and reprocessing of these data should be conducted.

Author Contributions: Conceptualization, X.-M.L.; methodology, X.-M.L.; formal analysis, all authors; investigation, all authors; resources, X.-M.L.; data curation, T.Z., B.H., and T.J.; original draft preparation, all authors; writing—review and editing, X.-M.L.; visualization, all authors; supervision, X.-M.L.; project administration, X.-M.L.; funding acquisition, X.-M.L.

Funding: This research was funded by the National Key Research and Development Program of China (2018YFC1407100), the National Natural Science Foundation of China (grant no. 41471309) and the National High-Resolution Project of China (grant no. Y20A14-9001-15/16).

Acknowledgments: It is acknowledged that the GF-3 SAR data were acquired from the ground segment of the Institute of Remote Sensing and Digital Earth, CAS and data portal of National Satellite Ocean Application Service (NSOAS). We specifically thank the ECMWF, NOAA NDBC and NCEP, and Remote Sensing Systems for providing the comparison datasets. The webpages where these data can be accessed are listed in the text.

Conflicts of Interest: The authors declare no conflict of interest.

Remote Sens. **2018**, *10*, 1929

References

1. Vesecky, J.F.; Stewart, R.H. The observation of ocean surface phenomena using imagery from the SEASAT synthetic aperture radar: An assessment. *J. Geophys. Res.* **1982**, *87*, 3397–3430. [CrossRef]
2. Liu, A.K.; Hsu, M.K. Internal wave study in the South China Sea using Synthetic Aperture Radar (SAR). *Int. J. Remote Sens.* **2004**, *25*, 1261–1264. [CrossRef]
3. Li, X.F.; Dong, C.P.; Clemente-Colon, P.; Pichel, W.G.; Friedman, K.S. Synthetic aperture radar observations of sea surface imprints of upstream atmospheric solitons generated by flow impeded by an island. *J. Geophys. Res.* **2004**, *109*, C02016. [CrossRef]
4. Reppucci, A.; Lehner, S.; Schulz-Stellenfleth, J.; Brusch, S. Tropical Cyclone Intensity Estimated from Wide-Swath SAR Images. *IEEE Trans. Geosci. Remote Sens.* **2010**, *48*, 1639–1649. [CrossRef]
5. Chapron, B.; Collard, F.; Ardhuin, F. Direct measurements of ocean surface velocity from space: Interpretation and validation. *J. Geophys. Res.* **2005**, *110*, C07008. [CrossRef]
6. Morena, L.C.; James, K.V.; Beck, J. An introduction to the RADARSAT-2 mission. *Can. J. Remote Sens.* **2004**, *30*, 221–234. [CrossRef]
7. Buckreuss, S.; Werninghaus, R.; Pitz, W. The German satellite mission TerraSAR-X. In Proceedings of the IEEE Radar Conference, Rome, Italy, 26–30 May 2008; pp. 1–5.
8. Covello, F.; Battazza, F.; Coletta, A.; Manoni, G.; Valentini, G. COSMO-SkyMed mission status: Three out of four satellites in orbit. In Proceedings of the IEEE International Geoscience and Remote Sensing Symposium (IGARSS), Cape Town, South Africa, 12–17 July 2009; pp. II-773–II-776.
9. Brusch, S.; Lehner, S.; Fritz, T.; Soccorsi, M.; Soloviev, A.; Schie, V.B. Ship Surveillance with TerraSAR-X. *IEEE Trans. Geosci. Remote Sens.* **2011**, *49*, 1092–1103. [CrossRef]
10. Pastina, D.; Fico, F.; Lombardo, P. Detection of ship targets in COSMO-SkyMed SAR images. In Proceedings of the IEEE RadarCon (RADAR), Kansas City, MO, USA, 23–27 May 2011; pp. 928–933.
11. Vachon, P.W.; Kabatoff, C.; Quinn, R. Operational ship detection in Canada using RADARSAT. In Proceedings of the IEEE International Geoscience and Remote Sensing Symposium (IGARSS), Quebec City, QC, Canada, 13–18 July 2014; pp. 998–1001.
12. Velotto, D.; Migliaccio, M.; Nunziata, F.; Lehner, S. Dual-polarized TerraSAR-X data for oil-spill observation. *IEEE Trans. Geosci. Remote Sens.* **2011**, *49*, 4751–4762. [CrossRef]
13. Zhang, B.; Perrie, W.; Li, X.F.; Pichel, W.G. Mapping sea surface oil slicks using RADARSAT-2 quad-polarization SAR image. *Geophys. Res. Lett.* **2011**, *38*, L10602. [CrossRef]
14. Migliaccio, M.; Nunziata, F.; Brown, C.E.; Holt, B.; Li, X.F.; Pichel, W.G.; Shimada, M. Polarimetric synthetic aperture radar utilized to track oil spills. *Trans. EOS* **2012**, *93*, 161–163. [CrossRef]
15. Choe, B.H.; Kim, D.J.; Hwang, J.H.; Oh, Y.; Moon, W.M. Detection of oyster habitat in tidal flats using multi-frequency polarimetric SAR data. *Estuar. Coast. Shelf Sci.* **2012**, *97*, 28–37. [CrossRef]
16. Geng, X.M.; Li, X.M.; Velotto, D.; Chen, K.S. Study of the polarimetric characteristics of mud flats in an intertidal zone using C- and X-band spaceborne SAR data. *Remote Sens. Environ.* **2016**, *176*, 56–68. [CrossRef]
17. Gade, M.; Wang, W.S.; Kemme, L. On the imaging of exposed intertidal flats by single- and dual-co-polarization Synthetic Aperture Radar. *Remote Sens. Environ.* **2018**, *205*, 315–328. [CrossRef]
18. Dierking, W.; Wesche, C. C-Band Radar Polarimetry is Useful for Detection of Icebergs in Sea Ice? *IEEE Trans. Geosci. Remote Sens.* **2014**, *52*, 25–37. [CrossRef]
19. Ressel, R.; Singha, S.; Lehner, S.; Rösel, A.; Spreen, G. Investigation into Different Polarimetric Features for Sea Ice Classification Using X-Band Synthetic Aperture Radar. *IEEE J. Sel. Top. Appl. Earth Observ. Remote Sens.* **2016**, *9*, 3131–3143. [CrossRef]
20. Johansson, M.A.; Brekke, C.; Spreen, G.; King, J.A. X-, C-, and L-band SAR signatures of newly formed sea ice in Arctic leads during winter and spring. *Remote Sens. Environ.* **2018**, *204*, 162–180. [CrossRef]
21. Ciappa, A.; Pietranera, L.; Coletta, A.; Jiang, X.W. Surface transport detected by pairs of COSMO-SkyMed ScanSAR images in the Qingdao region (Yellow Sea) during a macro-algal bloom in July 2008. *J. Marine Syst.* **2010**, *80*, 135–142. [CrossRef]
22. Cheng, Y.; Liu, B.; Li, X.; Nuziata, F.; Xu, Q.; Ding, X.; Migliaccio, M.; Pichel, W.G. Monitoring of Oil Spill Trajectories With COSMO-SkyMed X-Band SAR Images and Model Simulation. *IEEE Trans. Geosci. Remote Sens.* **2014**, *7*, 2895–2901. [CrossRef]

23. Ren, Y.; Li, X.M.; Gao, G.; Busche, T.E. Derivation of Sea Surface Tidal Current from Spaceborne SAR Constellation Data. *IEEE Trans. Geosci. Remote Sens.* **2017**, *55*, 3236–3247. [CrossRef]

24. Romeiser, R.; Suchandt, S.; Runge, H.; Steinbrecher, U.; Grunler, S. First Analysis of TerraSAR-X Along-Track InSAR-Derived Current Fields. *IEEE Trans. Geosci. Remote Sens.* **2010**, *48*, 820–829. [CrossRef]

25. Kerbaol, V.; Chapron, B.; Vachon, P.W. Analysis of ERS-1/2 synthetic aperture radar wave mode imagettes. *J. Geophys. Res.* **1998**, *103*, 7833–7846. [CrossRef]

26. Lehner, S.; Schulz-Stellenfleth, J.; Schattler, B.; Breit, H.; Horstmann, J. Wind and wave measurements using complex ERS-2 SAR wave mode data. *IEEE Trans. Geosci. Remote Sens.* **2000**, *38*, 2246–2257. [CrossRef]

27. Schulz-Stellenfleth, J.; Lehner, S.; Hoja, D. A parametric scheme for the retrieval of two-dimensional ocean wave spectra from synthetic aperture radar look cross spectra. *J. Geophys. Res.* **2005**, *110*, C05004. [CrossRef]

28. Schulz-Stellenfleth, J.; König, T.; Lehner, S. An empirical approach for the retrieval of integral ocean wave parameters from synthetic aperture radar data. *J. Geophys. Res.* **2007**, *112*, C03019. [CrossRef]

29. Li, X.M. A new insight from space into swell propagation and crossing in the global oceans. *Geophys. Res. Lett.* **2016**, *43*, 5202–5209. [CrossRef]

30. Stopa, J.E.; Mouche, A. Significant wave heights from Sentinel-1 SAR: Validation and applications. *J. Geophys. Res. Oceans* **2017**, *122*, 1827–1848. [CrossRef]

31. De Zan, F.; Guarnieri, A.M. TOPSAR: Terrain Observation by Progressive Scans. *IEEE Trans. Geosci. Remote Sens.* **2006**, *44*, 2352–2360. [CrossRef]

32. Romeiser, R.; Horstmann, J.; Caruso, M.J.; Graber, H.C. A Descalloping Postprocessor for ScanSAR Images of Ocean Scenes. *IEEE Trans. Geosci. Remote Sens.* **2013**, *51*, 3259–3272. [CrossRef]

33. Wang, H.; Yang, J.S.; Mouche, A.; Shao, W.Z.; Zhu, J.H.; Ren, L.; Xie, C.H. GF-3 SAR ocean wind retrieval: The first view and preliminary assessment. *Remote Sens.* **2017**, *9*, 694. [CrossRef]

34. Ren, L.; Yang, J.S.; Mouche, A.; Wang, H.; Wang, J.; Zheng, G.; Zhang, H.G. Preliminary Analysis of Chinese GF-3 SAR Quad-Polarization Measurements to Extract Winds in Each Polarization. *Remote Sens.* **2017**, *9*, 1215. [CrossRef]

35. Wang, H.; Wang, J.; Yang, J.S.; Ren, L.; Zhu, J.H.; Yuan, X.Z.; Xie, C.H. Empirical Algorithm for Significant Wave Height Retrieval from Wave Mode Data Provided by the Chinese Satellite Gaofen-3. *Remote Sens.* **2018**, *10*, 363. [CrossRef]

36. Shao, W.Z.; Sheng, Y.X.; Sun, J. Preliminary Assessment of Wind and Wave Retrieval from Chinese Gaofen-3 SAR Imagery. *Sensors* **2017**, *17*, 1705. [CrossRef] [PubMed]

37. Sato, A.; Yamaguchi, Y.; Singh, G.; Park, S.E. Four-component scattering power decomposition with extended volume scattering model. *IEEE Geosci. Remote Sens. Lett.* **2012**, *9*, 166–170. [CrossRef]

38. Li, Y.; Xiao, J.; Ding, L.; Wang, Z.; Song, W.; Fang, S.; Fan, S.; Li, R.; Zhang, X. Community structure and controlled factor of attached green algae on the Porphyra yezoensis aquaculture rafts in the Subei Shoal, China. *Acta Oceanol. Sin.* **2015**, *34*, 93–99. [CrossRef]

39. Zhou, M.J.; Liu, D.Y.; Anderson, D.M.; Valiela, I. Introduction to the Special Issue on green tides in the Yellow Sea. *Estuar. Coast. Shelf Sci.* **2015**, *163*, 3–8. [CrossRef]

40. Schneiderhan, T.; Lehner, S.; Schulz-Stellenfleth, J.; Horstmann, J. Comparison of offshore wind park sites using SAR wind measurement techniques. *Meteorol. Appl.* **2005**, *12*, 101–110. [CrossRef]

41. Christiansen, M.B.; Hasager, C.B. Wake effects of large offshore wind farms identified from satellite SAR. *Remote Sens. Environ.* **2005**, *98*, 251–268. [CrossRef]

42. Li, X.M.; Lehner, S. Observation of TerraSAR-X for Studies on Offshore Wind Turbine Wake in Near and Far Fields. *IEEE J. Sel. Top. Appl. Earth Observ. Remote Sens.* **2013**, *6*, 1757–1768. [CrossRef]

43. Li, X.M.; Chi, L.Q.; Chen, X.E.; Ren, Y.Z.; Lehner, S. SAR observation and numerical modeling of tidal current wakes at the East China Sea offshore wind farm. *J. Geophys. Res.* **2014**, *119*, 4958–4971. [CrossRef]

44. Vanhellemont, Q.; Ruddick, K. Turbid wakes associated with offshore wind turbines observed with Landsat 8. *Remote Sens. Environ.* **2014**, *145*, 105–115. [CrossRef]

45. Grashorn, S.; Stanev, E.V. Kármán vortex and turbulent wake generation by wind park piles. *Ocean Dyn.* **2016**, *66*, 1543–1557. [CrossRef]

46. Li, X.; Jackson, C.R.; Pichel, W.G. Internal solitary wave refraction at Dongsha Atoll, South China Sea. *Geophys. Res. Lett.* **2013**, *40*, 3128–3132. [CrossRef]

47. Jia, T.; Liang, J.J.; Li, X.M.; Sha, J. SAR observation and numerical simulation of internal solitary wave refraction and reconnection behind the Dongsha Atoll. *J. Geophys. Res. Oceans* **2017**, *123*, 74–89. [CrossRef]

48. Alpers, W. Theory of radar imaging of internal waves. *Nature* **1985**, *314*, 245–247. [CrossRef]
49. Li, X.; Zhao, Z.; Pichel, W.G. Internal solitary waves in the northwestern South China Sea inferred from satellite images. *Geophys. Res. Lett.* **2008**, *35*, 344–349. [CrossRef]
50. Vachon, W.P.; Wolfe, J. C-Band Cross-Polarization Wind Speed Retrieval. *IEEE Geosci. Remote Sens. Lett.* **2011**, *8*, 456–459. [CrossRef]
51. Zhang, B.; Perrie, W.; He, Y. Wind speed retrieval from RADARSAT-2 quad-polarisation images using a new polarisation ratio model. *J. Geophys. Res.* **2011**, *116*, C08008. [CrossRef]
52. Zadelhoff, G.J.; Stoffelen, A.; Vachon, P.W.; Wolfe, J.; Horstmann, J.; Belmonte Rivas, M. Retrieving hurricane wind speeds using cross polarization C-band measurements. *Atmos. Meas. Tech.* **2014**, *7*, 437–449. [CrossRef]
53. Mouche, A.; Chapron, B. Global C-band Envisat, RADARSAT-2 and Sentinel-1 SAR measurements in copolarization and cross-polarization. *J. Geophys. Res. Oceans* **2015**, *120*, 7195–7207. [CrossRef]
54. Hasselmann, K.; Hasselmann, S. On the nonlinear mapping of an ocean wave spectrum into a synthetic aperture radar image spectrum and its inversion. *J. Geophys. Res.* **1991**, *96*, 10713–10729. [CrossRef]
55. Hasselmann, S.; Brüning, C.; Hasselmann, K.; Heimbach, P. An improved algorithm for the retrieval of ocean wave spectra from synthetic aperture radar image spectra. *J. Geophys. Res.* **1996**, *101*, 16615–16629. [CrossRef]
56. He, Y.; Shen, H.; Perrie, W. Remote sensing of ocean waves by polarimetric SAR. *J. Atmos. Ocean. Tech.* **2006**, *23*, 1768–1773. [CrossRef]

MDPI

St. Alban-Anlage 66

4052 Basel

Switzerland

Tel. +41 61 683 77 34

Fax +41 61 302 89 18

www.mdpi.com

Remote Sensing Editorial Office

E-mail: remotesensing@mdpi.com

www.mdpi.com/journal/remotesensing

www.ingramcontent.com/pod-product-compliance
Lightning Source LLC
Chambersburg PA
CBHW051851210326
41597CB00033B/5851